The Complete Book of Home Inspection

The Complete Book of Home Inspection

Fourth Edition

Norman Becker, P.E.

New York Chicago San Francisco Lisbon London Madrid
Mexico City Milan New Dehli San Juan Seoul
Singapore Sydney Toronto

The McGraw-Hill Companies

Library of Congress Cataloging-in-Publication Data

Becker, Norman.
The complete book of home inspection / Norman Becker.—4th ed.
p. cm.
ISBN 978-0-07-170277-5 (alk. paper)
1. Dwellings—Inspection. I. Title.
TH4817.5.B43 2011
643′.12—dc22 2010024310

4 5 6 7 8 9 0 DOC/DOC 1 6 5 4 3

ISBN 978-0-07-170277-5
MHID 0-07-170277-6

Sponsoring Editor: Joy Bramble
Editing Supervisor: Stephen M. Smith
Production Supervisor: Richard C. Ruzycka
Acquisitions Coordinator: Alexis Richard
Project Manager: Aloysius Raj, Newgen Imaging Systems Pvt. Ltd.
Copy Editor: Anula Lydia, Newgen Imaging Systems Pvt. Ltd.
Proofreader: Helen Mules
Art Director, Cover: Jeff Weeks
Composition: Newgen Imaging Systems Pvt. Ltd.

Printed and bound by RR Donnelley.

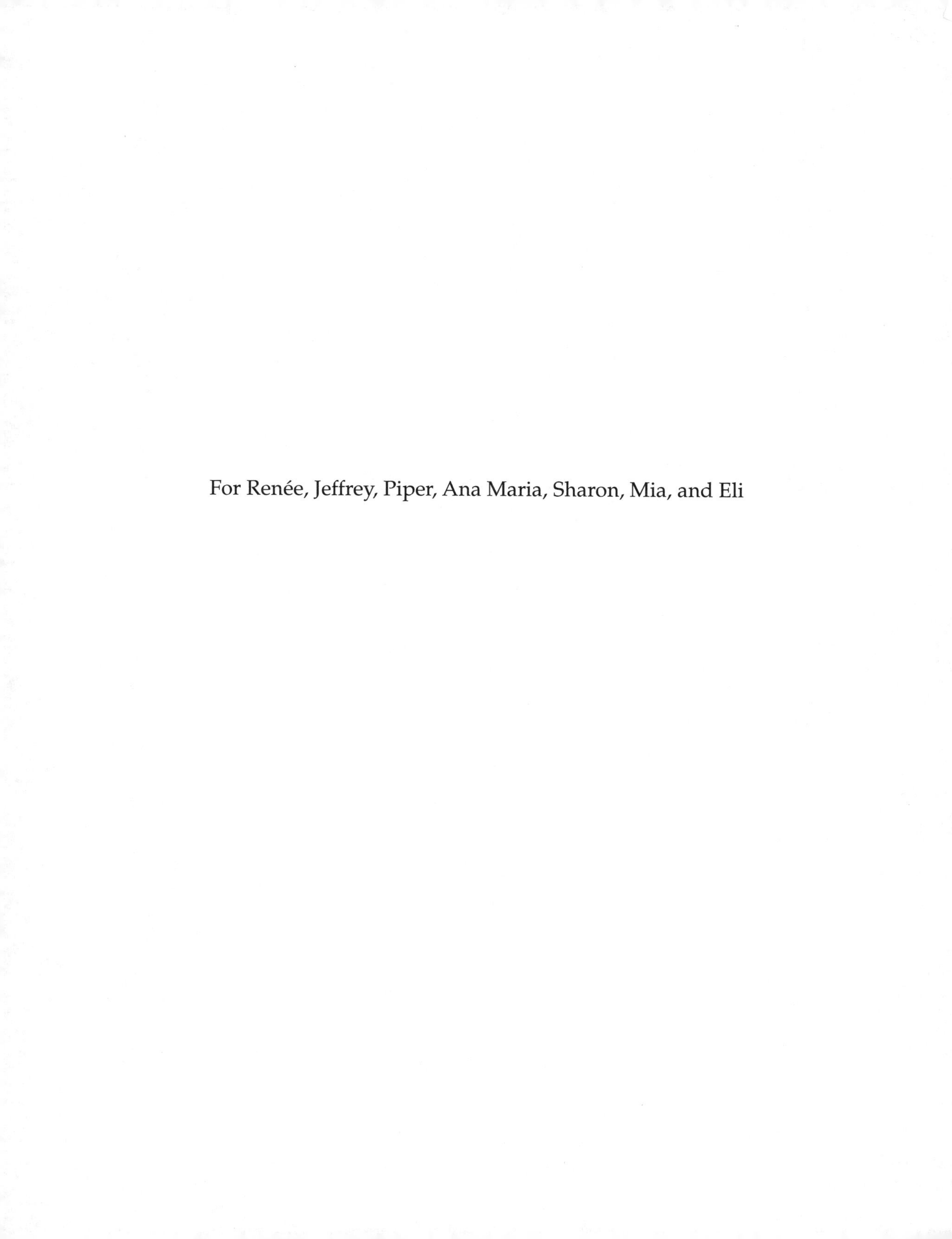

For Renée, Jeffrey, Piper, Ana Maria, Sharon, Mia, and Eli

About the Author

Norman Becker, P.E. (Hamburg, New Jersey), has more than 35 years' experience in home inspection. He has inspected homes of all ages, from newly constructed to prerevolutionary, from vacation homes to stately mansions. He is one of the founders of the American Society of Home Inspectors (ASHI) and has been qualified in court as an expert witness on the subject of home inspection. He also wrote the widely read "Homeowners Clinic" column for *Popular Mechanics* magazine for 24 years and is the author of the Popular Mechanics book *500 Simple Home Repair Solutions*. Mr. Becker is a licensed professional engineer in New York, New Jersey, and Florida.

Contents

Acknowledgments

I would like to thank Victor J. Faggella of Centurion Home Inspections in Mahopac, New York, for his suggestions and overall help and support in bringing this fourth edition to a timely finish. Thanks to David Stewart of Expert Home Inspections in Hamburg, New Jersey, for his helpful comments on New Jersey's licensing law. Also, many thanks to Douglas Hansen in San Rafael, California, for his help with the quiz at the end of this book, which is a test to see if you were asleep when you were reading the book.

Introduction

Over the years various new products and systems become available for house construction that are of interest to home buyers and owners as well as to home inspectors who have to examine and evaluate them. With each new edition of this book, I have tried to include those products and systems as well as items that may have inadvertently been omitted in the previous edition.

In this fourth edition, most of the chapters have been expanded to include additional items that should be checked during an inspection. Some of these topics are metal shingles, synthetic slate, fiber-cement siding, composite decking, free-standing decks, Superior Walls, conditioned space, Chinese wallboard, mold-resistant drywall, engineered lumber, laminate flooring, engineered hardwood flooring, bamboo flooring, air admittance valves, loop vents for island kitchens, hydro-air, geothermal heating and cooling, electronic low-water cutoffs, alternative septic systems (mound systems and alternating drainfields systems), PEX piping, recirculating hot water systems, tankless water heaters, and arc fault circuit interrupters. In addition, there is a new chapter titled "Green Home Technology" and an appendix that includes home inspector requirements.

By following the procedures outlined in this book, a home buyer will be able to look beyond the cosmetic and have a good idea as to the true condition of the house. The book is also helpful to an owner, who can use it to determine problems or potential problems in the house. There could be problems caused by deterioration as a result of aging or there could be safety or fire hazards. Very often the problems could be quite minor and could be corrected at little or no cost; however, if left unattended they could be quite costly to correct.

The section on home inspector requirements was included because I felt it was important to inform the public about the demanding requirements and the training needed to become a professional home inspector. It will also be of interest to the many engineers, architects, and contractors who would like to work as home inspectors. The book should also be quite helpful to a buyer who wants to hire the services of a home inspector to check the condition of a prospective home.

Norman Becker, P.E.

The
Complete Book of
Home Inspection

1

Tools and procedure

The inspection procedure outlined in this book is similar to the one that I use when inspecting residential structures. Because the various components of homes are basically the same, this procedure is valid regardless of the geographic location of the structure. It has been used on homes of all ages, from newly constructed to pre-Revolutionary, from vacation homes to stately mansions.

Tools needed

To inspect the house properly, you will need the following tools: a flashlight, to see in dark places (and you'll be surprised what you might find); a magnet, to determine whether plumbing pipes are iron; a marble, to note whether the floors are relatively level; an ice pick or screwdriver, to aid in looking for wood rot and termite infestation; a 6-foot stepladder, for those houses that do not provide direct access (built-in or pull-down steps) to the attic; an electrical tester, for checking the electrical ground connection and the electrical polarity, particularly in the kitchen and bathroom outlet receptacles; binoculars, to get a closer look at the roof and roof-mounted structures; a thermometer to check the temperature of the domestic hot water; and a compass, to determine the building's exposure. Knowledge of the exposure is helpful in evaluating the condition of various structural elements and components.

Since the first edition was published, a number of electronic tools have come on the market that are being used by professional home inspectors. The one tool that I would recommend for homeowners and home buyers is a battery-operated moisture meter that checks for water leaks without damaging the surface being tested. One manufacturer of moisture meters is Delmhorst Instrument Co. They can be reached at (800) 222–0638 for information on purchasing a meter. You may have read about infrared cameras that can be used for thermal imaging of a house. The cameras are used for detecting heat loss, insulation deficiencies, and specific areas of air infiltration, and for locating moisture problems. Handling of these cameras, which are quite expensive, requires special training so that the images can be accurately interpreted. Infrared cameras are not normally used during a home inspection. Most professional home inspectors do not have the camera, and those who do charge an additional fee for the service.

When performing the inspection, you should wear old clothes. Areas such as unfinished attics, basements, and crawl spaces are often quite dusty. The last items you need for inspection are a pencil and inspection worksheets. The worksheets are provided in the back of this book and should be completed as you perform the inspection. Later, you can use these worksheets to evaluate the true condition of the house and base your decision on facts rather than emotion.

Inspection procedure

A house, no matter how large or imposing, can be easily inspected if it is divided into its component parts, such as the exterior, interior, and electromechanical systems. The exterior and interior portions can be further subdivided. By approaching the inspection in a systematic order and using the worksheets provided as a guide, all the items of any consequence will be checked.

When driving up to the house, take a moment to notice the overall topography or shape of the land. Often, the topography in the immediate vicinity of the house is level; however, the overall topography might be inclined. Consequently, the possibility exists of subsurface water movement in the direction of the house. When you see inclined topography in the general area of the house, you should be alerted to the possibility of some water seepage into the lower level of the structure.

The exterior of the building should be inspected before the interior. This order is important because it provides you with an overall view of the structure that in turn can reveal the cause for some interior problems. Specifically, water seepage into a lower level can be the result of faulty gutters or downspouts, or improper grading (the ground immediately adjacent to the house slopes toward the house rather than away). A faulty roof can manifest itself in water stains one or two levels below the roof. Cracked and open exterior joints can allow the entry of water, which you will note as cosmetic damage to interior portions of the structure.

Exterior inspection

Before you start the exterior inspection, stand in front of the house and take a compass reading. The exposure for all four sides of the building should be marked on the worksheet (i.e., Front exposure—southerly; Right exposure—easterly, etc.). The exterior inspection is performed while walking around the house twice. The first time, you should look at the roof, gutters, chimney, vent stack, and anything else that is roof-mounted. The details of what to look for and how to inspect the various components of the house are discussed in the chapters that follow. During this first pass around the house, use binoculars so that you can get a closer view of the items on the upper portion of the structure. The binoculars should not have a magnification greater than 8X (eight times actual size). A more powerful pair of binoculars will tend to exaggerate hand movement, making it difficult to see details. The slightest hand movement will cause a blurred image. This problem, however, can be eliminated if you have a (more expensive) pair of 10X power binoculars with image stabilization.

After the first pass, the condition of those items inspected should be noted on the worksheet. If your first pass around the house is in the winter just after a snow, be careful where you walk. There may be a swimming pool behind the house that is covered over with snow. If you are concentrating on the house, the pool may not be that obvious and you might step right into it. (See FIG. 1-1.)

During the second pass, you look for many types of problems. Start at the front of the house and look at all the items that are either on the front of the structure or in the front yard. Examples of these items are paths, entry steps, exterior wall siding, windows, doors, decks, landscaping, fence, and so on. All of the items normally encountered during an inspection are discussed in detail in the following chapters. You should be concerned only with those items that apply to the house you are inspecting.

After inspecting the front of the house, apply the same technique to the left side, the rear, and the right side of the building. If any items of a suspicious nature require further investigation on the interior of the structure, make a note on the worksheet as a reminder. For example, if an elbow is missing from one of the downspouts and no splash plate on the ground deflects the effluent from the downspout away from the building, you should check for water seepage from that area into the lower level of the structure. Noting this fact on your worksheet helps you remember to check the interior wall opposite the downspout for signs of water seepage.

After going around the building the second time, you should be finished with the exterior inspection. Double-check your worksheet to see if you've recorded the condition of all the items inspected. At this point, you should inspect the garage. After the garage inspection, you are ready for the inside of the house.

Interior inspection

Enter the house through the front door. Try the doorbell to make sure it is operational. It's important to remember that you are looking at a house you are interested in buying. If all goes well, this will be your home, so don't be shy or feel embarrassed about doing things that any homeowner would do. As part of the inspection, you should open and close faucets on sinks, tubs, and showers; flush toilets; open and shut doors and windows; turn on the heating system and air-conditioning system by means of the thermostats; feel the airflow from heat/cooling registers; see if radiators

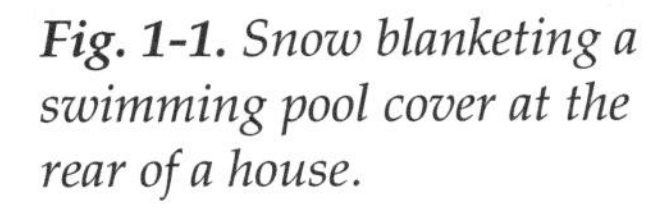

Fig. 1-1. *Snow blanketing a swimming pool cover at the rear of a house.*

get warm; and turn switches controlling lights and fans on and off.

Start the interior inspection at the uppermost portion of the building. If the house has an attic, that's where you start. To inspect the attic, you might need your ladder—check with the owner. Some homes don't have an attic, so begin this portion of the inspection with the rooms directly below the roof.

After the attic inspection, check all of the rooms on the level directly below. In some large homes you can easily miss a room. To avoid overlooking any rooms or items, begin your inspection at the entry to that level. If you start at the entry and walk either clockwise or counterclockwise, looking at each room in order, you will return to your starting point and will have inspected all of the rooms. However, if you jump around from one side to another, you can easily overlook a room or two. Again, there is no substitute for good procedure.

After all the rooms on one level have been checked, proceed to the next lower level, inspecting the connecting staircase along the way. Check all the rooms on this level in the same manner. After all the finished rooms have been checked, inspect any unfinished areas such as the basement and crawl space. This is the end of the interior inspection. At this point, all the rooms, halls, and staircases throughout the house have been checked. To complete the home inspection, you must now check the electromechanical systems.

Electromechanical systems

The systems and associated equipment included in this category are electrical, plumbing, domestic water heater, heating, air-conditioning, and swimming pool. The condition, operation, and adequacy of each system in your house must be checked as described in its respective chapter and recorded on the worksheet.

This final check concludes your home inspection. You have now looked at every item in the house of any consequence and should have recorded on the worksheets all problems and deficiencies. Some of the problems you uncover might require the services of a professional for further investigation. All situations requiring the services of a professional are indicated in the chapters that follow.

Also, after the physical inspection, you might want to have certain tests performed to determine whether the house has environmental problems, such as a high radon concentration, contaminated well water, a leaking buried fuel-oil tank, lead paint, or a mold buildup.

Look at your worksheet and test results and try to evaluate the major problems. Do not expect a perfect house. You will always find minor problems, and the costs for correcting these problems should not concern you. However, if you find many minor problems, the costs for correction can be significant.

Of main concern from a cost point of view are major problems, some of which are defined as follows:

- The need for structural rehabilitation to the foundation
- The need for re-siding the exterior walls
- Water penetration into the basement or lower level
- A malfunctioning or obsolete heating system
- The need for repiping the plumbing system
- The need for a new roof
- A malfunctioning air-conditioning system
- Inadequate electrical service
- Termite infestation
- The need for complete rehabilitation to:
  —paved areas

—deck
—detached garage
—retaining walls
- Environmental problems

If you find a major problem, have a contractor look at it and give you a written estimate on the cost for correction. At this point, you should be able to determine the true cost of buying the house—the purchase price plus the costs for upgrading substandard, deteriorated, or malfunctioning components.

Final inspection

On the day of, but prior to, the contract closing, you should take one final walk through the house and look at the walls, windows, doors, and plumbing fixtures for cracked and broken sections. Sometimes damage occurs when the seller's furniture is being moved out or through vandalism when the house is left vacant for a period of time. Specifically, look for physical changes that occurred between the time of the contract signing and the closing.

During your walk-through, check the operation of the electrical, plumbing, and heating systems. The central air-conditioning system and swimming pool equipment should also be checked if the weather permits. Check the operation of all the appliances that are considered part of the purchase. If any appliances or electromechanical systems are malfunctioning, list them on a sheet of paper, along with any items that have been badly damaged since the contract signing. This list should be taken to the closing and discussed with the seller. Very often, dollar adjustments are made to compensate for the cost of repairs.

2

Roofs

Every roof has two basic elements—the *deck* and the weather-resistant *covering*. The deck (also called *roof sheathing*) serves as a base for supporting the roof covering that protects the structure from the weather. A proper roof inspection includes an evaluation of both the roof covering and the deck. Even though a covering might be in good condition, the deck underneath might be soft, spongy, and structurally unsound. This condition can be caused by rot or delamination and is not necessarily noticeable in an exterior inspection.

Pitched roofs

The technique used for inspecting the roof differs depending on whether the roof is pitched or flat. Pitched roofs are checked during your initial pass around the house. Because of the hazards involved, I do not recommend that you climb onto a pitched roof. Begin your inspection by stepping far enough away from the house so that you are able to see all exposed sections of the roof as you circle the structure. The use of binoculars is recommended for this inspection to get a close-up view of the roof.

As you walk around the structure looking at the roof, make note of any uneven, sagging, or damaged sections. Unevenness in the roof might be the result of warped sections of deck or a poor installation of a second layer of shingles. This condition is usually not a problem. However, shingles in uneven areas are more vulnerable to damage and water intrusion. Make a note on your worksheet of the approximate locations of the uneven areas. During the attic inspection, you should check these areas for signs of leakage and to verify the cause of the unevenness.

Sagging sections in the roof, on the other hand, might be symptomatic of a structural problem or might reflect a problem that has been corrected. A sagging ridge beam or roof deck could indicate a structural failure, inadequate bracing, or inadequate spacing of wood-frame support members. The condition causing the sagging might have stabilized, so that no further corrective action is necessary. If you see a sagging ridge beam or sagging section of deck during your first pass around the house, have this condition evaluated by a professional.

Damaged sections can occur from falling tree limbs or swaying tree branches that overhang the roof. If you see a damaged section, record its location on your worksheet, since it must be repaired. Usually, patching the damaged area is all that is required.

Ventilation of the area directly below the roof deck is very important, especially in newer buildings where the deck is constructed of plywood panels rather than tongue-and-groove boards. If the area is inadequately ventilated, a moisture buildup can eventually cause the plywood sheathing to delaminate. This moisture problem is particularly acute in homes that have cathedral ceilings constructed in the following manner: The ceiling is plasterboard or an equivalent type of panel nailed directly to the roof rafters. Above the ceiling is insulation, and above this is the roof deck. Often there is a small air space between the insulation and the deck. When the moisture normally generated in the house by cooking, bathing, and so on reaches the area of the deck, there must be vent openings through which it can escape. Otherwise, rot and delamination can occur. A high percentage of the homes built with this type of cathedral ceiling have inadequately ventilated roof structures. Vent openings are needed near the top of each channel formed by the roof rafters and the ceiling and are also needed around the soffit. Often, only the soffit vents are installed. Vent openings for the top portion of the rafter channel can be provided through individual roof vents or a ridge vent. (See FIG. 2-1.) When the cathedral ceiling is the exposed roof-deck planks or panels, there is usually a rigid insulation on the top side between the deck and covering. This type of construction will not result in a deck having a problem with moisture accumulation, and therefore venting is not necessary.

If your house has a cathedral ceiling with no vent openings near the ridge, anticipate problems with the roof deck. You can tell if you have a deck problem by walking on the roof. If sections of the deck yield with each step and feel soft and spongy beneath your

Fig. 2-1. *Ridge vent. Note the ridge vent along the top of the roof. This low-profile ventilator helps circulate air through the area below the roof deck.*

feet, there are problems. Note that you should not attempt to walk on the roof if the pitch is steep or the shingles are a type that can easily be damaged, such as tile or slate. Also, if the roof is not readily accessible from a deck or an intermediate-level area, it is best to leave this part of the inspection to a professional. Even if the roof deck shows no signs of a problem, if the area is not adequately ventilated near the top of the roof, the installation of a ridge vent should be considered to prevent future problems.

Shingles

Pitched roofs are usually covered with shingles applied in an overlapping fashion. The shingles are not intended to be watertight; they protect the structure from rain intrusion by shedding water. The more common types of shingles are made of asphaltic material, wood, asbestos-cement, slate, and clay tiles. When inspecting the roof, pay particular attention to a slope that has a southerly or southwesterly exposure. These slopes receive a maximum sun exposure, and it is the sun's rays that cause the shingles to become brittle and age prematurely. Consequently, the shingles on these exposures will deteriorate more rapidly than the shingles on the other exposures. (See FIG. 2-2.)

Since shingles are intended only to shed water, any water that gets under them will leak into the interior of the structure. Shingles that are lifting, cracked, or broken are vulnerable to this type of water leakage. If you see this problem, it is an indication that some maintenance is needed. In areas where the winter temperature drops below freezing, roof leakage can occur as a result of an ice dam. Because of heat loss through the roof and heat from the sun, snow on a roof can start melting, even in freezing weather. As the water reaches the roof overhang, it often refreezes, forming an ice dam and blocking the melting snow from draining. As the snow continues to melt, the water backs up under the shingles and leaks into the interior. (See FIG. 2-3.)

Fig. 2-2. *The orientation of the house can affect the projected life of the roof shingles. The deteriorated shingles on the right slope have a southerly exposure, while the shingles on the left slope have a northerly exposure.*

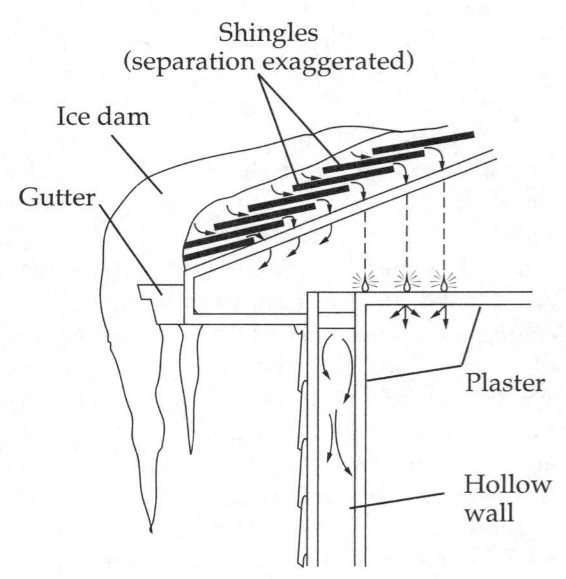

Fig. 2-3. *Ice dam at roof eaves. Because of the ice dam, water from melting snow backs up under the shingles and leaks into the house.*

Water leakage from this type of problem is not an indication of a faulty roof and should not be interpreted as a sign that roof repairs are necessary. It is an indication that adequate precautionary measures were not taken during the installation of the covering to eliminate or minimize the effects of an ice dam. The condition could have been reduced by the installation of eaves flashing. On existing roofs, the condition can be avoided by installing deicing cables along the edge of the roof and in the gutters and downspouts. This may not solve the problem completely, but it is somewhat effective. The deicing cables reduce the ice dam buildup by creating heated channels that allow water to drain into gutters and downspouts. The best method for minimizing an ice dam problem is to maintain what's called a *cold roof*. By overinsulating the attic floor and ventilating the attic profusely, the roof deck temperature will be lowered to the point where the snow won't melt.

When looking at a roof after all the snow has melted, you would never know whether there had been an ice dam and water leakage. Sometimes, however, you can see indications of a past problem—stained or warped sections of soffit trim or water stains on the ceilings of the rooms below, near the exterior walls. I have seen water stains on the ceilings of rooms two levels below the roof that were the result of water leakage because of an ice dam. Ice-dam problems will not necessarily occur every year; they depend on the severity of winter weather conditions.

Portions of the roof particularly vulnerable to leakage are the joints between the roof and roof-mounted structures, such as the chimney; the joint between the roof and a vertical sidewall; and the joint where two sloping sections of the roof intersect. The latter joint is commonly referred to as a *valley*. To protect the joints from water intrusion, they are normally covered with strips of a thin, impervious material called *flashing*. Sheet metal is usually used as a flashing material, with copper flashing as the top of the line; however, roll roofing strips are also used. Valley or sidewall flashing might not be visible, which depends on the type of joint construction. There are three basic types of valley construction: open, closed-cut, and woven valley. (See FIG. 2-4.)

When inspecting the roof, check the condition of the exposed flashing at the various joints. Loose, cracked, and deteriorated sections must be repaired. If there is leakage through any of these joints, it will usually be noted by water stains on the wood framing or roof sheathing in the

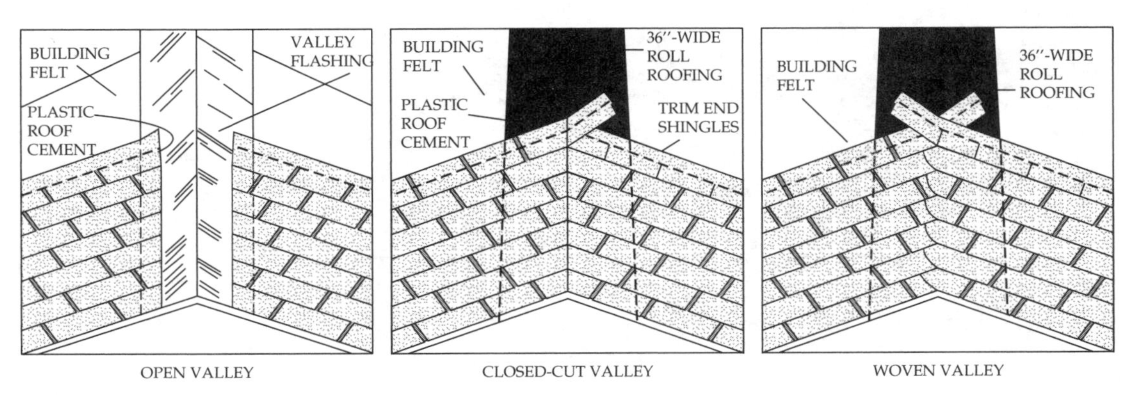

Fig. 2-4. *The three basic types of roof valley construction.*

attic or by stains in the ceilings of the interior rooms. Faulty joints are often resealed with an asphaltic cement rather than reflashed. The cement, however, is not as durable a seal as sheet-metal flashing, and the joint will often require periodic resealing.

Also check to see if the roof needs cleaning. Most pitched roofs need an occasional cleaning to remove an accumulation of debris—seed pods, twigs, pine needles, and leaves. Accumulated debris must be removed, especially from the spaces between the shingle tabs. If the debris is allowed to remain on the roof, it will retain moisture and promote the growth of moss and fungi, which is detrimental to the roofing. In addition, the litter can impede the runoff of rainwater, resulting in leaks.

Some asphalt shingle roofs develop a discoloration or what appear to be dirty streaks. The condition is often caused by wind-borne microscopic algae or mildew spores, which do not degrade or affect the performance of the shingles. It does, however, detract from the overall aesthetic appearance of the roof. The condition can usually be controlled by installing copper or zinc strips across the length of the roof and every few feet down the roof's slope.

Asphalt shingles The most common type of roof shingle used in this country is asphalt shingle, made by impregnating mats of fiberglass or organic felt materials such as rags, paper, and wood pulp, with asphalt and covering one surface with mineral granules. The *mat* is the vehicle for supporting the asphalt, which is a water-resistant material. The granules protect the shingle from damaging sun rays and provide color. When inspecting asphalt shingles, look for loss of granules, missing and torn sections with erosion of the mat. (See FIG. 2-5.) Some fiberglass-mat shingles have failed prematurely because of cracking, which can take the form of horizontal, vertical, and diagonal cracks across the shingles.

A particularly vulnerable location for leakage is the area between the shingle tabs. The granules in this area tend to come loose before those in other sections, exposing the mat to the weather. Although most roofs have a double and triple layer of shingles, a small section of the area between the shingle tabs has only one-shingle coverage. Thus, an eroded mat in this area is very vulnerable to water leakage. Loss of granules and erosion of the mat between shingle tabs is a deficiency that usually occurs on the roof slope with a

Fig. 2-5. *Deteriorating asphalt roof shingles. Note torn, missing, and brittle shingles with a loss of the granule covering, exposing the roofing mat.*

southerly or southwesterly exposure before other slopes. The condition is usually visible from the ground and can be clearly seen with the aid of your binoculars. When you see such a problem, you should anticipate early replacement of the roof shingles.

Most homes are designed to take three separate layers of shingles, although in some communities only two are allowed. When a new covering is needed on a structure that already has the maximum layers allowed, it is necessary to remove all the layers before installing the new shingles. When reroofing, it is more costly to remove existing layers of shingles than to install a new layer over existing shingles. Therefore, you should try to determine the number of layers. When the roof has an exposed edge, as in the case of a gable roof, look at the thickness of the layers. If you see two to three overlapping shingles, the roof covering is the first layer. In a hip type of roof, since there are no exposed edges, this type of determination cannot be made. In this case, try to find out the age of the house. Asphalt shingles have a projected life of seventeen to twenty-two years. The actual life span of the shingles will depend on the weight of the shingles, the type of mat, and the exposure. Asphalt shingles are classified by weight (pounds per roofing square); a roofing *square* is 100 square feet. Lightweight shingles, the least costly, weigh about 215 pounds per roofing square. Heavyweight shingles weigh about 350 pounds per square and have a longer life expectancy than lightweight ones.

If the covering is over seventeen years old, extended life for the shingles should not be anticipated. Even though the shingles might look all right (lying flat with no noticeable loss of granules or erosion), they are becoming brittle and vulnerable to wind damage. Also, these shingles will be more vulnerable to damage from someone walking on the roof when cleaning the gutters, installing a TV dish antenna, and so on. Often, these shingles will show signs of aging such as curling, cupping, cracking, and pitting. (See FIG. 2-6.) Such shingles are vulnerable to damage and will deteriorate rapidly. An exact estimate of the usable years or months remaining for the shingles is difficult. Some people do not replace an aging roof until there is a leakage problem. Others will replace it before any leakage occurs thus avoiding the cosmetic damage caused by leakage. The life span of an aging roof can be extended by patching and coating exposed cracks and eroded areas. However, even if you see no signs of leakage, the shingles on a roof that can take a second layer should be replaced before they become brittle, curl excessively, crack, and chip, as shown in FIG. 2-7. Because of the physical condition of these shingles, the surface of a second layer of shingles will be uneven, lumpy and aesthetically unattractive. In this case, for a nice even appearance the old shingles need to be stripped off the roof deck before the new shingles are installed.

Fig. 2-6. *Aging asphalt roof shingles. Note curling of the edges, with some pitting.*

Fig. 2-7. These old, weathered asphalt roof shingles are dry, brittle, cracked, chipped, and excessively curled.

If more than approximately one-third of the roof shingles show signs of advanced aging, I recommend reroofing. At this point, attempts to extend the life of the shingles are usually not economically justifiable.

Wood shingles and shakes In many parts of the country, wood shingles and wood shakes are used as a roof covering. The basic differences between the two are appearance and thickness. During the manufacturing process, shingles are sawed; shakes are split. Consequently, wood shingles have a relatively smooth surface, and shakes have a textured surface.

Because of the need for resistance to decay, most wood shingles and shakes (hereafter referred to as shingles) are made from cedar. They are also made from redwood and southern cypress. The shingles, although resistant to decay, are not immune to decay and will rot after prolonged exposure to moisture. (Rot-producing fungi are discussed in chapter 8.) The projected life expectancy for a wood-shingle roof is twenty-five to thirty years. As a wood-shingle roof ages, the shingles dry, crack, curl, and rot. As you walk around the house looking at the roof, be aware of aging shingles. Rotting shingles should be replaced. If you notice loose, damaged, or missing sections, repair is needed, even if you see no signs of water leakage. When approximately one-third of the shingles on a slope show signs of excessive aging (rotting, chipped, cracked, loose, missing, or curling), all the shingles on that slope should be replaced.

On the northern slope or on portions of the roof that are usually shaded, you might see moss growing in clusters between the joints of the shingles. It should be removed. The moss functions like a wick; the root system provides a direct path for water entry. In addition, as the moss cluster builds up, it might lift the shingles slightly, making them more vulnerable to water penetration, particularly during a driving rain.

Wood shingles have traditionally been installed on spaced sheathing, as opposed to solid sheathing, which acts as nailing strips.

This enables air to circulate on the underside of the shingles so that they can dry from both sides. The shingles are spaced between ⅛ and ¼ inch apart to allow for swelling during damp weather. Because of this space and the irregularities of some of the shingles (due to thickness and texture), daylight might be visible through portions of the roof from the attic. If during your inspection of the attic you see daylight through a wood-shingle roof, don't think that roof maintenance is necessary. If daylight is visible through the roof by means of an indirect path, maintenance is not required. On the other hand, if daylight is visible via a direct path, such as a crack, some maintenance is needed. Depending on the pitch of the roof, the shingles are two-, three-, or four-ply and are installed so that the joints between the shingles for the various plies do not line up. When daylight is visible via a direct path, cracks in the shingles line up with the joints. In this case, water can penetrate the roof, and maintenance is needed. On newer construction, you may find the roof deck consisting of solid sheathing. This is not a problem. In such a case, in order to provide the needed air circulation on the underside of the shingles, a product called Cedar Breather is probably installed between the shingles and the roof deck. Cedar Breather is a three-dimensional nylon matrix that is stiff enough to resist crushing, thereby allowing air movement to the underside of the shingles.

Asbestos-cement shingles Asbestos-cement shingles were manufactured by combining asbestos fibers with Portland cement under high pressure. Although the shingles are no longer manufactured, they can be found on many homes. Because of the asbestos ban in the mid-1970s, similar-type shingles are now manufactured using nonasbestos man-made fibers and cement. The shingles possess properties that make them highly suitable for exterior use. They are immune to rot, unaffected by exposure to salt air, and fireproof. One drawback is that they are weak in their resistance to impact and thus are vulnerable to cracking and chipping.

As you walk around the exterior of the building during the roof inspection, look for cracked, loose, chipped, and missing shingles. Note on your worksheet the areas that will require maintenance. Although asbestos-cement shingles individually last many years, an asbestos-cement-shingle roof should not be considered maintenance-free, and periodic repairs should be anticipated. Occasionally, as with a wood-shingle roof, clusters of moss might be found on the northern slope or slopes shaded by trees. If you see this condition, note it on your worksheet. The moss is a potential problem and should be removed.

Slate shingles Of all roof coverings, slate shingles are the most durable. If they are of good quality, they can last indefinitely (at least in excess of one hundred years). The slate roof over the Saxon Chapel at Stratford-on-Avon in England is over eleven hundred years old and, according to the Vermont Structural Slate Company, is in good condition. A slate roof, however, does not remain maintenance-free, even though the slates are of good quality. I have seen very few slate roofs that did not have some cracked, loose, chipped, or missing slates, a condition that requires some repair. These repairs are considered minor roof maintenance and should be anticipated on a periodic basis. Slate roofs are often patched with asphalt cement, which has a tendency to dry and crack and requires periodic application. When inspecting this type of roof from the exterior, look for cracked, loose, chipped, or missing shingles. If you find any, make a note on your worksheet.

Repairs to a slate roof, even minor ones, can be somewhat costly. Several roofers have told me that their fee reflects additional work above and beyond the required repair because they always anticipate accidental cracking

Fig. 2-8. Slate roof shingles. The ribbon slate is of inferior quality. Cracking often occurs along the ribbon after only 10 years.

of some of the slates during the repair. One difficulty you should be aware of is that when replacing a slate shingle, the roofer might not be able to match the color of the new slate to the existing weathered shingles.

Sometimes a poor-quality slate, *ribbon slate*, is used as a roof covering. (See FIG. 2-8.) The ribbons within the individual shingles are softer than the normal slate and will cause the shingles to crack along the ribbon. Often, cracking occurs along the ribbon after only ten years. Repairs to these shingles must then be made as needed.

When inspecting the roof, flaking slates may also be noted. Surface flaking is of no concern, since the shingles are at least 3⁄16 inch thick and are basically impervious to water. However, if any of the shingles are deteriorating as a result of excessive splitting and flaking (a condition brought about by winter freeze-thaw cycles), they should be replaced.

There are a number of synthetic slate look-alikes that are now on the market. These products have the durability, texture, and appearance of natural slate and are lighter in weight. The materials used for synthetic slates are quite varied. One product called FlexShake is made from recycled steel-reinforced rubber automobile tire treads. Others are made from recycled postindustrial rubber and plastic waste. There is also a synthetic slate made from ground natural slate, resin, and fiberglass, bonded under high pressure. Although synthetic slates are too new to comment on their life span, manufacturers estimate it at 40 to 60 years or more.

If the house you are inspecting is in the northern part of the country where snow might accumulate, look for snow guards on the lower portion of the roof. (See FIG. 2-9.) In particular, they should be located above doorways, sidewalks, or other areas where people will pass or gather. Snow guards are needed to prevent sliding masses of snow and ice from falling off the roof and damaging the gutters. It might interest you to know that the slate roofs on the buildings of the Harriman estate in New York had 35,000 copper-wire snow guards.

Clay tiles Clay tiles are available in many patterns. The most common are *Spanish*

Fig. 2-9. *Snow guards along the edge of the roof will help keep the snow from sliding off.*

and *Mission*. These tiles are made by shaping moist clay in molds and firing the various shapes. They are hard, durable, and fireproof. However, they are also brittle and can be easily damaged by falling tree limbs or climbing on the roof to make repairs. As with slate shingles, repair or replacement of individual tiles is more difficult and costly than that of asphalt shingles. Also, matching new tiles to the weathered tiles is usually a problem. When inspecting this type of roof, look for loose, broken, chipped, cracked, or missing tiles. If any of these conditions is found, it should be noted on the worksheet, as repairs are needed. Tiles can also deteriorate as a result of freeze-thaw cycles. You might find some cracked areas that have been sealed with asphalt cement. This condition is usually an indication of past problems. Since asphalt cement does dry and crack, periodic reapplication should be anticipated.

Check the joint (valley) between two sloping sections of the roof. (See FIG. 2-10.) If it's filled with asphalt cement, it's an indication of a problem condition that has been temporarily corrected. The flashing in that joint should be replaced. As with slate roofs, repairs, even minor ones, can be somewhat costly. This item should be noted on your worksheet.

Metal shingles Metal shingles are becoming more popular for residential roofing. They are available in several different shapes and are primarily made from painted or coated aluminum or steel panels although they are also made from copper. The panels are approximately 4 feet wide by 1 foot high and are formed to resemble wood shakes, slate, or Spanish, Roman, and Mediterranean tiles. The shingles are durable, lightweight, and fire resistant. Some have embedded stone chips for additional texture.

When inspecting the roof look for loose nails or loose shingle panels, both of which can result in leakage. Check to see if any of the joints between panels and valley sections are covered with roofing cement, which indicates

Fig. 2-10. Valley joint filled with a heavy layer of asphalt cement is an indication of a problem condition.

past and potential problems. From a cosmetic point of view, is the finish on the shingles fading or chalking? Record your findings on the worksheets.

In addition to shingles, flat sheet metal roofing has also been gaining in popularity for residential pitched roofs. Metal roof panels that come in widths of 12, 16, 18, and 36 inches are installed so that they run down the slope of the roof, essentially from ridge to eave. The joints between the panels are overlapped and interlocked to provide weathertightness. Do not climb up to the roof for an inspection. Metal roofs are slippery when wet, and depending on how the panels are supported they could be dented. The roof should be inspected from the ground with binoculars. If there are any doubts about its condition, have it inspected by a professional roofer.

Flat roofs

A roof that is perfectly level or slightly pitched is referred to as a *flat roof*. Since this type of roof is not visible from the ground, the inspection must be made from the roof itself. As with a pitched roof, a flat roof should be the first item inspected. Safe access to the roof is of prime importance. If the building is higher than one story, the roof should be accessible from the interior. Anything other than an interior means of access is a potential hazard and is considered a deficiency in the structure's design. If the roof you are inspecting is flat and is more than one story high with no interior access, it is best to have it inspected by a professional roofer.

Ventilation of the area directly below the roof deck is needed to minimize the moisture buildup in this area. An excessive moisture condition can result in deterioration of the roof deck, a lowering of the thermal resistance of the insulation, and, eventually, damage to the interior of the structure. During the cooler months, the moisture trapped in the area between the roof and the upper-level ceiling will condense, drip onto the insulation, and cause random water stains on the ceiling. Adequate ventilation of this area is also important in reducing the summer heat load on the rooms located immediately below the roof.

All too often, provisions for ventilation have been omitted by the builder. Therefore, when inspecting a flat roof, be sure to look for ventilation openings. The openings might be in the form of roof vents (vertical pipes protruding through the roof deck) or open areas in the side of the building just below the roof. Roof vents are often shielded from the rain by a cover and should not be confused with the plumbing vent stacks, which also protrude through the roof deck. The plumbing vent stack is connected to the house sewer line and is easily identified by the odor of the discharging gases. If no ventilation openings are noted, that fact should be marked on your work-sheet, and installation of ventilation openings should be considered.

A flat roof must have a watertight covering, rather than one that merely sheds water, to protect the area below the roof from water intrusion. The most common types of flat-roof coverings are built-up, single-ply, roll roofing, and metal. When inspecting a flat roof, look for ponded water. Unless the roof was specifically designed to hold standing water as an energy conservation measure, to reduce the heat load during the warm or hot months, its presence is considered a potential problem. Ponded water can become a breeding place for insects and can promote the growth of vegetation and fungi. The roots of plants growing on the roof can puncture an asphalt covering. The freezing of ponded water that has penetrated into the layers of a built-up roof can delaminate the roof covering. The temperature difference of the wet and dry areas on a randomly ponded roof results in differential expansion that might cause warping and cracking of the roof cover. If you see ponded water or signs of past ponding on the roof during your inspection, note the location on your worksheet as an area that should be drained.

A properly designed roof should have provisions for drainage. Two basic drainage designs are used in a flat roof—the perimeter system and the interior drainage system. In the perimeter system, water that drains from the interior portions of the roof collects in gutters or scuppers (openings in a parapet wall) located along the perimeter and then flows into downspouts or merely drips off the roof. In the interior drainage system, drains are located in the roof itself and are connected to downspouts that run through the interior of the structure. Look for one of these drainage systems as you walk around the roof. The interior roof drains are often clogged with debris and are sometimes set higher than the surrounding area, a condition that results in ponding. Look for cracks around the joints between the roof drain and the roof covering.

When inspecting a flat roof, you should also inspect all roof-mounted structures and projections such as skylights, hatch covers, chimneys, vent stacks, and so on. These items are discussed in chapter 3.

Built-up roofs (BURs)

This type of roof consists of bitumen (asphalt or coal-tar pitch) sandwiched between two to five layers of roofing felts and is usually covered with a mineral aggregate embedded in the top surface. The bitumen is the waterproofing agent, stabilized and reinforced by the roofing felts. The felts restrain the bitumen from flowing in hot weather and help resist cracking in cold weather. The aggregate surfacing generally consists of gravel, slag, or crushed rock. Its purpose is to protect the bitumen from the damaging effects of the sun's infrared and ultraviolet rays. These rays, through a combination of heat and photochemical oxidation, accelerate the aging of bitumen, resulting in premature brittleness and cracking. Depending on the number of layers in the builtup roof and the quality of construction, the projected life can vary from ten to twenty years.

An aggregate-covered built-up roof is difficult to inspect, principally because of the mineral-aggregate topping. Nevertheless, look for telltale signs of past, current, and potential problems. Walk around the roof systematically so that you cover the entire roof area. Look for patched areas, cracking, blistering, surface erosion, alligatoring, and wrinkling. In particular, look for cracks at the joints of roof projections and roof-mounted structures.

Usually, this type of roof has a flanged metallic strip along the perimeter, a *gravel stop*. It provides a finished edge for the built-up roofing and also prevents loose aggregates from washing off the roof. Look for cracks in the joint between the gravel stop and the roof. All cracks in the roof cover should be sealed. Record the location of any cracked or patched areas on your worksheet so that you can check the ceiling of the rooms below these areas for water stains.

Not all builtup roofs have a top surface covered with aggregates. A smooth-surface roof treatment has a weathering surface that is an asphalt-saturated roofing felt. Since the top surface is exposed, defective or deteriorated sections are visible. To maintain the weathering surface in reasonably good condition, the surface should be covered with a protective coating such as an asphalt-based emulsion every three to six years. Excessive application of the emulsion will result in alligator-type surface cracking, which is a potential problem. (See FIG. 2-11.)

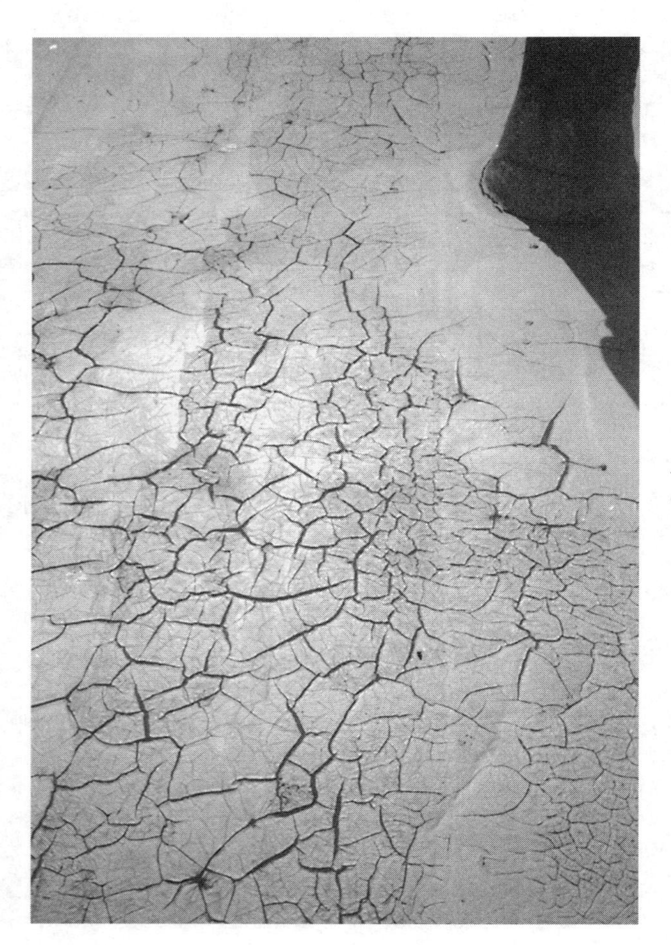

Fig. 2-11. *Roof surface with alligator-type cracks.*

Incidentally, when walking on the roof, be careful not to step on any *blisters*—weak spots in the roof covering, usually the result of air or water vapor being trapped between the layers of roofing felts. Depending on a blister's brittleness, the weight of someone walking on the blister might cause the roofing felt to crack.

Single-ply membrane

In the 1970s, the notion that a single ply of rubber, plastic, or modified bitumen material could provide the same weather protection as a built-up roof gained acceptance. Initially, single-ply membranes were used on commercial and apartment buildings; today, they are also used on residential structures with flat roof decks.

Several methods are used for installing a single-ply membrane. The membrane can be loose-laid over the roof deck and attached only at the perimeters. It is then held in place everywhere else by the weight of gravel or concrete pavers placed on top. Another method is to adhere the membrane to the roof deck fully with a self-adhesive or an adhesive applied at the jobsite. Single-plies can also be installed as a partially adhered system by using mechanical fasteners or applying an adhesive at predetermined intervals.

When inspecting a single-ply roof membrane, look specifically for open joints at seams that are not properly adhered; check for membrane erosion and punctures; and look for cracks, splits, and open sections around the perimeter walls and roof projections. As with a BUR, if the membrane is covered with gravel or paving blocks, a visual inspection of the weathering surface is not practical. In this case, pay particular attention to the condition of the ceilings in the rooms directly below the roof.

Roll roofing

This type of covering is often found on older structures in urban areas and over porch roofs. It is also found on pitched roofs of low-cost homes and outbuildings. It consists basically of an asphalt-saturated roofing felt applied directly over the roof deck and provides only single- or two-ply coverage. Quite often, the outer surface is coated with hot tar to seal joints and small cracks. Roll roofing should be inspected for blistered, cracked, eroded, weathered, torn, and aging sections. (See FIG. 2-12.) Be careful not to walk on blistered sections. Sometimes the lapped joints between the strips curl and lift, making the joints vulnerable to water seepage. Check the joints between the roof and the parapet wall and other roof projections. These joints are vulnerable to cracking and periodically require resealing. (See FIG. 2-13.) If any of the above items are noted, they should be recorded on your worksheet, since they indicate the need for some maintenance.

Metal roofing

In older homes, the roof that covers porches, garages, or an exterior side projection such as a sunroom is usually made of galvanized iron (sheet iron coated with zinc), *terne* metal (steel-coated with a mixture of lead and tin), or copper. A galvanized roof with a heavy zinc coating will last many years without painting. However, a lightly galvanized roof will rust and must be kept well painted. A terne metal roof, if kept well painted, can last for over forty years. A copper roof will also last for many years and does not require painting. Occasionally, a sheet-metal roof covering might be found on the main roof of a Victorian structure. When you are inspecting a sheet-metal roof, look for cracks or open joints at the soldered seams; because of the large amount of expansion that

Fig. 2-12. Weathered, cracked, and aging roof membrane.

Fig. 2-13. Cracked and open joint between the roof and the parapet wall.

takes place, the seam is a vulnerable joint for leakage. Also, look for exposed corroding sections of metal; they must be scraped and coated with paint or a bituminous compound such as asphalt cement. In some cases, problem conditions that have occurred over the years have necessitated coating the entire roof with a bituminous compound as a means of sealing corroded areas. Be aware that this type of roof coating requires periodic maintenance.

Checkpoint summary

Pitched roofs

- ❍ Visually inspect all portions (slopes) of the roof.
- ❍ Note any sagging, uneven, damaged, or patched sections.
- ❍ Look for overhanging tree limbs or branches that can cause damage.
- ❍ Is the area directly below the roof deck ventilated?
- ❍ Is the roof in need of a cleaning?

Asphalt shingles

- ❍ Look for curling, cracked, torn, or missing shingles.

- ❍ Are shingles losing their stone granules?
- ❍ Look for eroded sections in the slot between the shingle tabs.
- ❍ Pay particular attention to slopes with a southerly or southwesterly exposure.
- ❍ How old are the roof shingles?
- ❍ How many layers of shingles?
- ❍ If roof has been recently reshingled, is a guarantee/warranty available?

Wood shingles, shakes

- ❍ Look for rotting, loose, cracked, chipped, or missing sections.
- ❍ Pay particular attention to the slopes with heavy shade.
- ❍ Any signs of moss?

Slate, asbestos-cement, clay tiles

- ❍ Look for missing, cracked, chipped, flaking, or loose sections.
- ❍ Sections patched with asphalt cement?
- ❍ Are there snow guards along the lower edge of the tile roof?
- ❍ Are the valley joints filled with asphalt cement?

Metal shingles

- ❍ Check for loose nails and loose shingle panels.
- ❍ Are any joints or valley sections covered with roofing cement?
- ❍ Are there any chalking, faded, or dented sections?

Flat roofs

- ❍ Safe access available?
- ❍ Cracked, blistered, eroded, split, punctured, or torn sections?
- ❍ Look for open joints and seams.
- ❍ Areas of ponding water or low points where water will accumulate?
- ❍ Drainage system functional?
- ❍ Area below the roof deck adequately ventilated?

Built-up roofs (BURs)

- ❍ Are there areas with missing aggregate?
- ❍ Look for blisters in the membrane.
- ❍ Look for patched areas, surface erosion, and alligatoring.
- ❍ During the interior inspection, check the ceilings of the upper-level rooms for water stains.

Roll roofing (tar paper)

- ❍ Sections drying, eroding, or blistered?
- ❍ Look for open seams.

3

Roof-mounted structures and projections

When inspecting a roof, you should also inspect all roof-mounted structures and projections. Specifically, look at chimneys, plumbing vent stacks, roof vents, roof hatch, skylights, TV antennas, and gutters and downspouts. Except for the last two items, an area of concern is the joint between the item and the roof. This joint, although normally sealed with flashing, is vulnerable to water leakage. With years of weathering, the flashing can develop cracks, pinholes, or breaks, resulting in periodic leaks. To correct this type of water leakage, rather than reflashing the joints, the homeowner might often seal the joints with asphalt cement. While this is an effective correction, the asphalt cement will eventually become brittle and crack, and the joints will require periodic resealing.

Chimneys

Chimneys are used to vent smoke and combustion gases from heating units and fireplaces. If, during your exterior inspection, you do not find a chimney, it does not necessarily mean that the house lacks a heating system. In all probability, the house is heated electrically.

Chimneys are normally constructed of masonry (brick, concrete, stone) or are prefabricated from metal or a cement-asbestos material. Masonry chimneys are usually supported by their own footings, which in northern communities extend below the frost line. These chimneys are not dependent on the main structure for support. When inspecting a masonry chimney that extends up the side of a building, if you see open joints between the chimney and the sidewall, it is an indication of some settlement and is often not a problem. (See FIG. 3-1.) However, the open areas should be sealed to prevent water intrusion. On the other hand, if the chimney is no longer vertical, it indicates excessive settlement because of the lack of a footing, or one that's improperly designed, and it should be checked by a professional. Depending on the degree of settlement, the flashing at the joint between the chimney and the roof might need repair. The movement of the chimney could result in open and loose sections of flashing. (See FIG. 3-2.)

Fig. 3-2. *Open and loose sections of flashing caused by movement of the chimney.*

Fig. 3-1. *Uneven settlement of chimney. Note the open joint between the chimney and the sidewall.*

If your house has a brick chimney, look at the area above the roofline to see that it is vertical. Over the years, the mortar joints can weaken on one side and cause the chimney to lean. (See FIG. 3-3.) A leaning chimney represents a potential safety hazard and should be checked by a professional mason to determine whether or not corrective measures are needed, such as either bracing or rebuilding from the roofline up. If you ask the owner about the chimney, you'll probably learn that it has been in that condition for at least fifteen years. That fact does not mean that corrective action is not needed. A leaning chimney is an indication of weakening mortar joints and should be considered of questionable structural integrity.

Sometimes brick chimneys on relatively new houses might be coated with white mineral deposits called *efflorescence,* a condition often caused by the absorption of water by the bricks. The minerals in the bricks dissolve in the water and then surface when the water

Fig. 3-3. *The portion of the chimney that extends above the roof is leaning, a potentially hazardous condition.*

evaporates. Although efflorescence is quite common in new brickwork, it can easily be scrubbed or washed off with a dilute solution of muriatic acid. Recurrence can usually be controlled by covering the bricks with a vapor permeable water repellant. Efflorescence on a brick chimney that has been up for many years usually means that water is getting inside the chimney, through cracks in the joints on top of the chimney or cracks in the bricks or mortar joints. If you see heavy efflorescence on an older chimney, make a note on your worksheet. The top of a masonry chimney should have a cement finish that slopes from the flue to the edge of the chimney. The purpose of this finish is to deflect rain and protect the joints between the flue and the chimney. This cement finish is vulnerable to cracking, and periodic resealing of this area should be anticipated.

The top of the chimney must extend above the roofline to prevent downdrafts caused by the turbulence of the wind as it sweeps past nearby obstructions or over sloping roofs. The top of the chimney should extend at least 3 feet past the highest point where it passes through the roof, and 2 feet higher than any roof part within 10 feet measured horizontally (FIG. 3-4). When a chimney is 10 horizontal feet beyond a roof ridge, builders often terminate the top of the chimney below the ridge. This is not quality construction, but it satisfies the code. In this case, downdrafts can still occur as a result of air currents formed when the wind hits the side of the building. The downdrafts can affect the efficiency of the heating system or result in backsmoking of the fireplace. This type of problem can usually be controlled by installing a concrete or stone cap about 8 inches above the top of the flue.

Older homes very often have unlined chimney flues. Although these chimneys might operate satisfactorily, they are a potential hazard. Over the years, the corrosive gases can have a deteriorating effect on the mortar joints. (See FIG. 3-5.) If an unlined chimney in your house is connected to a fireplace, and if you intend to use the fireplace, you should have a flue liner installed down the existing flue. This

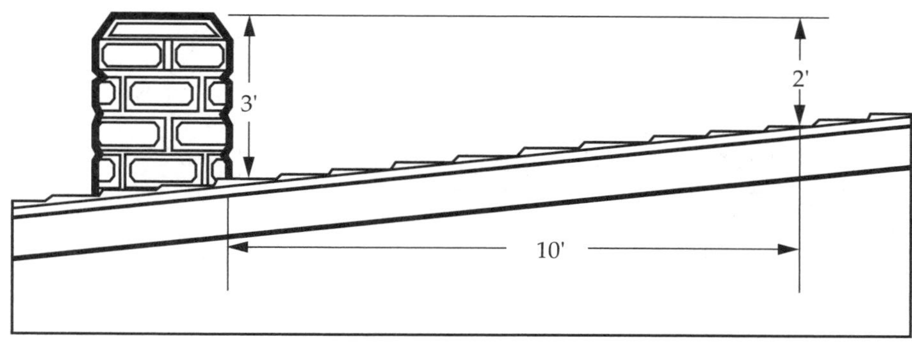

Fig. 3-4. Chimneys should extend at least 3 feet past the highest point where they pass through the roof, and 2 feet higher than any roof part within 10 feet measured horizontally.

Fig. 3-5. Deteriorated chimney with no flue liner. Chimney should be rebuilt and flue should be lined.

is an important fire safety measure and should be recorded on your worksheet.

In many homes when the flue damper in the firebox has deteriorated, it is replaced with a chimney-top damper. (See FIG. 3-6.) This spring-loaded damper is mounted over the flue opening at the top of the chimney. The device has a stainless-steel wire that runs from the damper down the flue into the opening of the fireplace. If you see a chimney-top damper during your inspection, don't forget to check its operation when inspecting the fireplace.

All masonry chimneys should be inspected for cracked, loose, chipped, deteriorating, and missing sections of brick and mortar joints. Some masonry chimneys have a stucco finish. In this case, look for cracked, chipped,

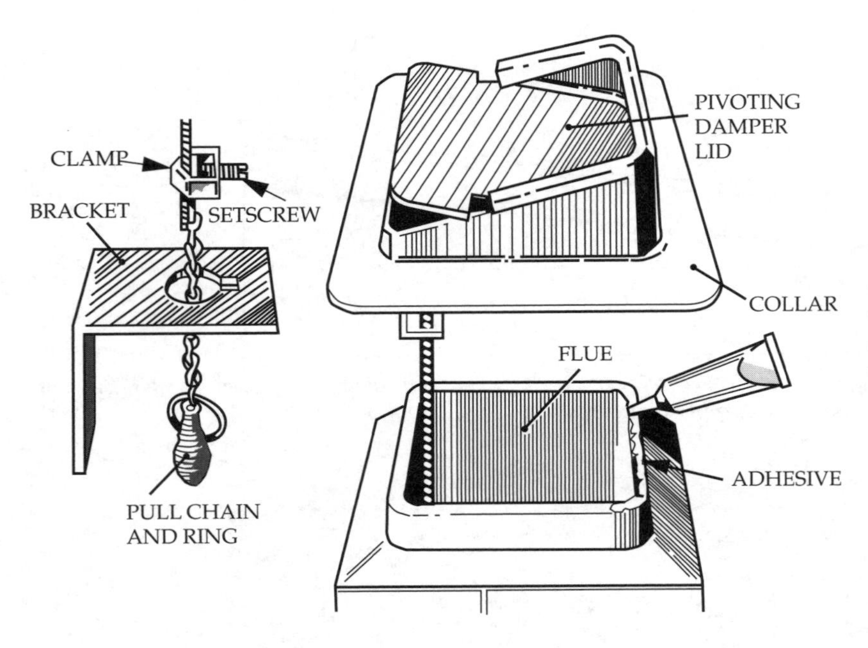

Fig. 3-6. *Spring-loaded chimney-top damper covers flue opening at top of chimney.*

and loose sections of stucco. Brick chimneys on older homes are sometimes covered with an asphalt-type coating, especially above the roofline. This technique is often used to prolong the life of the chimney when there are many cracked mortar joints and deteriorating bricks. It is considered makeshift. If you see any of the above items, they should be recorded on your worksheet.

Is there a *cricket* (also known as a *saddle*) behind the chimney? (See FIG. 3-7.) There should be one if the chimney is located along the slope of a roof and is more than 2 feet wide. The cricket prevents debris or snow and ice from piling up behind the chimney. This can cause rain or melting snow to back up under the shingles and leak into the house. The cricket also deflects water running down the roof around the chimney.

For the most part, very little maintenance is needed for prefabricated chimneys. Metal chimneys have a tendency to rust and should be checked for corrosion holes. If a metal chimney has no rain cover, note it on the worksheet; a cover is recommended.

Vent stacks

The vent stack is part of the plumbing system. Its purpose is to permit adequate circulation of air in all parts of the sanitary drainage system and to allow a means for sewer gas to vent harmlessly to the atmosphere. All homes should have at least one vent stack. (See discussion of air admittance valves in the Violations section of chapter 9.) The absence of a vent stack usually indicates that the plumbing system is not properly vented, almost always a violation of the plumbing code. As you inspect the roof, look for a pipe that projects through the roof and terminates about 8 inches above the roofline. In newer homes, the vent stack is usually fairly obvious. In older homes, however, the vent stack might be missing or at least not visible when looking at the roof. On occasion, I have found vent stacks that terminate in the attic. This condition is a violation of the

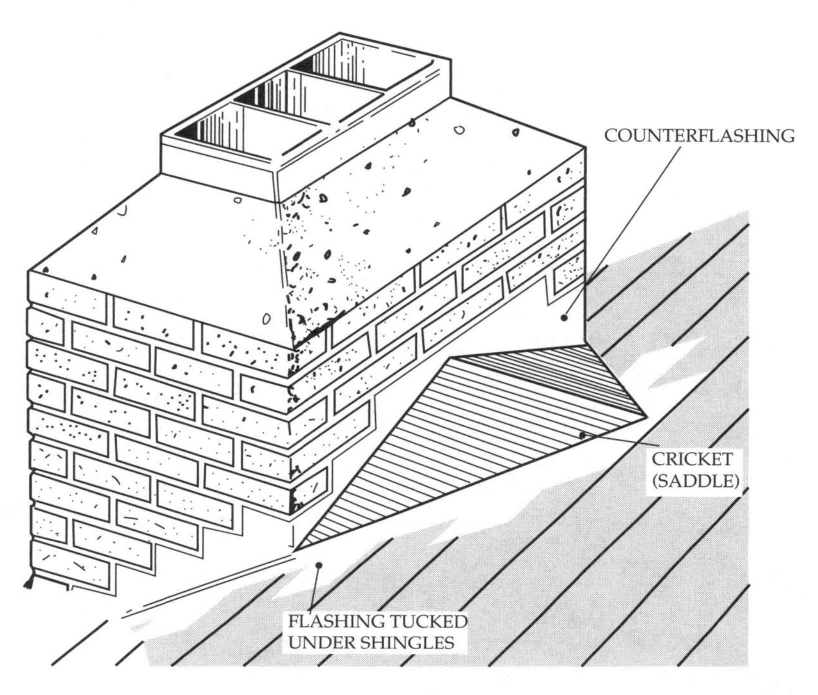

Fig. 3-7. Cricket (saddle) prevents debris, snow, and ice from accumulating behind the chimney, causing roof leaks.

plumbing code; the vent stack must extend above the roofline so that the escaping gases can discharge to the outside.

In some older homes, when kitchens or bathrooms are renovated, vent stacks are often run up along the outside of the building to a point above the roofline. (See FIG. 3-8.) If these homes are located in the northern part of the United States, this type of installation is undesirable. Because of the moisture in the escaping gas, during cold weather it could freeze over and eventually block the pipe. Sometimes you find vent stacks that terminate near a window. (See FIG. 3-9.) This installation is also undesirable because the discharging sewer gas might seep into the house when the window is open.

For vent stacks that terminate above the roof, very often you'll see a black ring at the base of the stack adjacent to the flashing. This black ring is asphalt cement that has been used to seal the joint. This joint is vulnerable to leakage and should be periodically checked. On one inspection, I found water dripping from a ceiling tile in a suspended ceiling located two levels below the roof. When I lifted the tile, water came cascading down. After the initial shock and after all the water had spilled away, it became obvious that the water was not caused by a leak in the above laundry room but had collected during the previous day's driving rain when water had leaked into the structure through the small crack between the vent stack and the flashing. This area should be periodically checked and sealed as needed.

Roof vents

These vents can be found on pitched and flat roofs and are available in both round and square hood styles. Normally, there are no problems with these types of vents. However, the joint between the vent and the roof is vulnerable to water leakage. Even though the joint might look okay from the roof side, it is

Fig. 3-8. Exterior-mounted plumbing vent stack.

best to check the vent openings from the attic periodically. If there is leakage, it will be noted by water stains on the roof deck in those areas. Water leakage around the joints can be easily corrected with asphalt cement or a suitable caulking.

Roof hatch

Access to a flat roof from the interior is provided through a roof hatch. Don't be surprised if you find a roof hatch on a house with a pitched roof. In some older and larger structures, they were installed as a means of easy access. Roof-hatch covers should be checked to determine whether they are operational; they should be. I have found that in about 30 percent of the cases, the hatch covers could not be opened. In all probability, during reroofing or maintenance, asphalt cement accidentally or intentionally sealed the hatch-cover frame. If this condition exists, it should be corrected. Usually, the hatch cover is constructed of wood and is covered with either sheet-metal or asphalt roll-roofing. Of particular concern is the integrity of the waterproofing cover. There should not be any cracked or open joints. The wood framing should not have any cracked or broken sections.

Skylights

Skylights are installed on a structure to provide daylighting and in some cases ventilation. The newer styles are prefabricated, with aluminum frames and plastic or glass panels or domes. Skylights can be found on pitched or flat roofs. These units should be checked for cracked or broken panes and signs of leakage. The leakage can be checked by inspecting the interior area below the skylight. If there are leaks around the skylight, you'll note water stains on the wall or finished ceiling in those areas. The skylights found on older flat-roof homes should be carefully checked for corroding frames and cracked and broken panes. Quite often the frames have corroded through and provide no support for the panes. This condition should be marked on your worksheet, because these skylights require complete rehabilitation.

TV antennas

Although roof-mounted TV antennas were quite common when the first edition of this book was written, for the most part they have been replaced by cable television or dish antennas. Nevertheless, they can still be found on many houses. Some TV antennas are strapped to the chimney for support. This means of

Fig. 3-9. Plumbing vent stack terminating near window. If the window is open, the discharging sewer gases can be blown into the house.

bracing the antenna is undesirable because repeated twisting action on the antenna from strong winds can cause stresses on the chimney that can in turn result in cracked mortar joints. When you see a TV antenna strapped to the chimney, look carefully at the joints. Sometimes the antenna is supported on the roof by means of guy wires anchored to the roof deck. When inspecting the roof, check the area around the anchors for deterioration of the roof covering. Sometimes the guy wires are strapped to vent stacks. This practice is undesirable because in a strong wind the movement of the antenna can cause the joint at the base of the vent stack to crack. I have also run into a situation where a homeowner inserted the TV antenna mast down into a vent stack to secure the antenna. If you see this situation, make a note of it so that you can remove the antenna after you take possession of the house. Finally, look for a ground wire that connects the antenna to a metal rod embedded in the ground. According to the National Electrical Code, this wire should not have any intervening splices or connections. The TV ground wire does not protect the house from lightning. Its purpose is to protect the television set in the event of a lightning surge.

Lightning protection

Although there are about 90 million lightning strikes each year in the United States, most homes don't have a lightning protection system. If a house has one, you will notice pointed metal rods projecting up above the high points of the building. These rods are connected to stranded cables, which are then connected to at least two grounded conductors. Of particular concern is whether the connections are properly bonded and if the system is properly grounded. Because of the destructive power of a lightning strike, the integrity of the system should be checked by a company specializing in lightning protection.

Gutters and downspouts

Gutters and downspouts are installed on a structure to control and direct rain runoff from the roof. The absence of gutters might result in water seepage into the basement or crawl space, rotting sections of wood trim, damage to foundation plantings, and the erosion of topsoil. Whether they are masonry-constructed or have long overhanging eaves, most residential structures that are not in the snowbelt would benefit from gutters. In the snowbelt, gutters are considered more an inconvenience than a help because the snow and ice often tear them from the supports and maintenance is constantly necessary. If you do not see gutters on the structure during your inspection, indicate their absence on your worksheet. There are basically two types of gutters, built-in and exterior-mounted.

Built-in gutters

These gutters are essentially extensions of the roof framing with waterways built into the roof surface over the edges. The gutter channel might be lined with asphalt roll-roofing or some other type of impermeable material. These channels require periodic maintenance, such as applications of an asphalt-type cement. Leakage through these channels can often be detected by water stains in the soffit below the leaks or by water stains in the interior. Leaks in this type of gutter often result in rotting sections of trim. If stains or rotting sections are noted, they should be indicated on the worksheet. Built-in gutters are seldom used on modern residential structures.

Exterior-mounted gutters

These gutters can be made from copper, galvanized iron, wood, plastic, or, most commonly, aluminum. Copper gutters are considered the top of the line. They are expensive, virtually corrosion-resistant, and have a projected life in excess of forty years. However, as these gutters age, they corrode and develop tiny holes in the bottom portion of the gutter channel. Depending on the gutters' height, the holes might not be visible to the naked eye from the ground. However, by standing directly below the gutters and looking straight up, the sky is often visible through any corrosion holes.

The joints on copper gutters are always soldered. In some parts of the country, the joints on galvanized gutters are also soldered rather than clipped. When the gutters are painted, you can tell the difference between them by using a magnet. The magnet will not stick to a copper gutter. Sometimes leaks develop around the soldered joints. For the most part, because of their cost, copper gutters and downspouts are no longer used when replacing faulty gutters and downspouts. A number of years ago I inspected a church in Ossining, New York, that had a wet basement. It turned out that most of the copper gutters and downspouts had been stolen from the structure by someone in search of a quick dollar because there was a copper shortage at that time.

Galvanized gutters have been used on many homes because of their low initial cost. However, they rust easily and require periodic maintenance such as patching corrosion holes and repainting. The inside portion of the gutter channel should also be painted. This area is often overlooked by the homeowner.

Aluminum gutters are quite popular because they do not have the corrosion problems of galvanized gutters. Older aluminum gutters on occasion leak around the seam, a condition requiring resealing. Leakage from seams can be noted by discoloration at the joint or some water stains or erosion on the area directly below the joint. Aluminum gutters can now be manufactured to almost any length, producing *seamless gutters*. Leaks in this type of gutter usually occur at corner joints or the joint around the downspout.

Wood gutters are usually made of Douglas fir or red cedar and have a tendency to crack and rot at the various end joints and seams along their length. (See FIG. 3-10.) The joints around the end sections, particularly those where the connection is made to the downspout, deteriorate more rapidly than other portions and should be checked for rot and cracking. Wood gutters should be painted every few years and the inside channel coated with an asphalt-type roof paint.

Fig. 3-10. Wood gutters. Note the cracked and rotting corner joint.

Plastic gutters, although relatively maintenance-free, have not received wide acceptance. They are found occasionally, but not necessarily, on homes with vinyl siding.

When inspecting the roof with binoculars, you should check the gutters to see if there are any loose support straps or spikes that should be resecured. (See FIG. 3-11.) In addition, an overall view will show whether any gutters are pitched incorrectly or are sagging and should be reset. Sagging is a condition occasionally found on those homes with slate roofs that do not have snow guards. All types of gutters have a tendency to become cluttered with leaves, twigs, seed pods, and mineral granules from roof shingles. They should be cleaned

***Fig. 3-11.** Loose gutter spike. Should be reset and resecured.*

Fig. 3-12. Gutters must be cleaned periodically to be effective. Note the weeds growing out of this gutter.

at least twice a year—once in the spring after the trees have bloomed and once in the fall after the leaves have fallen. Gutter screens are available to help prevent larger items, such as leaves and twigs, from cluttering the gutter channel. However, often a homeowner installs the screens and then forgets about cleaning the gutters. The gutters still require cleaning, although at less frequent intervals. Sometimes you can tell from the ground whether a gutter channel requires cleaning. Figure 3-12 shows weeds growing out of the gutter. Obviously, this gutter has not been cleaned for some time.

Fig. 3-13. Water stains on the exterior siding are the result of a missing downspout.

Downspouts

Downspouts are normally constructed of aluminum, copper, or galvanized iron. Copper and galvanized downspouts that have aged often have corrosion holes in the elbow sections. In some cases, the copper corrodes so that cracked sections are paper thin and can be stripped away easily. If you see this condition, it indicates that those sections should be replaced. Some downspouts have loose and open seams along their length that interfere with their effectiveness.

If you do not specifically look for the downspouts, you might not realize that some of them are missing. Figure 3-13 shows water stains on the asbestos-cement shingles, the result of a missing downspout. In this case, it is obvious that a downspout is missing; however, if the house is inspected during dry weather, you might not see any stains, and this item can be easily overlooked. Check the assembly of the downspout at the various joints. The lower portion of the downspout should be outside the upper portion. Otherwise, water will leak around the joint. Figure 3-14 shows an incorrect downspout assembly. On occasion some downspouts come loose from their connection to the gutters and should be resecured. Loose support straps around downspouts should be resecured.

The water from downspouts must go somewhere; it can be piped away underground to suitable drainage or discharged onto the ground at the base. Any water discharging from the base of the downspouts must be directed away from the structure. Otherwise, it can accumulate around the foundation and can eventually enter the lower level. To help deflect this water away from the structure, there should be an elbow at the base of the downspout and a splash plate below on the ground. Effluent discharging around the foundation should be indicated on your worksheet.

Sometimes the downspouts terminate directly into the ground. It is usually not possible during an inspection to determine whether the downspouts are connected to free-flowing drain tiles or to dry wells. Free-flowing tiles are more desirable; dry wells can become clogged (see FIG. 3-15) or will become

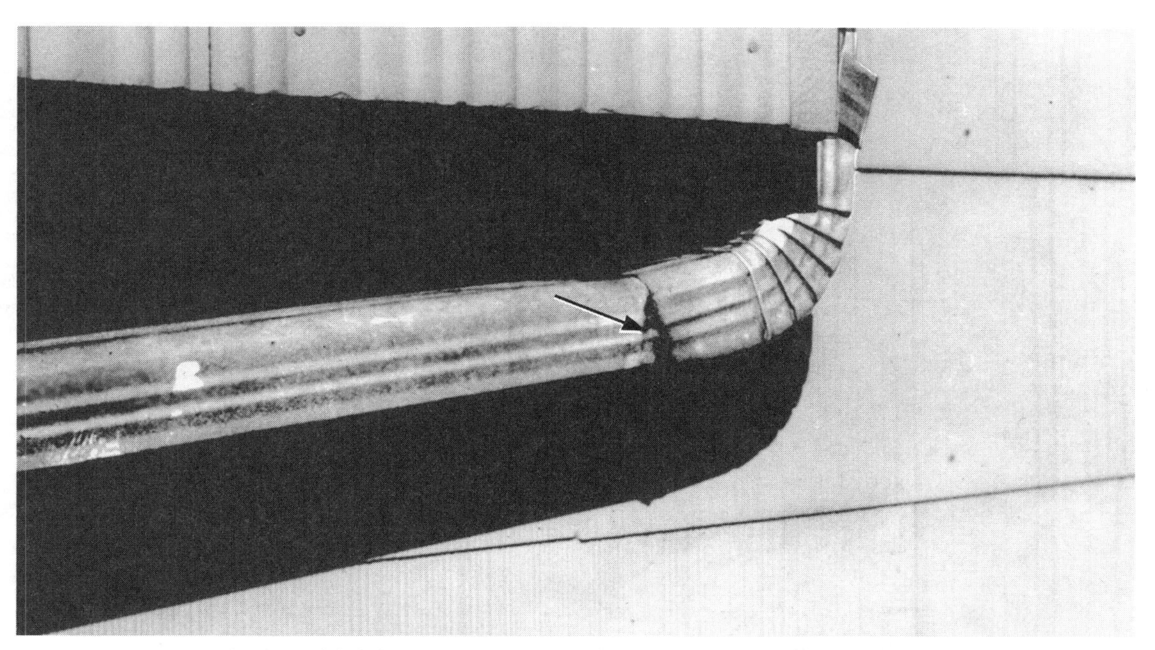

Fig. 3-14. *Incorrectly assembled downspout joint; the lower section is inside the upper section. It should be reversed; otherwise, water will leak out of the joint.*

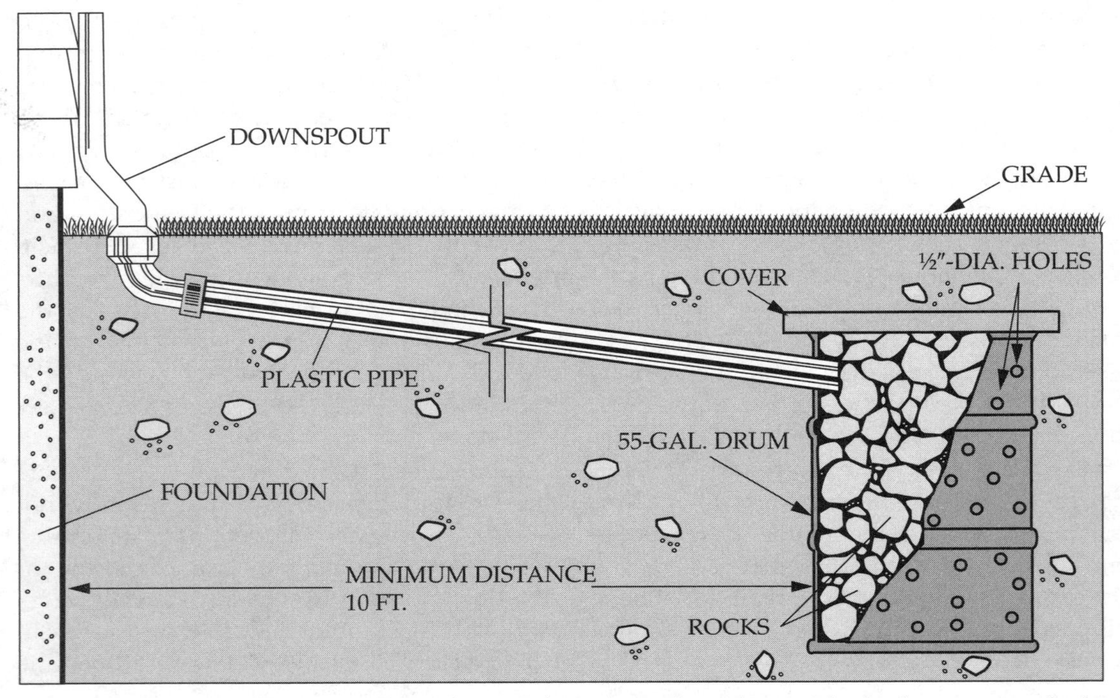

Fig. 3-15. *Over the years dry wells can become clogged, causing rainwater in the downspout to back up.*

ineffective if the level of the subsurface groundwater (water table) is high. Occasionally, some free-flowing tiles are visible at the street curb. If there are outlets at the curb, they must be kept clear.

Checkpoint summary

Chimneys

Masonry type (brick, stone, concrete block)

- ❍ Inspect for cracked, loose, chipped, eroding, or missing sections of masonry.
- ❍ Check mortar joints for cracked, loose, and deteriorating sections.
- ❍ For stucco-finished chimneys, cracked, chipped, missing, or loose sections of stucco?
- ❍ Chimney vertical?
- ❍ Open joints between the chimney and the sidewall?
- ❍ If the roof is flat, does the chimney extend 3 feet above the roofline?
- ❍ If the roof is pitched, does the chimney extend 2 feet above the roof ridge?
- ❍ If possible, check for cracked or missing sections of the chimney cap.
- ❍ If possible, check the chimney flashing for holes, tears, or loose sections. Check these vulnerable areas for leakage again during your attic inspection.
- ❍ If possible, check to see if the chimney flue is lined.
- ❍ Is there a chimney top damper?
- ❍ Is there a cricket (saddle) behind the chimney?

Metal type (prefabricated)

- ❍ Check for corrosion holes, rusting, or missing sections.
- ❍ Rain cover present?
- ❍ Note condition of flashing.

Vent stacks

- ❍ Plumbing vent stacks visible? (Their absence can be verified during attic inspection.)
- ❍ Black rings (asphalt cement) at the base joint of the vent stacks?
- ❍ Note any questionable roof joints and check further during attic inspection.
- ❍ Note vent stacks that
  —terminate near windows
  —run up an exterior side of the house (in northern climates)
  —have TV antennas strapped to them.

Roof vents, hatches, skylights, TV antennas, dish antennas, lightning protection

- ❍ Check all roof joints associated with vents, hatches, and skylights.
- ❍ Note questionable sections and verify tightness of joints during attic inspection.
- ❍ Does structure contain a roof hatch?
- ❍ Can you open it?
- ❍ Cracks or open joints in the cover?
- ❍ Check skylights for corroding frames and cracked or broken panes.
- ❍ Inspect ceiling area below for signs of leakage.
- ❍ Check TV antenna mast and guy wire connections to roof.
- ❍ Is antenna adequately grounded?
- ❍ Are lightning rods secure?

Gutters and downspouts

Exterior-mounted gutters

- ❍ Check for missing sections of gutters.
- ❍ Note type of material: copper, aluminum, galvanized iron, wood.
- ❍ Note gutter sections incorrectly pitched.
- ❍ Check metal gutters for corrosion holes, sagging sections, loose support straps or spikes, and leaking sections.
- ❍ Check wood gutters for cracked sections and areas of rot, particularly at connections and end sections.

Built-in gutters

- ❍ Check for areas with rotting trim.
- ❍ Check for signs of leakage.
- ❍ Note signs of seepage (stains) in soffit trim below gutters.
- ❍ Wherever possible, check the condition of the gutter channel.

Downspouts

- ❍ Note type of material: copper, aluminum, galvanized iron.
- ❍ Check for missing sections and improper joining.
- ❍ Inspect for loose straps, open seams, and corrosion holes at elbows.
- ❍ Do downspouts have elbows at base and extensions and splash plates (where required) to direct roof rain runoff away from the house?
- ❍ If downspouts terminate in the ground, try to find out whether they are connected to dry wells or to free-flowing outlets.

4

Paved areas around the structure

As you walk around the house on the second pass, inspect the paths, steps, patio, and driveway. The problems normally encountered with these items usually do not require immediate correction. Nevertheless, a tripping hazard might exist, cosmetic maintenance might be needed, or a condition might make the lower level vulnerable to water penetration. If you see problems in these areas, record them on your worksheet for early correction.

Sidewalks

Sidewalks are usually constructed of concrete, asphalt, or stone embedded in the ground. Not all communities have sidewalks. If a sidewalk is in front of or on the sides of the house, look for cracked or uneven sections because they are a potential tripping hazard. Even though the sidewalk might be on property owned by the town, its maintenance is usually the homeowner's responsibility. If someone should be hurt because of a cracked or uneven section, you would be vulnerable to a lawsuit. Very often uneven sections are caused by the roots of trees growing nearby. Correcting the condition would require chopping that section of root away and resetting the sidewalk. You should be aware of the fact that trees adjacent to the sidewalk are often the property of the local municipality, and prior to altering the roots, permission should be obtained.

Street-level/ driveway-level steps

Steps might lead from the sidewalk or driveway to the front path. These steps should be inspected for cracked, chipped, broken, or uneven sections. Look specifically for dimensional variations in the step risers. The riser (vertical distance) for the top step or the bottom step is often a different size than those for the other steps. This difference is a tripping hazard because it interrupts the natural rhythm of ascending and descending the steps. If there are more than two steps, there should be a handrail as a safety precaution.

Front and side paths

Two general types of paths are used with residential structures. One is a ribbon type, generally constructed of concrete or asphalt; the other is a sectional type, generally inlaid with material such as stones, bricks, sections of tree trunks, or precast concrete blocks. The sectional type normally requires periodic maintenance because of the tendency toward uneven settlement and weed growth between the sections. Sometimes the sections are loose and uneven and present a tripping hazard. Occasionally, the sections are set in mortar. Look for loose, cracked, and chipped mortar joints that require repointing. When inspecting the ribbon-type path, look for cracked, uneven, and broken sections. Very often damage occurs because the base below the path was not properly prepared during construction. Look particularly for settled sections that are sloping toward the house. These areas will direct rain runoff toward the house, so that water accumulates around the foundation. This water can enter the lower level. In this case, the path should be repositioned so that it slopes away from the structure or is completely rehabilitated.

While inspecting the paths, look for small, abrupt changes in the elevation. Occasionally, I have found a single step in the middle of a path. This step is a potential tripping hazard because it often goes unnoticed. A single step in a path, except at an entrance, should be avoided. If a slight elevation change is necessary and there is a step, it should be converted to a ramp, or shrubs should be planted at the step to call attention to the elevation change. An outdoor light should also be placed here. (See FIG. 4-1.) You might find a path partially blocked by overgrown shrubs; consequently it is no longer functional. In this case, if a path is needed, the shrubs should be pruned or the path repositioned.

Entry steps

Entry steps can be made of stone, concrete, brick, metal, or wood. As a safety precaution, when more than two steps are necessary, at least one handrail should be installed especially if the house is located in an area where the temperature drops below freezing.

Fig. 4-1. *A single step in the middle of a path is a potential hazard. The shrubs on both sides of this path call attention to the step. However, there should also be an outdoor light in the area.*

In those areas, the steps can be coated with a layer of ice after a freezing rain. (See FIG. 4-2.) Look specifically for differences in the vertical distance between the steps (risers). (See FIG. 4-3.) Any dimensional variations in the risers are potential tripping hazards. Some steps are designed so that the vertical distance between the treads is open (open risers). Although this might be aesthetically pleasing, it is a potential tripping hazard.

If the entry door opens onto the entry steps, a landing platform rather than a step tread is necessary at the doorway. The turnaround area of a single step tread is not considered adequate to operate a door safely. The platform should provide sufficient space to allow adequate standing room while opening the door, which swings over the landing. (See FIG. 4-4.)

When inspecting the steps, look specifically for cracked, broken, rotting, chipped, and loose sections. The treads should be level.

Fig. 4-3. *Note that the vertical distances between the steps vary. These steps are a potential tripping hazard. The bottom riser needs masonry repair.*

Fig. 4-2. *Handrail is needed for steps, especially when they could be covered with ice.*

Uneven sections are a tripping hazard. If the steps are masonry-constructed, look at the step foundation walls for cracked, broken, and chipped sections. Any wood *stringers* (the side portions of the steps that support the treads) should be resting on a concrete pad rather than on the soil. Check the base of wooden stringers for rot by probing the area with a screwdriver. If the screwdriver penetrates the wood easily, the stringer should be replaced. Metal handrails are often used with exterior steps. Very often, these handrails are corroding and have deep pockets of rust. In this case, the handrails should be scraped, primed, and repainted.

Patio

There are probably as many types and styles of patios as there are houses. The more common types are concrete slab, stone set in mortar, and brick or paving blocks set in the ground. In all cases, a particular concern is whether any tripping hazards can result

Fig. 4-4. *The turnaround area of the top step is too narrow to operate the door safely.*

from cracked, broken, or uneven sections. Some patios have a grid pattern consisting of wood embedded in the ground around sections of brick or concrete. This wood should be pressure-treated to protect it from rot. However, often the wood is not pressure-treated, so that after a few years it rots and requires replacement. Concrete-slab patios should have expansion joints and control joints for cracking. Very often these items are omitted, and the patio cracks in a random fashion that is aesthetically undesirable. Look at the patio for signs of uneven settlement. If a patio is adjacent to the house and has settled so that it is sloping toward the house, water can accumulate around the joint between the patio and the foundation wall. This water will eventually accumulate around the foundation and if the house has a basement or crawl space, it could enter those areas. This condition should be indicated on your worksheet for later correction and as a reminder for you to check the area of the basement adjacent to the patio.

Driveway

Driveways are normally constructed of asphalt, concrete, gravel, and, more recently, individual paving blocks. They also might just be a clearing with no covering. The latter is not particularly desirable from a cosmetic point of view and because of the ease with which ruts can develop. Gravel driveways on occasion also develop ruts and require periodic replacement of the gravel. Concrete and asphalt driveways should be inspected for cracked, broken, or settled areas that require rehabilitation. Extensive cracking, such as alligator cracks, is usually an indication of poor

Fig. 4-5. Extensively cracked and deteriorating driveway.

drainage of the subbase or poor construction. (See FIG. 4-5.) In this case, the cracks cannot be sealed effectively. The entire area should be patched. If a large portion of the driveway has this type of cracking, resurfacing is required.

If the base below a paving block driveway has not been prepared properly, sections will settle causing water to pond during a rain. (See FIG. 4-6.) Although the water will drain into the joints between the blocks, this item should be recorded on your worksheet so that it can be corrected at a later date.

Sometimes the driveway is on an incline, so that subsurface water flows below the driveway, undermining the subbase. In this case, prior to resurfacing, adequate provision for drainage of the subbase should be made.

When standing at the front of the driveway, look to see if it pitches directly down toward the house. If it does, the garage and any other portion of the lower level are vulnerable to flooding unless a large operational drain at the base of the driveway intercepts the surface water. Usually a channel drain that runs across the width of the driveway is needed to control such water runoff. The drain, however, should have a free-flowing outlet. If the drain is connected to a dry well, it might be ineffective during those months when there is a high water table; one corrective procedure would be

Fig. 4-6. Water ponding on a paving block driveway.

to connect the drain to a sump pit and remove the water by means of a sump pump.

When a house is located on a street that is inclined, the curb cut for the driveway should not be feathered into the street. Instead there should be a small ridge on the driveway at the joint between the driveway and the street. This ridge will prevent water that normally accumulates around the curb from flowing onto the driveway. This is normally not a problem in newer structures. However, in older houses the ridge tends to deteriorate, allowing water to overflow. This condition should be corrected.

In some northern areas, raised and inclined driveways have scratch marks that are caused by studded snow tires. This indicates that those driveways will be difficult to negotiate during some winter months; 100 pounds of salt and sand should be stored in the garage or near the driveway.

The minimum width for a driveway is 8 feet, although 9 feet is preferred. If the driveway is used both as an areaway for the car and a walkway in place of a path, it should be at least 10 feet. Anything less will make walking quite difficult when a car is parked in the driveway. Note whether the driveway discharges into a heavily trafficked street. If it does, an area in the driveway should function as a turnaround to allow a driver to head onto the street rather than back out onto the street. Also, look for overgrown trees and shrubs at the end of the driveway that might obstruct the driver's view when entering the street.

Checkpoint summary

Sidewalks, paths

- ❍ Inspect for cracked, missing, eroding, and uneven sections.
- ❍ Check for areas that might present a tripping hazard.
- ❍ Check slope of all paved paths adjacent to the house for improperly pitched sections.

Street-level/driveway-level steps

- ❍ Inspect for cracked, chipped, broken, or uneven sections.
- ❍ Check for missing handrails or railings.

Entry steps (masonry, wood)

- ❍ Inspect for cracked, broken, loose, or deteriorating sections.
- ❍ Note potential tripping hazards such as variations in riser heights and narrow treads.
- ❍ Check (probe) for rot in wood stringers, step treads, and handrails.
- ❍ Are wood stringers supported on concrete pads, or are they resting on the earth?
- ❍ Check for handrails.
- ❍ Inspect metal handrails for rusting, loose, and broken sections.
- ❍ Inspect wooden handrails for cracked, broken, loose, and rotting sections.
- ❍ Does entry door open onto a step or landing?

Patio

- ❍ Check for cracked, broken, eroding, and uneven areas.
- ❍ Inspect for uneven and settled sections adjacent to the house that can allow water to accumulate around the house foundation.
- ❍ Check for rot and insect damage to embedded wood sections.

Driveway

- ❍ Inspect for cracked, broken, eroding, or settled areas. Note extensively cracked and deteriorated areas for future rehabilitation.
- ❍ Check slope of driveway: level, raised, or inclined?
- ❍ For an inclined driveway, is there an adequate drain at the base?
- ❍ Does drain discharge to a drywall or to a free-flowing outlet?
- ❍ Is driveway width adequate (8 feet minimum, 9 feet preferred)?

5

Walls, windows, and doors

As you walk around the house inspecting the paved areas, you should also inspect the exterior walls, windows, trim, and doors. Before actually inspecting these items, however, look at the overall wall area for indications of past or current structural problems. Are the window and door lines square? Are any portions of the walls sagging or bulging? Are the walls and corner sections vertical? A problem condition should be recorded on your worksheet for further investigation. If by the end of the house inspection you cannot determine the cause, you should have the condition checked by a professional.

While inspecting the exterior walls, also note for further investigation any pipe or hood projecting through a wall or basement window. These items are usually not problem conditions, but it is useful to understand their function. The hood is often covering the discharge end of an exhaust fan or clothes-dryer duct, and the pipe may be the discharge line for a sump pump or condensate line from an air-conditioning system. If the pipe is connected to a sump pump, it might indicate a past or current water-seepage problem. Sump pumps and water seepage are discussed in detail in chapter 11.

Exterior walls

The exterior walls in most residential structures will be either wood frame or masonry, sometimes a combination of the two. The latter is commonly called a *veneer wall*. The exterior walls rest directly on the foundation and are *bearing* (load-supporting) walls. They support the roof, floors, and vertical loads imposed by other building components. The outer covering of the exterior walls provides protection from the weather and, if properly installed, minimizes the flow of air, moisture, and heat into or out of the structure.

When the walls are wood frame, the vertical framing members (*studs*) support all the vertical loads, and the outer finish covering (generally called *siding*) provides weather protection. Insulation is normally located in the spaces between the studs. In masonry walls, the masonry (clay tile, brick, stone, concrete block, etc.) provides both the structural support and the weather barrier. A masonry-veneer wall is a wood-frame wall with masonry used in place of the siding. Although the masonry in a veneer wall is not used for supporting the vertical loads, it does support its own weight.

Basically, a wood-frame exterior wall consists of 2-by-4-inch or 2-by-6-inch studs covered on the interior side by materials such as plaster, Sheetrock, plasterboard, wood, or hardboard panels (as described in chapter 10) and on the exterior side by sheathing, sheathing paper, and the finish siding. (See FIG. 5-1.) In some parts of the country where rot and termite activity are problems, metal studs are now being used in place of wood studs.

Sheathing is installed over the studs to provide bracing and minimize air infiltration. Depending on the type of sheathing, it can also be used to form a surface onto which the exterior finish can be nailed. Wood boards, plywood, fiberboard, and plasterboard are often used for wall sheathing. Fiberboard sheathing adds a small amount of insulation to the overall exterior wall; however, it should not be used as a nailing base for the direct attachment of the exterior siding. Rather, the siding should be nailed either to the studding through the sheathing or to wood nailing strips that have been attached to the sheathing. In many communities, when the exterior siding is capable of supplying adequate bracing and weather protection (as with exterior plywood panels), the sheathing is often omitted.

The purpose of the sheathing paper is to resist the direct entry of water during a driving

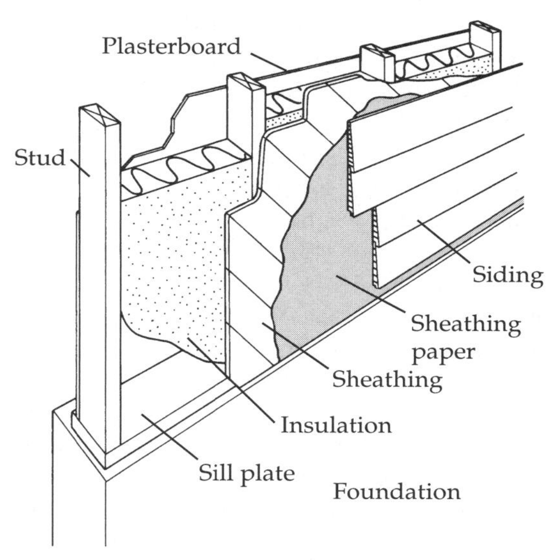

Fig. 5-1. *Components of a wood-framed exterior wall.*

rain. Sheathing paper (an asphalt-saturated felt) is water-resistant but not vapor-resistant and allows water vapor (which often builds up in the voids of the frame wall) to escape rather than condense and cause problems. Sheathing paper can also be very effective in reducing air infiltration.

Exterior siding

There are many types of exterior siding. If the siding is of good quality and has been maintained, it could last as long as the house. Normally, the type of maintenance required (other than painting) is the repair of cracked, broken, loose, rotting, or missing sections. When maintenance is needed, it is usually on a small section rather than the entire wall. However, some homeowners neglect the siding and allow it to deteriorate to a point where complete re-siding is necessary. Several clients have asked me about the condition of the old siding on a house that had been re-sided. Actually, the degree of deterioration of the old siding does not matter as long as the new siding is properly installed and provides the needed weather protection.

When inspecting the siding, pay particular attention to the sections that are facing south or southwesterly. These areas receive the maximum exposure to the sun and are more vulnerable to weather deterioration. The bottom of exterior siding should not be close to, or in contact with, the ground. Because of the dampness associated with the ground, the bottom of the siding should be at least 8 inches above the finished grade. Otherwise, the wood siding or the wood nailing boards for nonwood siding will be vulnerable to rot and termite infestation.

On occasion I have found vines growing up an exterior wall, in some cases reaching the roof. Although this might be aesthetically pleasing, the vines are undesirable. They can cover a multitude of problems and can cause problems. The vines can conceal termite shelter tubes (see chapter 8) or cracked portions of the siding. They can widen cracks, damage mortar joints, and loosen shingles. In addition, the dampness associated with the vines can promote rot and cause paint to blister and peel. If you find vines growing up an exterior wall, you should consider removing them.

Wood siding Wood siding is a broad classification that includes shingles, shakes, boards (applied vertically and horizontally), plywood panels, and hardboard. When inspecting wood siding, pay particular attention to exterior corner joints and the joints between the siding and window frames and doorframes. In addition, the area where the siding joins a dissimilar material (such as masonry or metal) is vulnerable to water penetration during a driving rain. These areas should be checked for weathertightness and rot.

On wood siding you might find dark, blotchy sections. This condition is generally caused by spores of fungi or mildew and often occurs in shaded areas. It is not a concern because it does not cause the siding to decay. However, it is unsightly. On painted surfaces, it can often be removed by washing with bleach and water. New paint on such areas should contain a mildew-inhibitor additive. On unpainted wood surfaces, this condition can usually be controlled by coating the siding with a penetrating preservative containing a mildewcide. You might also find brown and black discolorations on the siding (FIG. 5-2). This staining is caused by rusting of the nails used to secure the siding. The discolorations could have been avoided if aluminum or galvanized (rust-resistant) nails had been used. Eliminating this condition is somewhat difficult and usually not cost-justified.

Wood shingles/shakes Most shingles and shakes (hereafter referred to as shingles) used for exterior sidewall application are made from cedar or redwood. They are basically the same ones that are used for roofing (as described in chapter 2). However, their application is somewhat different. Because vertical walls present fewer water-penetration problems than roofs, the shingles on walls can be installed with a greater weather exposure than those on roofs. In addition, roofs generally have a three-ply layer of shingles, whereas exterior walls have only a two-ply layer. For full weather protection, the butt joints between the wall shingles for the upper ply should not line up with the vertical joints for the lower ply. Otherwise, water can penetrate the wall during a driving rain.

Since the shingles are decay-resistant, they do not have to be painted for weather protection. However, new shingles that replace deteriorated unpainted weathered shingles will not match the remaining shingles. Many shingles are painted to achieve a color decor. After a number of years, the paint begins to peel and flake. Consequently, once the shingles are painted, they will require periodic repainting for cosmetic purposes.

When inspecting the shingles, look for cracked, loose, chipped, rotting, and missing sections. Check for open joints around window and door trim. (See FIG. 5-3.) Open joints allow water penetration during a rain. Inspect for warped shingles; they generally will be on the sidewall with the southerly or southwesterly exposure. Also look at the quality of the shingles. You might find that the top portions of some are paper thin and can crack or chip very easily. These shingles are of a lower quality and are intended for use as an undercourse or installation with less shingle-length exposure. This type of shingled sidewall has a short projected life, and periodic repairs, with eventual residing, should be anticipated. Try to lift a few shingles gently. They should not lift up. If they do, it is an indication that they were improperly nailed. Shingles that are nailed directly to fiberboard sheathing rather than to wooden nailing strips (attached to the sheathing) will lift up under gentle pressure. Any of the above conditions should be recorded on your worksheet for future correction.

Wood boards Wood-board siding can be applied horizontally or vertically. Horizontal siding tends to make a house appear lower and longer; vertical siding tends to make a house appear taller (and is popular on one-story houses). The wood used for board siding should be free of knots. Otherwise, over a period of time, shrinkage can cause the knotty

Fig. 5-2. *Discolorations of wood shingles caused by rust stains from iron nails.*

Fig. 5-3. *Open shingle joints allow rainwater penetration.*

cores to drop out, leaving the siding with holes that are vulnerable to water penetration.

With the exception of redwood and cedar boards, most wood siding is painted for protection against weathering and decay. When inspecting the siding, look at the condition of the paint. Are there any bare, peeling, flaking, or blistered sections? If there are, paint touchup or repainting may be needed. Blistered and peeling paint is often caused by moisture in the painted wood, although it can also result from a poor paint job. Determination of the exact cause for blistering and peeling paint cannot be made during a single inspection. Nevertheless, this condition should not affect your thinking about the house. If it is caused by moisture in the wood, it can be corrected after you move in, and usually at minimal expense.

Wood-board siding should be inspected for cracked, loose, and rotting sections. In addition, look for loose or missing knots. All holes should be patched with a wood filler. In vertical siding, check the joints between the vertical sections for weathertightness. In both vertical and horizontal siding, pay particular attention to the outside corner joints. These joints are vulnerable to water penetration during a rain, and any open joints must be sealed.

Plywood panels Plywood panels are also used for siding. They are made from exterior-type plywood in which the veneer layers are bonded together with a waterproof glue. The exterior facing of the panel comes in a variety of surface textures and grooves. The panels are 4 feet wide by 8, 9, or 10 feet long. The thickness of the panel will depend on the depth of the grooves and will generally vary between ⅜ and ⅝ inch. Plywood panels are usually installed in a vertical position with the vertical joints over studs. This minimizes the number of horizontal joints, which is desirable since plywood panels are often applied directly to the studs rather than over sheathing. Because of their vulnerability to water penetration, any horizontal joints should be shiplapped (not visible to the inspector) or protected by metal flashing (visible). When inspecting plywood siding, look for loose, warped, cracked, delaminated, and rotting sections. Also check for open and nonweathertight joints. If the panel siding is painted, check the finish for peeling and flaking paint and blistered sections.

Hardboard siding Hardboard siding is made by bonding (under heat and pressure) wood fibers that have been ground almost to a pulp. The siding is dense and tough, and has a fairly good dimensional stability, although not as good as that of plywood. Hardboard, like plywood, is available in a wide range of textures and surface treatments. It is available in 4-foot-wide panels and 9- and 12-inch-wide planks. When inspecting hardboard siding, look for cracked, chipped, broken, deteriorated, and loose sections. Also check horizontal and vertical joints for weathertightness.

Aluminum siding Aluminum siding is often used in new construction and is often used when re-siding the exterior walls. The siding comes in planks that are either smooth or embossed with a wood-grain texture (to resemble painted wood boards), also as shingles and vertical panels. Aluminum siding is relatively maintenance-free. It is noncorrosive and termite-proof, and will not rot. The siding surface is generally covered with a baked enamel-paint finish that can stand up for many years before it fades, becomes dull, and needs a coat of paint. If the siding is scratched, bare aluminum is exposed. However, since the aluminum does not corrode, the scratch is only of cosmetic concern and can easily be corrected with touch-up paint. One problem with the siding is that it can be dented if struck hard enough—as with a baseball or stone thrown from a power mower. Many communities require that aluminum siding be grounded electrically as a precaution against electrical shock.

Aluminum siding is available with or without insulation backer boards. The insulation is generally a rigid foam (such as polystyrene) or fiberboard. Although the backer boards are only about ⅜ inch thick, they are quite effective as an insulator for a house that has no insulation in the exterior walls. Because of increasing energy costs, even a house with insulation in the exterior walls benefits from the additional insulation. The backer boards reduce heat loss in the winter and heat gain in the summer. They also increase the strength and rigidity of the siding. However, insulation-backed siding (and tight siding jobs) can cause moisture to accumulate within the exterior walls of houses that have no vapor barriers on the inside surface. (Insulation and vapor barriers are discussed in chapter 19.) You can usually tell whether the aluminum siding has an insulation backer board by pressing on it. If the siding is relatively firm, it has a backer board, but if it yields and bends under the pressure, it has no insulation board. Another method is to tap the siding. If it has no insulation, you will get a hollow sound.

When inspecting aluminum siding, look for loose, missing, and dented sections. Check the exterior joints for open sections and weathertightness. In those areas where electrical grounding of the siding is required, you should look for an electrical ground connection, a wire that runs from the siding to the inlet water pipe or a rod or pipe that has been driven into the ground. (See chapter 12.) You can find out whether an electrical ground connection is required by checking with the local municipal building department.

Vinyl siding Vinyl siding is very much like aluminum siding in size, shape, application, and appearance. Quite often close examination of the siding is needed to tell the difference between the two. The coloring in vinyl siding is embedded in the material and is the same throughout its thickness. Since the coloring in aluminum siding is only on the surface, an end cut or scratch in the aluminum reveals the silvery color of the bare metal. To tell whether the siding is vinyl or aluminum, look at an end cut or joint.

Vinyl siding is usually installed with an insulation backer board behind each sheet. In addition to insulation value, the board adds rigidity and strength. Vinyl siding normally does not dent from impact; it merely flexes and springs back to its original shape. However, during very cold weather, the siding becomes brittle, and a hard blow could crack or shatter it. When inspecting vinyl siding, check for cracked and broken sections and loose and sagging sections with open joints.

Vinyl siding expands and contracts as the temperature changes. When the siding is improperly nailed, this movement usually results in waviness and blisters in the vinyl panels. If you see this type of unevenness in the vinyl panels, record the fact on your worksheets for future correction.

Asbestos-cement shingles As with roofing shingles, asbestos-cement siding shingles were manufactured by combining asbestos fibers with Portland cement under pressure. These shingles are currently called *mineral-fiber shingles*. Although the shingles are no longer manufactured, they can be found on many homes in a variety of textures and colors. Since the shingles are unaffected by the weather and are immune to rot and termite activity, they require very little maintenance. However, the shingles are brittle and can be damaged and cracked by impact. The lower courses of the shingles are most vulnerable to damage. Usually damaged shingles are replaced rather than repaired. Depending on the weathered condition of the existing shingles, the new shingles may not match the older ones. (See FIG. 5-4.) When inspecting asbestos shingle siding, look for cracked, chipped, broken, loose, and missing shingles. Shingles that have slipped out of place were usually improperly nailed, or the nails were not rust-resistant and deteriorated. If the condition is caused by the latter, additional shingles will slip out of place in the future, and maintenance should be anticipated.

Asbestos-cement shingles were generally installed with sheathing-paper-backer strips behind the vertical shingle joints. These backer strips provided additional protection against water penetration. You might find sections of backer strips that have slipped out of place or are hanging loose between the shingles. Since the shingles were normally installed over sheathing paper, which is waterproof, replacing the loose backer strips is usually not necessary.

Fig. 5-4. Missing and patched (mismatched) asbestos-cement shingles.

Fiber-cement siding Fiber-cement siding is made from a mixture of cement, sand, cellulose fibers, and other components such as water, waxes, and resins. The material then goes through a multistep high-pressure curing process in which the mixture is subjected to highly pressurized steam that presses the material into a variety of siding patterns. There are several companies that manufacture fiber-cement siding; the most popular is James Hardie Building Products. Hardie fiber-cement siding products that you will find on houses are HardiePlank lap siding, shingles, and vertical siding.

Fiber-cement siding is popular because it is fire-resistant and will not rot, warp, or delaminate. It is also resistant to termites, and because it is normally manufactured into a much thicker and more durable product than its vinyl counterparts (according to Hardie, its siding is 5 times thicker than vinyl siding and weighs about 1.5 times as much as wood), it resists cracking and impact damage. The main concern with fiber-cement siding is not the product, but improper installation, which can result in skewed trim, exposed nails, and cracked, chipped, loose, or damaged sections. If you see any problems, record them on your worksheets for future correction.

Asphalt siding Asphalt siding is made by impregnating an organic felt material or glass-fiber mat with asphalt. The siding is available as shingles or as a roll. The exterior surface of the roll material is coated and embossed so that from a distance it looks like bricks. As the siding ages, it becomes dry and brittle, and cracks easily. For the most part, asphalt materials are no longer used for siding or residing residential structures. However, they can be found on existing buildings. When inspecting asphalt siding look for cracked, chipped, and eroded sections. Also check for open and lifting joints and loose, torn, and missing sections. If you find any areas in need of repair or replacement, record them on your worksheet.

Stucco A stucco finish on an exterior wall is basically a concrete sheet that has been built up in layers. It is usually made from a mixture of cement, lime, sand, and water. Stucco is weather-resistant, immune to termite and fungus attack, rigid, and durable—qualities that are desirable for an exterior wall finish. In addition, it can be applied to curved or irregularly shaped surfaces and to wood-frame walls that have been prepared with backing (sheathing) paper and metal lath. The backing paper is needed to resist water penetration through open joints or cracks that might develop in the stucco. The metal lath provides the means for bonding the stucco mix.

Stucco is generally applied in two coats on a masonry wall and three coats on a wood-frame wall. The minimum thickness for a three-coat wall is ⅞ inch; for a two-coat wall, ⅝ inch. The top layer of stucco is the finish coat and can be relatively smooth or have a rough texture. It can be prepared in a wide range of colors or painted.

Because stucco is a rigid material, cracks can develop as a result of a slight movement of the house. (See FIG. 5-5.) Movement occurs from foundation settlement and from wind forces. You generally find more cracks in stucco on a wood-frame house than on a solid masonry house. Shrinkage of the wood-framing members creates stresses in the stucco that often result in cracks. Once a crack develops, water can penetrate into the wall during a driving rain and can cause problems. In time, the portion of the metal lath around the crack rusts and deteriorates, and, depending on the condition of the backing paper, the wood framing and sheathing might rot. In addition, in cold climates, accumulated water behind the stucco can freeze, causing further deterioration as a result of frost action.

Fig. 5-5. *Cracked stucco wall. If wall was covered with vines, these cracks would be concealed.*

All cracks should be sealed. Hairline cracks and cracks up to $\frac{1}{16}$ inch generally can be sealed by coating them with a cement-based paint. The only difficulty is in matching the color of the wall. Larger cracks can be sealed by filling them with a mortar mix. Broken and loose sections of stucco must be rehabilitated by a skilled craftsman. When inspecting a stucco wall, look for chipped, cracked, loose, and broken sections. If you find areas in need of repair, record their location on your worksheet. Stucco does not require painting. But if it has been painted, check the condition of the paint. Once a stucco wall has been painted, periodic repainting will be required for cosmetic purposes, although at less frequent intervals than wood.

Synthetic stucco Synthetic stucco, commonly known as an exterior insulation and finish system (EIFS), is an exterior wall siding that consists of four primary components:

- Foam insulation boards attached to the exterior wall sheathing
- A base coat that is applied to the insulation board
- A fiberglass reinforcing mesh embedded in the base coat
- A finish coat applied over the fiber glass mesh

Although the EIFS may resemble stucco in appearance, the two siding systems are quite different, especially with regard to controlling water intrusion. The traditional stucco system anticipates eventual water penetration through open joints and cracks. To prevent rot and deterioration of the wood framing and sheathing in the exterior walls, it uses a housewrap or building paper behind the stucco surface to carry any water that accumulates in that area down and out the bottom of the wall.

The EIFS, on the other hand, was not designed for water intrusion. It was considered a surface barrier system because it resisted water penetration at its outer surface. It was assumed that moisture would not penetrate the surface and reach the wall sheathing and/or framing. The system was not originally intended to drain water that got behind the EIFS cladding. In practice, however, water did penetrate the wall, not through the surface but through jambs and sills of window frames, and at the joints between the exterior walls and door, window, deck, and roof intersections. Water that penetrated the wall could not easily escape. It was trapped between the EIFS cladding and the sheathing.

Over a period of time the water was absorbed by the sheathing and framing, which increased their moisture content to a level above saturation, causing rot and structural damage. Depending on weather conditions and the quality of construction, significant damage due to moisture intrusion could occur. Damage from water intrusion has been found in the exterior walls of houses that are only 3 to 5 years old.

The industry, recognizing the problems that resulted because of the lack of drainage in the wall, modified the system to include the installation of drainage channels and building paper between the foam insulation and the sheathing. In appearance the new system is similar to the old one, and with a visual inspection you cannot tell the difference between the drainable and nondrainable EIFS. Not only is the location of water entry often difficult to see, but moisture damage to the sheathing and framing behind the exterior wall cladding cannot be readily detected by visual inspection.

If the exterior wall cladding is an EIFS, you can do a preliminary inspection by checking for cracked and open joints at the interfaces between the EIFS and dissimilar materials such as windows, doors, and wall penetrations. If any are found that require caulking, record their location on your worksheet. Also, if you have a noninvasive moisture meter, check those areas for water intrusion. However, because of the nature of the potential problems and the high cost for correction, it is recommended that you have the exterior walls inspected and evaluated by a professional.

Veneer wall A veneer wall is a wood-frame wall with an attached masonry facing. Unlike exterior siding, which is held in position by being fastened to the sheathing or studs, the masonry rests on top of the foundation wall and supports its own weight. It is attached to the wood backing by corrosion-resistant metal ties. The ties are considered the weakest point in this type of construction. If the ties have deteriorated or are not properly attached, the masonry facing can pull away from the wood frame.

The masonry—usually clay brick, concrete brick, or split stone—is normally positioned so that there is a 1-inch air space between the veneer wall and the wood backing. (See FIG. 5-6.) Small holes (*weep holes*) are usually installed at the base of a veneer wall. These holes allow water that might accumulate in the air space to drain. When the masonry facing is brick, the weep holes are generally formed by eliminating the mortar in a vertical joint.

Some bricks absorb more water than others. Bricks with high absorption should not be used for the exterior facing of walls, especially in colder climates. Unfortunately, they are used occasionally. Alternate freezing and thawing of bricks that have absorbed water will cause them to deteriorate. Depending on how much water has been absorbed, the interior walls might become damp, a condition that can usually be detected during the interior inspection. This condition can often be controlled by coating the bricks with a silicone sealant.

When inspecting a veneer wall, look for chipped, cracked, loose, deteriorating, and missing bricks or stones. In addition, check for cracked, chipped, and deteriorating mortar joints. Pay particular attention to the mortar joints. Occasionally, because of excessive shrinkage of the wood framing or slight foundation settlement, you might find large open cracks, especially around window frames and doorframes. Also look for loose and bulging sections of veneer wall. If you find any of the above items, record their location on your worksheet for later correction.

Masonry wall

Unlike a wood-frame wall where the structural support and weather barrier are provided by two separate components, studs and siding, the masonry units in a masonry wall (clay tile, brick, stone, or concrete block) provide both the support and the weather protection. Because of the low thermal resistance of masonry, a masonry wall allows greater heat loss than a wood-frame wall.

To reduce the heat loss, insulation can be added by applying a rigid foam insulation board to the interior side. In addition to providing insulation, the board can also be

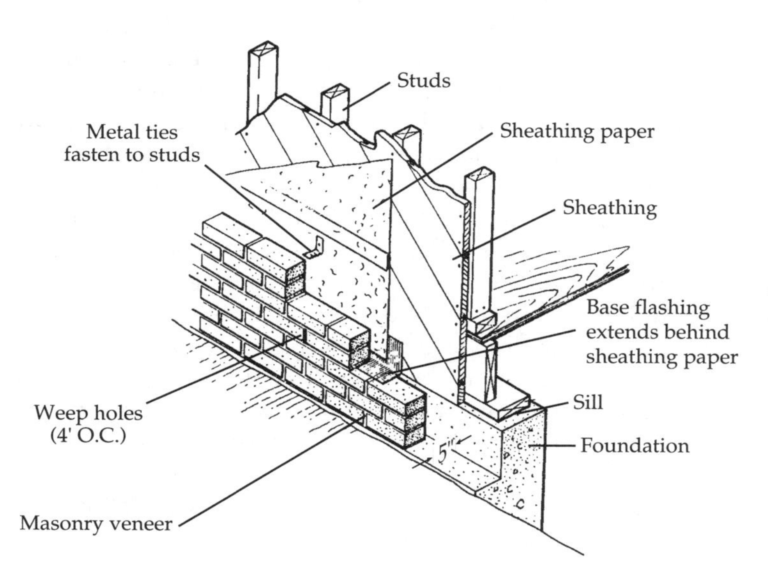

Fig. 5-6. *Components of a brick-veneered exterior wall.*

used as a base for plastering. Another approach is to apply furring strips to the inside wall; the furring strips create an airspace into which insulation can be placed prior to installing the finishing wall panel. In some cases, however, the interior side of the masonry wall is left completely exposed and serves as a decorative element or a base for direct plastering. This is quite wasteful from an energy-conservation point of view.

Because of the rigidity of masonry walls, differential movement within the wall might cause serious cracking. Wall movement might be the result of unequal foundation settlement, or expansion and contraction from temperature and humidity changes. Many cracks are not of structural concern, although they should be sealed to eliminate the possibility of water penetration. If you have any doubt about the severity of a crack, have the condition checked by a professional.

A common problem with masonry walls is *efflorescence* on the exterior surface. Efflorescence is a deposit of soluble salts that were originally within the masonry, usually brought to the surface by water in the wall. When the water evaporates, the salts are deposited on the surface. Efflorescence generally can be removed by scrubbing with a stiff brush or washing with a dilute solution of muriatic acid. However, if the condition is a recurring problem, it is an indication that water is penetrating the wall through cracks or faulty joints or flashing.

When inspecting a masonry wall, pay particular attention to the joints around window and doorframes. All joints should be weathertight. Are there any cracks around the corners of window or door openings? These are areas of high stress concentration and are vulnerable to cracking. Cracked and chipped mortar joints and deteriorated masonry should be indicated on your worksheet for later repair. If you notice bulging sections in the exterior walls or large cracked sections, have the condition checked professionally, since it might indicate structural problems.

Trim

All portions of the exterior finish, other than the wall covering, are generally classified as exterior trim. This includes the moldings and sills around windows and doors, fascia boards, soffits, louvers, shutters, and decorative columns. (See FIG. 5-7.) Trim does not serve a structural function. It is used as finishing around openings and to protect joints, edges, and ends. Most exterior trim is made of wood or wood products, although aluminum and vinyl trim have become quite popular. Many older, traditionally designed homes have decorative sheet-metal cornices, which are considered part of the trim. The problem with sheet-metal trim is that if it is not maintained and kept adequately painted, it will rust and deteriorate.

Wood trim that is exposed to the weather should be decay-resistant so that it does not rot. (See the section on rot in chapter 8.) Some types of preformed trim are factory-treated with a water-repellent preservative to make them water- and decay-resistant. When the trim is cut to size during construction, the ends or miter joints must be treated to make them water-resistant. All too often they are not treated, and the joints, which readily absorb water, begin to rot. When inspecting wood trim, pay particular attention to the joints that are vulnerable to decay. A house with a wide roof overhang at the eaves and gables provides greater weather protection of the sidewalls and trim than one with no roof projection beyond the walls. All nontreated wood continually exposed to moisture is prone to decay. The trim around the edge of the roof is particularly vulnerable. Although the Asphalt Roofing Manufacturers Association recommends the installation of a metal drip edge along the eaves of a roof deck,

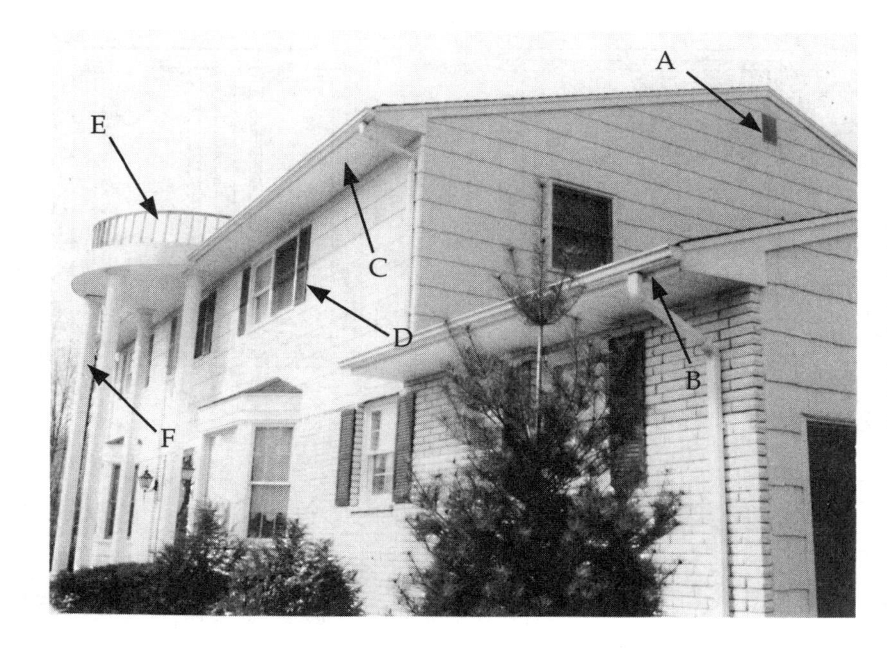

Fig. 5-7. Exterior trim on a house: A—gable louvers, B—fascia, C— soffit, D—shutters, E— widow's walk balustrade, F—decorative columns.

in practice it is often omitted. The drip edge is designed to allow water runoff to drip free of the underlying trim. Without it, water tends to curl back under the shingles, wetting the edge of the roof sheathing and trim.

Wood trim should be inspected for cracked, loose, missing, and rotting sections. If the trim is painted, is the paint peeling and flaking in sections? Does the trim need repainting for weather protection? In older Tudor-style houses with timbers embedded in the stucco siding, inspect the timbers at the stucco joints for decay, especially if the joint is horizontal. Over the years, the joints tend to open slightly, allowing water to penetrate. With nonwood trim, check for loose, missing, and deteriorated sections.

Windows

The windows should be checked during the exterior and the interior inspections. The overall condition of the windows should be checked during the exterior inspection; the operation of the windows should be checked during the interior inspection. (See the section on windows in chapter 10.)

Many types of windows are used in residential structures. The most common types, as shown in FIG. 5-8, are double-hung, horizontal sliding, casement, awning, jalousie, and fixed-pane.

The *double-hung window* is the most common window unit in older and newer homes. It consists of upper and lower sashes that slide vertically past each other. The sashes are held in a fixed position within the window frame by a friction fit, counterweights, or spring balances. In some windows, the sashes are removable for ease of maintenance such as cleaning or painting. One variation of the double-hung window is the single-hung window; the upper sash is fixed, and the lower sash is movable.

The sashes in a *horizontal sliding window* slide horizontally on separate tracks. The most

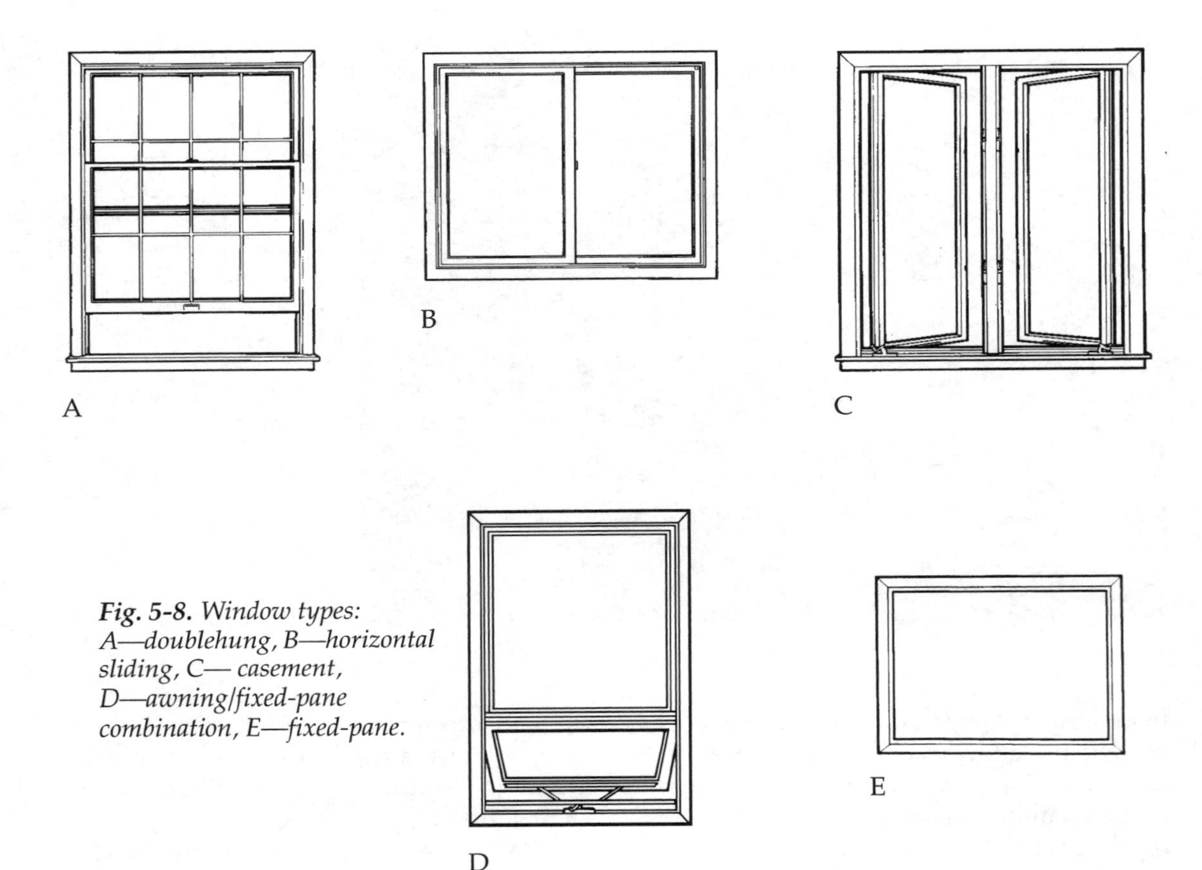

Fig. 5-8. Window types: A—doublehung, B—horizontal sliding, C— casement, D—awning/fixed-pane combination, E—fixed-pane.

common design consists of two sashes, both of which are movable. However, sometimes one sash is fixed. In most of these windows, the sash can be removed for cleaning.

Casement windows consist of two or more sashes hinged at the side and mounted so that they swing outward. The sashes are opened and closed by a cranking mechanism, a push bar mounted on the frame, or a handle fastened to the sash. Because the sash opens outward, a storm sash or screen must be attached to the inside of the window.

Awning windows have one or more sashes hinged at the top and mounted so that they swing out at the bottom. They are opened by push bars or cranking mechanisms similar to those on casement windows. As with casement windows, screens and storm sashes are mounted on the inside.

Jalousie windows are basically adjustable louvers. The louvers are glass slats (several inches wide) held by an aluminum frame at each end. The frames are interconnected by levers so that the slats open and close in unison, like venetian blinds. Jalousie windows are crank-operated and provide good ventilation. However, because of the many glass slats, they are troublesome to wash and are not weathertight. Even with a storm sash, cold-air leakage occurs around the windows. In northern climates, jalousies are usually limited to use on porches and breezeways.

Unlike the preceding windows, which are movable and provide ventilation, *fixed-pane windows* are stationary and are used only to provide daylight and outdoor views. They can be used alone or in combination with sliding, double-hung, or swinging windows to achieve a custom design.

With the exception of the jalousie windows, available only in aluminum, all of the windows are available in wood, metal (steel or aluminum), and vinyl-clad frames. If you are not sure whether the metal window frames are steel or aluminum, you can always tell the difference by using a magnet. A magnet sticks to a steel frame but not to an aluminum one.

Metal frames get quite cold during the winter months. Consequently, some of the water vapor in the air inside the house tends to condense on the frames. To reduce this condensation, several manufacturers produce metal-framed windows with a thermal barrier that prevents the outside frame from touching the inside frame. This substantially reduces the condensation but does not eliminate it. Since wood has a greater thermal resistance than metal, wood window frames normally do not get cold enough for condensation to form on their inside surface.

Although wood windows provide a better insulation than metal windows, they do have a tendency to swell or shrink with changes in moisture. A wood sash that absorbs moisture will expand and bind in the frame, so that it does not operate freely. Wood windows should be treated (by the manufacturer) with a water-repellent preservative to resist decay and moisture absorption. In the past, not all windows were so treated; thus rotting sections and binding sashes are occasionally found.

The glass used in windows must be of sufficiently high quality to minimize distortion. Years ago, because of the difficulty of manufacturing large distortion-free panes, window glass was available only in small sheets. Consequently, to fill a large opening such as a window sash, small panes of glass were used and held in position by framing strips called *muntins*. Today, even though large windowpanes are available, muntins are still used to create a special architectural effect. Some window manufacturers provide preassembled wood or plastic dividers that simulate muntins. The dividers merely overlay the large windowpane and snap in and out of the sash. Some companies also install muntins between the panes of double-glazed windows during the manufacturing process.

Windowpanes are also available with single and double glazing. A double-glazed window, a thermal (or insulated) pane, reduces heat loss or gain through an equivalently sized single-pane window by about 50 percent. It also reduces condensation on the inside surface. With some windows, triple glazing is also available. See the section on windows in chapter 10 for a discussion of insulated windows and associated problems.

Inspection

When inspecting windows, look for cracked, broken, and missing panes. Check the joints between the sash and the window frames to see whether they are filled with paint. If they are, the windows might not open, and minor maintenance will be needed. The condition of the joints between the glass pane and the sash should be inspected. Are any panes loose? Most windowpanes are secured to the sash by bedding and sealing with a glazing compound such as putty. If putty is used, are any sections cracked, loose, chipped, or missing? Some panes are secured to the sash by wood strips (trim). Check the strips for cracked, loose, and missing sections. Note any of the above items on your worksheet for later correction.

Look at the overall condition of the window frames and sash. Windows exposed to the elements are vulnerable to weathering deterioration. Are any wood sections cracked and rotted? Steel windows can rust. Depending on the extent of the deterioration, some windows require simple repair; others, replacement. All the windows generally are not visible or accessible during your exterior inspection, especially the windows on upper levels. The exteriors of those windows, and their operation, should be checked during your interior inspection.

Exterior doors

The two basic types of exterior wood doors are flush doors and stile-and-rail doors. Flush doors are made by bonding face panels to solid or hollow cores. Stile-and-rail doors (also referred to as *panel* doors) are solid doors that consist of vertical and horizontal members (called *stiles* and *rails*, respectively) that enclose wood or glass inserts. (See FIG. 5-9.)

Fig. 5-9. *Exterior doors.*

Most *flush doors* have hardwood veneer face panels, although hardboard and softwood panels are also available. Some doors are made with cutouts for windows or louvers. Flush doors are used as interior and exterior doors. As exterior doors, they should be made with waterproof adhesives rather than water-resistant adhesives. You can often tell whether the proper adhesive was used by looking at the top edge. In many quality doors, you will find a small red plastic plug in the edge. This indicates that the door was bonded with a waterproof adhesive that is suitable for exterior use. When an improper adhesive is used, the exterior face panel eventually begins to delaminate and peel.

A solid-core flush door provides greater heat and sound insulation, fire resistance, and dimensional stability than a hollow-core door. The solid core of a flush door might be made of wood blocks or a composition material that has been formed into a rigid slab. A hollow core is generally made of wood or wood derivatives (cardboard) that have been formed into a honeycomb or parallel strips. When resistance to heat, sound, and fire are not important factors, the hollow-core flush door is sometimes used as an exterior door. This is not considered quality construction. When used on the exterior, the door should be treated with a water-repellent preservative and bonded with a waterproof adhesive. Unfortunately, this is not always the case.

One variation of the flush door is the metal-clad insulated entrance door. This door is available in a selection of surface styles and is becoming more popular. Basically the door consists of metal face panels with an insulating core. Some doors are provided with a thermal break to separate the interior parts of the frame and door panel from the exterior parts, thereby minimizing condensation during the winter months. Because of the insulating characteristics of this type of door, the need for a storm door is essentially eliminated.

Although *stile-and-rail doors* are not used as commonly as flush doors, they are available in a greater variety of designs. In addition to wood, the stile-and-rail doors are available in steel and fiberglass. Because of the number of panels and joints, the stile-and-rail door is not as effective an insulator as a flush door. Also, one or more panels in a wood door might crack as a result of shrinkage. If you stand inside of the house looking at the door, cracked sections are very noticeable. Daylight is visible through the cracks.

Some exterior doors are not considered secure because of the location of the glass panes relative to the door lock. If the door lock can be reached by breaking a glass pane in the door or in side panels, an auxiliary lock is recommended. This lock should be positioned where it is not accessible from the outside. Since you normally do not know whether you were given all the keys to the exterior door locks, it is recommended that after you take possession of the house, you replace all of the door locks or at least have the locks rekeyed.

The condition of the exterior doors should be checked during your exterior inspection, and the operation of the doors should be checked during your interior inspection. Outside, inspect a wood door for cracked, chipped, and delaminating sections. Inside, check the doors for cracks (visible daylight) and ease of operation. Check a steel door for dents and scratches, which will rust if not painted. A fiberglass door can crack under severe impact and should be checked. Does the door open and close easily, or does it bind? Also check for weatherstripping around the exterior joints. Weatherstripping is desirable, since it minimizes air infiltration. If you find any problems with the doors, record them on your worksheet.

Storm windows, screens, and storm doors

Storm windows

Windows are a major source of heat loss. On a per-square-foot basis, more heat is lost through windows than any other area. In fact, the heat lost through a single-pane window is approximately fourteen times greater than that lost through a well-insulated wall of comparable size. However, with storm windows, the heat loss can be reduced by about 50 percent. There are three basic types of storm windows: storm sashes, storm panels, and combination units.

A *storm sash* is a removable sash, usually made of wood, containing a fixed-pane window. The storm sash fits over the window on the outside or inside, depending on how the window opens. Storm sashes are most commonly found on the outside of older double-hung windows. They are not desirable for year-round use because they cannot be opened to admit breezes during the warmer months. Storm windows are generally taken off in early spring and reinstalled in late fall, a task that can be somewhat awkward and time-consuming.

A *storm panel* looks like a storm sash, but it is usually mounted in a narrow metal frame and is attached to the movable window sash rather than being fitted over the entire window opening. Since storm panels are attached directly to the movable sash, they do not interfere with the operation of the window and need not be removed during the warmer months.

Combination units refer to storm and screen sashes combined in a single frame. The unit is mounted over the outside of the window and is therefore effective in reducing air infiltration around the window joints in addition to reducing heat loss. Combination storm and screen windows are available in two- and three-track units. With two-track units, the outside track contains the storm sash in the upper half and a screen sash in the lower half. With the screen in position, the upper storm sash cannot move. The inside track contains the lower storm sash, which can slide up and down. In a triple-track unit, there is a separate track for each of the two storm panes and for the screen. Combination storm and screen units are generally found on double-hung and horizontal sliding windows. Since these units do not interfere with the operation of the movable sash and can also be opened to provide ventilation, they are not normally removed once installed.

Combination units are available in aluminum, steel, or solid vinyl. The aluminum frame with a baked enamel finish looks like a vinyl frame. You can tell the difference by looking at an edge or joint. The aluminum edge will have a silvery color; the vinyl edge will be the same color as the frame. Over the years, many aluminum frames with a mill finish (plain aluminum) show the effects of weathering, such as pitting, corrosion, and a degraded appearance. An anodized or baked enamel finish will offer greater protection against weathering. Combination storm and screen windows made of steel have a tendency to rust and require periodic painting.

Screens

Most window screens for residential structures are either mounted on a wood sash or rimmed with a metal or plastic frame. The wood-framed screen is usually used in conjunction with a storm sash; the metal- or plastic-framed screen is used in the combination storm and screen unit. Metal- or plastic-framed screens are also used for casement and awning-type windows. However, for these windows, the screen must be equipped with panels that provide access to the cranks or push bars.

Another type of screen (not very common) is the roll-up screen. This screen is mounted on the inside of the window and is similar in operation to a roll-up shade. When the screen is not being used, it can be rolled up and hidden from view. The sides of the screen move in metal tracks to prevent insects from flying in around the edges. As the screen ages, the joint between the screen and the track tends to open and become less effective.

Storm doors

In colder climates, unless the exterior doors are insulated, they are often used with storm doors to reduce heat loss and cold-air infiltration around the joints. Storm doors are generally lightly constructed wood or metal stile-and-rail-type–doors with a glass-panel insert. On many of these doors, the glass insert is interchangeable with a screen panel so that they can function as both storm and screen doors. Because storm doors are constantly being opened, they are not as effective as storm windows in reducing heat loss. Nevertheless, they are effective from an overall energy-conservation point of view.

Inspection

When inspecting the house, look for storm windows. If you do not see any or the inspection is being performed during warmer weather when storm sashes are normally not installed, ask the owner whether the house has storm sashes for all the fixed and movable windows. If no storm windows or only a few are available, record the fact on your worksheet. Installing a complete set of storm windows can be quite costly.

Storm windows should be inspected for cracked, broken, and missing panes. When the storm sash is wood, check for cracked, broken, and rotting sections. On combination units, look specifically at the corner joints. These joints should be tight so that no cold air leaks into the unit. Check the overall condition of the frames. Are any sections loose, broken, rusted, or corroded? Are any of the panes loose in the sash? Look for torn sections and holes. In metal- and plastic-rimmed screens, the screening is normally held in position by a spline that has been forced into a groove around the frame. Periodically, I find splines hanging loose within the frame of the combination unit. Loose splines are an indication that the joints in the associated screens must be resecured.

Storm doors should be checked for ease of operation and overall condition. Are sections cracked, broken, loose, rotted, or corroded? Is the glass panel loose, cracked, broken, or missing? If you find any problems with the storm windows, screens, or storm doors, record them on your worksheet as a reminder for later correction.

Caulking

As you walk around the house inspecting the walls, windows, trim, and doors, look for cracked and open joints. All exterior joints should be caulked (sealed) so that they are watertight and airtight. If they are not adequately caulked, wind-driven rain can enter and cause wood members to rot, metal ties to rust, and masonry sections to crack and chip. In addition, cold air can infiltrate the house, resulting in higher heating costs. A vulnerable joint for cracking is one that joins two dissimilar materials; for example, the joint between a brick facing and a non-masonry sidewall. Dissimilar materials usually have different expansion and contraction characteristics, often causing the joints between them to crack and open. These joints should be sealed with a non-shrinking flexible caulking compound.

There are several types of caulking compounds. The four most popular types are oil base, acrylic latex base, butyl-rubber base, and silicone base. *Oil base* caulking compounds are the cheapest and will readily bond to most surfaces—wood, masonry, and metal. However, they are not very durable; they tend to dry and crack after a short time. Joints sealed with this type of caulk require periodic inspection and maintenance. *Acrylic latex* caulking compounds are medium-priced, durable, and flexible, and should last for many years. *Butyl-rubber* caulks are medium-priced, durable, and paintable; however, they exhibit high shrinkage, a characteristic that is acceptable when caulking narrow cracks and inside corners. *Silicone-base* caulks are the most expensive. They are very pliable and are good for sealing joints subject to movement. No one caulking compound is ideally suited for every application. However, based on cost, durability, and ease of application, acrylic-latex caulks are considered by many to be the best choice.

When inspecting the exterior joints, check the condition of the caulking. Look for cracked, chipped, crumbly, and missing caulking compound. The location of joints that need recaulking should be recorded on your worksheet. Recaulking is a relatively simple task and can be done after you move into the house.

Checkpoint summary

General considerations

- ❍ Inspect exterior walls for sagging, bulging sections, and for corners that are not vertical.
- ❍ Check for window frames and door frames that are not square.
- ❍ Structural problems whose cause cannot be determined should be evaluated by a professional.
- ❍ Note wall locations that have pipe or hood projections.
- ❍ Determine their usage (i.e., sump-pump discharge, condensate line, dryer vent, etc.).
- ❍ Check for vines growing up the exterior walls.

Exterior walls

Wood siding (shingles, shakes, boards, plywood panels, hardboard)

- ❍ Check bottom course of siding for sections in contact with, or in close proximity to, the ground (less than 8 inches).
- ❍ Check wood shingles/shakes for open joints, cracked, chipped, loose, or missing sections.
- ❍ Note areas of rot or discolorations.
- ❍ Check for peeling and flaking paint and warped shingles, particularly on sidewalls with a southerly or southwesterly exposure.
- ❍ Inspect for poor-quality shingles and shingles that have been improperly nailed.
- ❍ Check wood boards for open joints, cracked and rotting sections, loose or missing knots, peeling paint, and blistered sections.
- ❍ Inspect plywood panels for open joints, loose, warped, cracked, delaminated, or rotting sections.
- ❍ Check hardboard siding for cracked, deteriorated, or loose sections.

Aluminum/vinyl siding

- ❍ Check aluminum siding for loose, missing, torn, or dented sections.
- ❍ Check joints for open sections and weathertightness.
- ❍ Does siding contain insulation backer boards?
- ❍ Check siding for an electrical ground connection. (This requirement can be verified with the local building department.)
- ❍ Check vinyl siding for open joints, loose, cracked, or sagging sections.
- ❍ Check vinyl panels for waviness and blisters.

Asbestos-cement shingles/asphalt siding

❍ Check asbestos-cement shingles for loose or missing sections; cracked, chipped, and broken areas.

❍ Inspect asphalt siding for open or lifting joints; missing, loose, torn, cracked, chipped, or eroding sections.

Fiber-cement siding

❍ Check for skewed trim, exposed nails, and cracked, chipped, loose, or damaged sections.

Stucco-cement–finished walls

❍ Check for bulging, missing, loose, cracked, or chipped sections. Note areas in need of rehabilitation.

❍ If stucco is painted, check condition.

Synthetic stucco (EIFS)

❍ Check for cracked and open joints at the interface between the EIFS and windows, doors, wall penetrations, and so on.

❍ Are there indications of moisture in those areas?

Veneer and masonry walls

❍ Inspect for loose or bulging sections and large open cracks, particularly around door and window frames.

❍ Check for cracked, chipped, or missing sections of brick or stone.

❍ Inspect mortar joints for deterioration, cracked or loose sections.

❍ Check exterior surfaces on masonry walls for signs of water seepage (efflorescence).

Trim

❍ Check trim for cracked, loose, missing, or rotting sections.

❍ Inspect for areas of bare wood, blistered and peeling paint.

❍ Check nonwood trim for cracked, torn, missing, or loose sections.

Windows

❍ Check for cracked, broken, or missing panes.

❍ Are any of the windows painted shut?

❍ Are the panes properly secured to the sashes?

❍ Check the condition of the window frames and sashes.

Exterior doors

❍ Check wood door for cracked, chipped, or delaminating sections.

❍ Check metal door for dents or scratches.

❍ Check fiberglass door for cracks.

❍ Check for weatherstripping around exterior joints.

Storm windows, screens, and storm doors

❍ Check for missing units and/or partial installations.

❍ Inspect storm windows for loose, cracked, broken, or missing panes.

❍ Inspect wood units for cracked, broken, or rotting sections.

❍ Inspect combination units for loose, broken, rusting, or corroded sections.

❍ Inspect screens for torn sections and holes.

❍ Inspect doors for ease of operation; missing glass; cracked, broken, rotting, or corroded sections.

Caulking

❍ Check joints for cracked, chipped, crumbly, missing, or loose areas of caulking compound.

6

Lot and landscaping

In addition to all of the items mentioned in the previous chapters, you should inspect the drainage around the house and the landscaping. Any retaining walls, decks, or fences also should be inspected.

Drainage

As housing developments and shopping centers sprout up in the countryside, they affect the drainage characteristics of the surrounding areas. Normally, in undeveloped areas most of the water falling to the earth soaks into the ground. The remainder flows over the surface into lakes, rivers, and streams, or accumulates in low-level areas, forming ponds. In built-up areas, thousands of acres that had been soaking up rain have been rendered impervious to water because of buildings and paved areas. The surface-water runoff in these areas might be two to ten times more than it was when the land was undeveloped.

In built-up areas, surface water usually flows into storm drains (catch basins) that in turn discharge into rivers and streams. In some cases established housing developments have been inundated with surface water after a heavy rain because their storm-drainage facilities were not adequate for the increased water flow resulting from adjacent new housing developments. In most cases, the increased runoff results in the rivers and streams swelling, although they are usually contained within their banks. However, after a heavy prolonged rain, a river or stream can overflow its banks and flood the surrounding area. Many people do not realize that even a small creek, which might be a trickle when they see it, can become a raging, destructive torrent following an excessively heavy rain.

The area normally flooded when a river or stream overflows its banks is called a *flood plain*. Between 5 and 10 percent of the land in the United States is in a flood plain. Much of this land is level and from outward appearances seems to be desirable. As land in urban and suburban areas became more and more scarce, builders constructed homes directly on the flood plains of streams and other waterways. These homes are all vulnerable to flooding. (See FIG. 6-1.) In many parts of the country, some owners do not realize that their homes were built in a flood-prone area, and they probably will not realize it until it is too late.

If you have doubts about whether the house is located in a flood plain, you should check with the local town or county engineer. If the engineer is not available, often the local highway superintendent can tell you whether the area periodically floods. In many communities, federal flood insurance is available for those homes located in a flood plain. If you are considering such a home, you should consider purchasing flood insurance.

Surface runoff is of concern to the homeowner because it can result in soil erosion, ponded water, and water in the basement or crawl space. Soil erosion occurs whenever water flows over bare earth. Soil particles are loosened and are carried away by the flow. Water seeks its own level and will therefore flow from a higher elevation to a lower elevation. The paths the water takes

Fig. 6-1. *Backyard flooding of a house located in a flood plain.*

when flowing to a lower level are called *natural drainageways*. Areas particularly vulnerable to erosion are steep banks and drainageways. (See FIG. 6-2.)

The basic principle for preventing or minimizing erosion is to have the ground covered as much as possible with growing vegetation such as grass, trees, bushes, shrubs, and even weeds. If the vegetation does not root and keeps washing out, a substitute cover such as gravel, stones, or mulch can be used. This cover is not as effective, but it does reduce the erosion. Sometimes the banks are too steep for a ground cover of any kind; they must be stabilized by terracing or retaining walls.

In addition to a ground cover, erosion can be reduced by slowing down the water flow. For example, if there is a concentrated surface runoff along a natural drainageway, the water can be diverted to a man-made channel or ridge that follows a level contour. This spreads out the water and slows the flow so that the water does not scour and erode the soil.

The effects of surface runoff can be minimized by reshaping the ground surface. This can be done by terracing and/or regrading the lot into gentle slopes with diversionary ridges and swales. A *swale* is a depression in the ground that like a ridge will intercept surface runoff and redirect it to an area where the water will not cause damage. When a house is located on a sloping lot, a swale or ridge should be in the portion of the lot that slopes toward the house. This type of a diversion will prevent surface water from accumulating around the house. Otherwise, the surface runoff might seep into the basement or crawl space. (See chapter 11.)

Many building lots have low, level areas that will tend to accumulate water after a rain or from surface runoff. When the soil is slow-draining, as with clay and silt, the water will

Fig. 6-2. *Bank erosion on a steep slope. Correcting this condition will require a retaining wall with adequate drainage.*

pond rather than soak into the ground. The ponded areas retain the water until it evaporates or eventually seeps into the ground.

Depending on the location of the pond, the accumulated water might not be a problem. If the pond is over the leaching field of a septic system or is in an area that normally has a lot of foot traffic or is used by children, corrective action is necessary. The problem can often be corrected by bringing in fill and regrading the area. When regrading is not practical, the area can be drained by laying a line of perforated drainpipe through the affected area and directing one end of the pipe to a low spot. The pipe is generally encased in a bed of gravel or broken stones. If conditions warrant it, a concrete-block catch basin with radial spokes of perforated pipe can be installed at the low point. (See FIG. 6-3.) Water collected in the pipes and catch basin can then be directed to another area. If there are no other low areas to which the ponded water can be redirected using a pipe with gravity flow, the water can be directed to a sump pit and pumped to the desired location.

Fig. 6-3. *Concrete block catch basin with several perforated inlet pipes and a solid outlet pipe that drains to a suitable location. Basins of this type are used to drain low-level lawns that are collection ponds for rainwater.*

Groundwater, water table

Water that soaks into the ground eventually percolates downward under the influence of gravity until it reaches an impervious layer it cannot penetrate. After the water reaches the impervious layer, it begins to move in a lateral direction. This underground flow is known as *groundwater*, the top surface of which is commonly called the *water table*. The level of the water table will vary with the amount of rainfall. Consequently, the water table might be several feet higher after a prolonged rainy period than during a prolonged spell of dry weather.

A high water table can result in a flooded basement or failure of the septic system's leaching field. In many parts of the United States, the seasonal high water table is only 2 to 5 feet below the surface. In those areas, houses should be built on a slab or over a crawl space rather than over a full basement. Unfortunately, houses with full basements have been built in areas where the water table (during the wet season) is above the level of the basement floor. This invariably results in water penetration into the basement. Depending on the soil, even when the water table is several feet below the basement floor, some water might seep into the area as a result of the capillary rise of groundwater. (See chapter 11.)

Homes located in areas where the seasonal high water table is only a few feet below the ground surface should not have septic systems for waste disposal. Ideally, they should be connected to a sewer system. For proper operation of a septic system, the water table during the wet season should be at least 4 feet below the bottom of the leaching field or seepage pit. The operation of a septic system is explained in chapter 13.

Basically, the top surface of flowing streams, rivers, lakes, and oceans is the water table. Consequently, the terrain that gradually slopes into the waterway has a high water table. Homes built in these areas are not only vulnerable to problems associated with a high water table but are also vulnerable to flooding. The water table tends to follow the general contour of the land and in some areas might intersect the ground surface, thus forming marshy wetlands. To a homeowner, these wetlands are quite undesirable, because not only are they costly to drain, but they are also a breeding place for insects.

Excessive grading or reshaping of the ground surface (such as cutting out the side of a hill to locate a house, FIG. 6-4) can change the natural drainage patterns and cause groundwater to seep to the surface. I have inspected many such houses and have found (in the early spring or after an excessively heavy rain) water oozing out of the cut side of the hill. This water, if not redirected away from the house, can work its way into the basement or crawl space.

In areas with a seasonal high water table, if the topography is such that the land slopes toward one side of the house, in addition to surface-water runoff, subsurface water will flow toward the house. This water, if allowed to accumulate around the foundation, can seep into the basement or crawl space. This condition can usually be controlled by installing a *curtain drain* in the hillside parallel to the house to divert the water away from the house.

A curtain drain consists of a perforated drainpipe installed in a trench that is filled with gravel and covered with soil. The trench normally extends several feet beyond the house, with one end leading to a suitable disposal area. Incidentally, the perforations in the pipe should be facing downward, not upward as is popularly believed. As the subsurface water level rises, it enters the holes along the length of the pipe. Since water always takes the path of least resistance, once

Fig. 6-4. *House located on the cutout side of a hill. The building site requires special provisions to minimize erosion and drainage problems.*

inside the pipe it flows to the outlet, which must be located away from the house and must be unobstructed. The outlet, however, should have an animal screen to prevent a small animal (such as a rabbit) from entering, becoming lodged, and blocking the flow.

In areas with a seasonal high water table or a potential for surface water to accumulate around the foundation, it is advisable to have foundation footing drains (perforated drainpipes) installed parallel and adjacent to the foundation footing. (See FIG. 6-5.) As with curtain drains, footing drains are installed with the holes facing downward. The purpose of the footing drain is to channel the water that accumulates around the foundation to another location. Footing and curtain drains either must have a free-flowing outlet or discharge into a sump pit where the accumulated water can be pumped to the desired location.

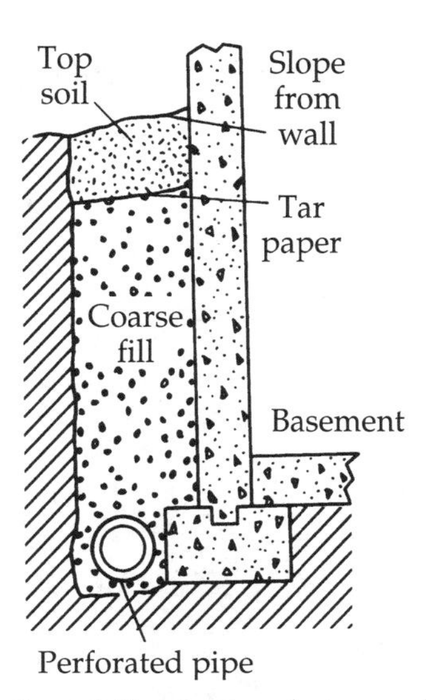

Fig. 6-5. Foundation footing drain, used for channeling water that accumulates around the foundation to another location.

Inspection

The drainage inspection should begin as you are driving to the house. When you approach the house, notice the overall topography. Is it level or inclined? If it is inclined, is it a gentle slope or a steep slope? With inclined topography, you should be concerned about the possibility of surface and subsurface water movement toward the house. If the house is located near the bottom of an inclined street, is a storm drain (catch basin) in the street at the low point? There should be, especially if the street is paved. Otherwise, after a rain or a snowmelt, water accumulates at the low area and depending on the amount, can flood the adjacent yards and driveways. Did you notice a waterway (stream, brook, etc.) on the street as you approached the house? If you did, the house might be located in a flood plain.

When you arrive at the house, notice whether the land between the house and the street is above or below the street level. If the land slopes downward from the street to the house, the house is vulnerable to drainage problems. The surface water, if not properly controlled, can accumulate around the foundation or can pond on the lawn or over the entry path. If the house is inspected when it is not raining, you might not see any problems. However, based on the slope and overall grading of the land around the house, you can at least determine the potential for a problem.

As you walk around the house, notice whether the ground immediately adjacent is graded so that it slopes away from the house on all sides. It should be. Otherwise, surface water can run directly to the foundation (see FIG. 6-6), seep down along the foundation walls, and accumulate at the lower section. This usually results in water seepage into the basement or crawl space. A rule of thumb for grading this area is a drop of about 1 inch per foot. The lawn should slope away from the house for at least 10 feet and should be pitched so that there is approximately a 10-inch drop over that distance. (See FIG. 6-7.)

Fig. 6-6. Improper grading of the land adjacent to the house. The lawn pitches toward the door, resulting in surface water ponding in front of the entry area.

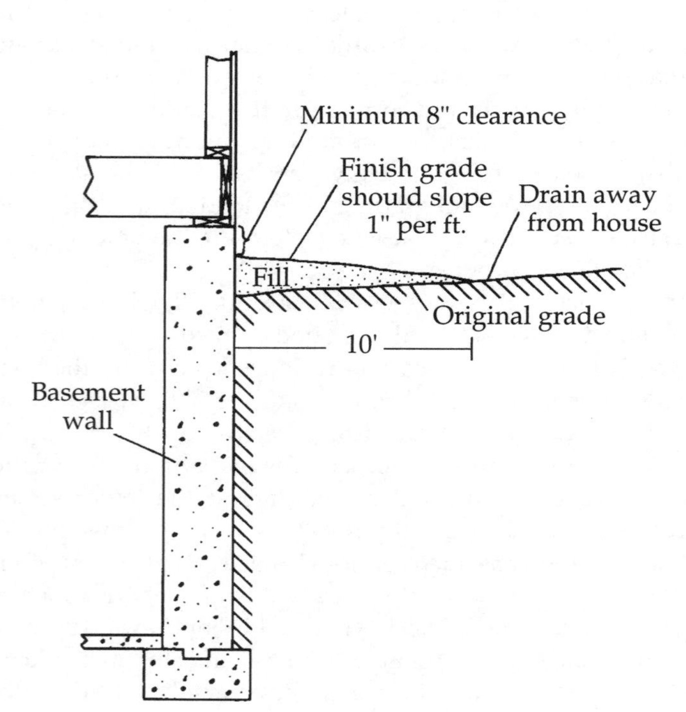

Fig. 6-7. Finish grade of the ground adjacent to the foundation, sloped for proper drainage.

Because of normal soil settlement and compaction, clogged gutter or faulty downspout, and foot traffic (especially when the ground is wet), the slope of the ground adjacent to the house usually changes with time. Consequently, the area around the foundation should be periodically checked for proper grading.

When checking the grading around the foundation, you might occasionally see a pipe protruding through the foundation wall or from a basement window. Record the fact on your worksheet for further investigation when you inspect the basement. Usually, this pipe is connected to a sump pump and is used for discharging the water that accumulates in the sump pit. The end of the pipe should be extended away from the house so that the water does not accumulate around the foundation. Sometimes the pipe terminates just beyond the foundation wall, which negates the advantage of a sump pump. The discharging water accumulates around and under the foundation and reenters the sump pit, only to be pumped out again. (See FIG. 6-8.)

If you find a stream on the property, you should realize the potential for flooding. As discussed previously, flooding can occur because of increased surface runoff resulting from a prolonged heavy rain. In addition, if the stream channel is blocked by fallen trees, tree limbs, sediment, or trash, flooding can result. Depending on the location of the stream relative to the house, occasional overflowing of the stream banks might or might not be a problem.

As you walk around, look at the overall landscape. If the topography is sloping, does a natural drainageway direct surface runoff away from the house, or does the lot need a swale or ridge? If there is an abrupt change in the grading, is there a need for a retaining wall? Are any areas eroding to the extent that corrective action is necessary? Are low or level areas vulnerable to ponding? Ponded water on the lawn does not necessarily indicate a drainage problem. It might be caused by a malfunctioning septic system, a faulty sewer hookup, or a break in the main water-supply pipe. If you find ponded water during your inspection, try to determine the cause.

Fig. 6-8. *Sump pump discharge pipes. Left—The pipe terminates just beyond the foundation wall. Right—The discharge pipe has been extended so that effluent discharges away from the foundation.*

If there are footing drains around the foundation or curtain drains in a hillside, they will not be visible during your inspection. For these drains to function properly, they must have a free-flowing outlet. (See FIG. 6-9.) Ask the seller if there are drains, and if so, ask for the location of the outlets. Unfortunately, most homeowners do not know whether there are footing or curtain drains. When a house is sold, this type of information is usually not discussed, although it should be. Consequently, the location of the drain outlet is lost for all future owners. If you are lucky enough to have a seller who knows the location of the drain outlet, you should inspect the opening to make sure that it is not obstructed. If the footing drain discharges into a stream channel, an overflowing stream could be a problem. It all depends on the elevation of the footing drain relative to the level when the stream is running full. If the stream is above the level of the footing drain, then water will back up into the drain around the foundation and could cause a water seepage problem. Look for evidence of seepage when inspecting the basement or crawl space. In new homes, the drain outlet is sometimes inadvertently blocked when the lawn is landscaped. If you are buying a new house, ask the builder to show you the location of the drain outlet.

Fig. 6-9. *Free-flowing outlet of curtain drain. Note water discharging from pipe.*

Retaining walls

Retaining walls are mostly used for stabilizing and controlling erosion on steep banks. In some cases, they are used in conjunction with terracing of rear or side yards to provide a level area for recreation. In either case, they must be designed to withstand the lateral pressures being exerted on them by the soil.

Retaining walls are normally built with construction timbers, railroad ties, stone, concrete, or concrete blocks. Some concrete and concrete-block walls have stone or brick veneer facing. On occasion, you might find a *gabion* retaining wall—steel baskets filled with stones. (See FIG. 6-10.) As gabions age, the steel baskets tend to corrode and deteriorate, especially on the side facing the embankment. Over the years, however, soil sediment usually fills the voids between the stones and tends to hold the wall in place.

Stone retaining walls are often referred to as dry or wet, according to whether mortar

Fig. 6-10. *Gabion retaining wall.*

was used between the stones. A *dry* retaining wall is one that has been constructed without mortar. It depends on the weight and friction of one stone upon another for stability. Frost heaving is not a problem with this type of wall. The stones are not bonded together and will therefore be raised and lowered together by the frost. Consequently, the bottom course of the wall is usually only about 6 inches below grade rather than below the frost line.

A *wet* wall is one with mortar between the stones. The mortar secures one stone to another and thereby achieves a monolithic wall with greater stability. Because this wall is integral, frost heaving will cause cracking. To prevent frost heaving, the bottom of the wall must be below the frost line. A wet wall offers greater solidity, and therefore loose soil will not wash out or run through the voids. Also, a wet wall is less of a hazard because no loose stones can be kicked out of place or fall off the top.

When constructing a retaining wall, provisions must be made for draining the water that normally accumulates behind the wall. Otherwise, a hydrostatic pressure buildup can cause structural failure of the wall. Drainage should be provided by installing a continuous perforated drainpipe at the lower portion of the wall and backfilling the area with broken stones or gravel. The pipe should be directed so that the effluent flows to a suitable location away from the wall. In many monolithic retaining walls, the perforated pipe is replaced by *weep holes*, holes in the wall that run from the front to the gravel backfill. The weep holes allow the water that accumulates in the gravel to drain out through the wall. Weep holes should be placed at a 5- to 10-foot spacing and should be 4 inches in diameter (the same diameter as the drainpipes).

Unfortunately, retaining walls are often built without adequate drainage provisions. The gravel backfill and drainpipe might be omitted, or the weep holes might be too few or too small to be effective. The weep holes must be kept clear so that the water behind the wall can be adequately drained.

A retaining wall built with construction timbers or railroad ties should be anchored into the hillside to provide the resistance to overcome the lateral forces exerted on it. If the wall is not tied back into the earth, it can bow, buckle, or heave and eventually collapse. Anchoring of the wall is achieved by using *tiebacks* and *dead men*. A tieback is a construction timber that has been placed perpendicular to the wall. The front end is flush with the wall and fastened to it with large spikes. The rear end is fastened to a dead man, a small section of timber perpendicular to the tieback and parallel to the wall. (See FIG. 6-11.) When the area around the wall anchor is backfilled with soil, a force is developed on the anchor that resists the lateral force on the wall. Because of the open joints between the railroad ties, weep holes are not needed.

Many railroad-tie or timber retaining walls are not constructed with anchors. You can tell whether anchors were used by looking at the wall. If tiebacks were used, end sections will be visible in the face of the wall. (See FIG. 6-12.) However, from a visual inspection, you cannot tell the length of the tiebacks or whether dead men have been installed. Your inspection, therefore, should concern itself with the condition of the wall rather than its construction, unless it is a new wall. If it is newly constructed, you should inquire about a guarantee.

Inspection

When inspecting a retaining wall, look at its overall condition. With a dry stone wall, look for missing and loose stones and crumbled sections. This type of wall is relatively easy to repair and generally does require periodic maintenance. With a wet stone wall, check

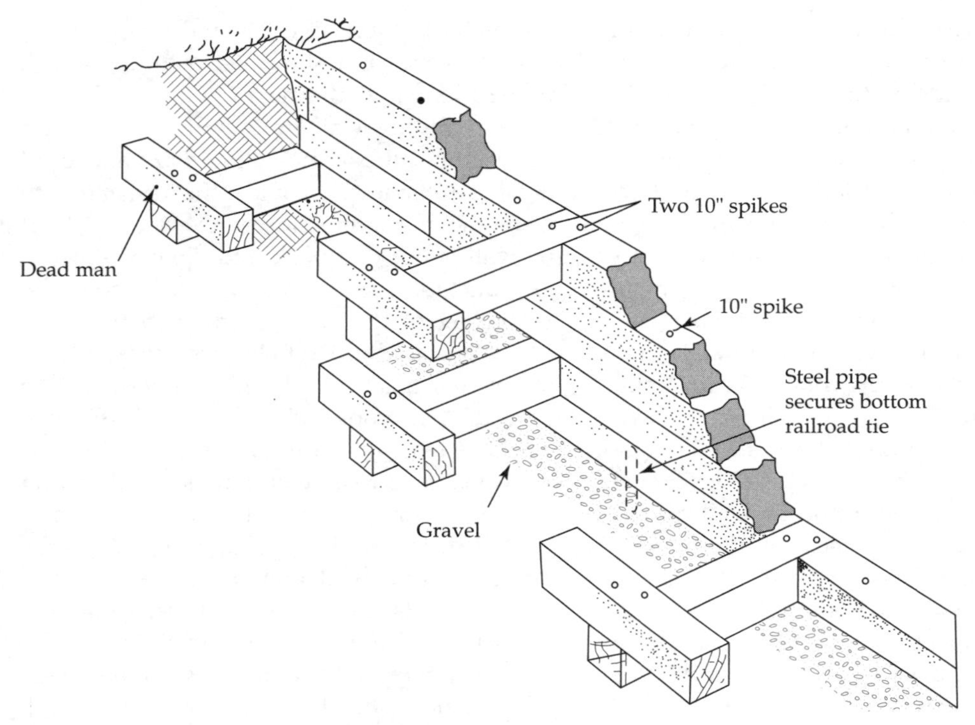

Fig. 6-11. Construction timber/railroad tie retaining wall.

Fig. 6-12. Timber retaining wall. Note the end sections, indicating tiebacks used for anchoring the wall.

the mortar joints for cracked, loose, and deteriorated sections. Are there weep holes in the wall? If so, are they adequately sized and unobstructed? Concrete and concrete-block walls should also have weep holes. Are there any cracked and heaved sections? Wood-constructed retaining walls should be checked for cracked, rotting, loose, and heaved sections. Some retaining walls are completely covered with vines. Try to push the vines aside so that you can inspect the wall. Quite often I have found cracked sections in the wall through which the vines were growing. While inspecting the wall, look for loose, heaved, and deteriorated sections.

All retaining walls should be vertical or inclined slightly toward the embankment. They should not be leaning forward. When they are, it is an indication that they could not withstand the lateral forces being exerted on them by the terraced or sloped earth behind. Once a wall cracks and heaves, the pressure that caused the condition is relieved, and the wall might stay in the leaning position for many years. (See FIG. 6-13.) However, additional forces might cause the wall to continue to heave and eventually to collapse. If the heaving is excessive or by the wall's collapsing someone can get hurt, the wall must be rehabilitated or braced. If you have any questions about what is excessive heaving, have the wall inspected by a professional.

In recent years, manufactured concrete blocks have been used for retaining walls for residential construction (see FIG. 6-14). The blocks come in different sizes and shapes depending on the manufacturer. Often these walls are 5 feet high or higher, and many are constructed in tiers. In most states, if the retaining wall is taller than 4 feet, it must be designed or approved by a qualified, licensed professional engineer. In order to prevent failure and to reduce the earth's pressure behind the retaining walls, they are reinforced with horizontal layers of a geosynthetic polymer mesh, which extends back into the

Fig. 6-13. *Cracked and heaved retaining wall.*

Fig. 6-14. Concrete block retaining wall.

embankment. Since the length of the mesh will not be visible, from an inspection point of view all you can do is to check the wall for tilting, cracking, or bowing. If any are noted, it should be recorded on your worksheet for later inspection by a professional engineer.

Landscaping

As you walk around the house, inspect the landscaping in the front, rear, and side yards. Specifically, look at the lawn, shrubs, and trees. A nicely landscaped area greatly enhances the beauty and value of the house. However, do not jump to a hasty conclusion about the house based on the landscaping. I have seen many neglected houses with beautiful landscaping and many well-maintained houses with poor landscaping.

Lawn

A lawn serves two purposes. It adds to the aesthetic beauty of the property, and, more important, it prevents erosion and washout of the topsoil. If you find that a large portion of the lawn consists of crabgrass and other weeds, do not be distressed. With a planned program of weed control, seeding, and fertilizing, you can upgrade the lawn so that it can be the "showcase of the neighborhood" within a few years, and at not too great an expense.

If you find holes or sunken sections in the lawn, they should be filled in, since they represent a potential tripping hazard. Occasionally, sunken sections are caused by the collapse of rotted, decayed, or deteriorated construction debris buried on the site years before. In new homes, all construction debris should be removed from the site rather than buried there.

In some parts of the country, moles are a problem in lawn maintenance. They burrow in the ground near the surface in search of food (grubs, caterpillars, and insects) and in the process create soft ridges (mole hills) that spoil the lawn's appearance. If you see ridges over portions of the lawn and they feel soft when you walk over them, suspect moles. This condition can generally be controlled through soil treatment and should be discussed with the proprietor of a local nursery.

Some lawns have steep sloping areas that from a maintenance point of view are quite difficult to mow, even when mowing across the slope. Because of the danger involved, sit-down riding mowers should not be used when cutting the grass on a steep slope. These mowers have been known to topple over and severely or fatally injure the driver. In some homes, the steep sections of the lawn have been replaced by terraced areas with steps that lead from one level to another. If there are terrace steps on the lawn, you should inspect them for cracked, loose, missing, and deteriorated sections. Also, check for uneven treads and dimensional variations in the risers (a tripping hazard). In addition, if there are more than two steps, a handrail is recommended.

If you are planning to buy a newly constructed house, find out whether it will be your responsibility or the builder's to establish a new lawn. If it is yours, it can be quite expensive, depending on the size of the lawn and whether topsoil needs to be added. Over 100 tons of topsoil is needed to cover an area one-third of an acre to a depth of 2 inches.

Shrubs

When new homes are landscaped, the shrubs are often intentionally planted very close to one another to produce an immediately pleasing effect. Many homeowners do not plant shrubs with the future appearance in mind. Consequently, as the shrubs grow and fill out, they tend to crowd one another, losing their individuality. Eventually, they become unsightly, with portions dying off due to lack of sunlight. In addition, the growing shrubs often block walkways and produce so much shade that the area around the house is always damp, a condition conducive to the growth of decay fungi and mildew. By extensive pruning of the shrubs and transplanting others (if you want to save them), the area can often be completely rejuvenated and restored to its original beauty. For some people, their dream house is one that is covered with ivy. (See FIG. 6-15.) Actually, vines of any type growing up the outside walls of a house are quite undesirable. As vines grow they can cause problems. They can lift wall

Fig. 6-15. *Vines growing up and partially covering exterior shingle wall.*

shingles and roof shingles. They can also grow into mortar joints and crush downspouts. (See FIG. 6-16.) If you see vines on the house, you should consider their removal.

Trees

If there are trees on the property, they should be checked to see if any are dead or have any large dead branches. All dead trees should be taken down. Because they are vulnerable to insect damage and decay, they are a potential hazard, especially if they are located near the house. Large dead branches are also a hazard. On a windy day, they can break off the trees and fall to the ground or, worse, onto the house.

After a deciduous tree has lost its leaves, it can be somewhat difficult to determine whether it is dead or has any dead branches. However, if you see any limbs with the bark peeled off, you can assume that those branches are dead. (See FIG. 6-17.) If you have any doubt, after you move into the house, you should have the trees checked by a professional, or wait until spring and summer when all the trees are in full bloom.

Fig. 6-16. *Downspout crushed by vines.*

If you find any dead trees or dead branches, record their location on your worksheet for later removal. Depending on the size and location of the dead tree, its removal can be somewhat costly. This type of work should be performed only by a professional who is insured in the event that the tree causes damage when it falls to the ground. In addition to dead branches, all limbs that are overhanging or resting on the roof should be pruned back. Otherwise, they might eventually damage the roof.

If you are buying a newly constructed house with trees on the property, you should be aware that the roots of some of the trees might have been damaged during construction. This could occur as a result of heavy equipment (tractors or trucks) being driven too close to the tree. Trees that have had root damage during construction do not necessarily show any immediate effects. However, within a year or two and depending on the degree of damage, the trees may die. If care is taken during construction, this problem can be avoided. Your best bet is to buy a house from a quality builder.

Decks

There are many types and styles of decks. However, from an inspection point of view, your main concern should be safety rather than appearance. When inspecting a deck, unless it is a rooftop or cantilever type, you should begin with the supports on the underside. If the deck is more than a few feet above the ground, it will generally be supported by wood or metal columns (posts). Unless wood

Fig. 6-17. Large, dead branches are a potential safety hazard and should be removed.

posts have been pressure-treated, they should not be in direct contact with the soil. Untreated wood in contact with the ground is vulnerable to rot and termite activity. Also, the dampness normally associated with the soil can promote rust deterioration of a metal post or connector.

Each post should rest on a concrete pad that has a footing below the frost line. Otherwise, the footing is subject to frost heave. Probe the base sections of the posts with a screwdriver to determine whether there is deterioration. If the screwdriver can penetrate the post beyond the surface, a problem exists that should be corrected. In some cases, the post might require replacement. Push the post to see if it moves. It should not. Occasionally I find columns that are loose and not adequately supporting the deck. (See FIG. 6-18.) The condition usually results from uneven settlement of the support footings and inadequate fastening at the top or bottom of the column. A loose post is a potential hazard and must be resecured as soon as possible.

When the deck is less than a few feet above the ground, it is usually supported by masonry piers. Inspect the piers for cracked, broken, loose, or deteriorated sections. They must be repaired. With some "ground-hugging" decks, usually less than a foot above the ground, this inspection might not be possible. Most of the support piers are not visible.

When one side of the deck is attached to the house, there are usually no support posts below that section. Consequently, if the joint between the house and the deck should weaken, there is a potential for the deck to collapse. Check the joint between the deck and the house to see if it is securely fastened. Is it pulling away from the house? It shouldn't be. This is a potential problem and should be called out on your worksheet for additional bracing. In some cases, the deck is fastened to the house with undersized or too few nails. I know of one community where this type of installation resulted in two decks collapsing. Because of these failures, the town passed an ordinance that requires using lag bolts rather than nails to secure the deck to the house. (See FIG. 6-19.)

Next, check the joist supports at the portion of the deck attached to the house. Since there are no posts, there will not be a girder to support the joists. In this case, the joists should be supported by metal brackets fastened to the

Fig. 6-18. Inadequately supported deck. Column is loose and can easily be knocked over.

header or by being toenailed into the header with a ledger below them. The former method is preferred because the ledger used is often skimpy. I have seen many decks where the joists were toenailed into the header but the ledger was never installed. If you find this type of installation, you should install angle brackets to support the joists or at the very least install a ledger as a precautionary measure. Check the condition of any and all metal brackets, connectors, nails, and screws. Over time they can corrode and rust, either because of the weather or because of the chemicals used in the deck's pressure-treated wood.

In addition, depending on the size of the deck, there might be a need for diagonal bracing. This provides additional rigidity to the deck and can be achieved by placing a 2-inch-by-6-inch board between two diagonal corners and nailing it to the underside of each joist.

When deck planks are installed, there should be a space of about ¼ inch between them. This space allows rainwater and melting snow to drain. In some cases, the planks are butted up against one another so that there is virtually no space between them other than the crack of the joint. Water entering this crack does not readily drain and instead promotes decay. If you can reach the underside of the deck planks and the top portion of the joists, probe them for rot. If there are rotting sections and the decay is advanced, you might see the fruiting bodies

Fig. 6-19. Lag bolts are used to secure the deck to the house. Note that joist is resting on a ledger board for support.

of the decay fungi. (This is discussed in detail in the section on rot in chapter 8.) If steps lead to the deck, the treads, stringers, and handrails should be inspected for decay. As with the deck support posts, the stringers should rest on a concrete pad rather than soil.

After inspecting the underside of the deck and the steps, you should inspect the top portion. When the deck is more than 30 inches above the ground, there should be a guardrail around the perimeter as a safety precaution. Naturally, the higher the deck, the greater the need for a rail. Issues such as wobbly railings, loose stairs, and ledgers that appear to be pulling away from the home are all causes for concern. Metal connectors, nails, and screws can corrode over time. Look for rust and other signs of corrosion that can weaken the structure of your deck. Check the rail to see if it is loose and wobbly. If it is, record it on your worksheet to be resecured. If the deck is to be used by small children, additional protection is needed to block the open area between the railing, railing posts, and deck planks. When balusters are used for this purpose, they should be spaced 4 to 5 inches apart.

The guardrails and deck planks should be inspected for cracked, rotting, and loose sections. (See FIG. 6-20.) Depending on the quality of the wood and the upkeep, a deck need not deteriorate to a point where it requires complete rehabilitation. Repairs or replacement of deteriorated sections should be performed as needed. If you find sections of the deck that are in need of repair, you should indicate those areas on your worksheet.

Fig. 6-20. *Cracked and rotting deck planks. Planks should be replaced as needed.*

Free-standing decks

Free-standing decks are independent self-supporting structures. If constructed properly they are as safe and secure as attached decks and in some cases more secure. A number of attached decks have collapsed because the joint between the deck and the house has pulled away, because either it was improperly or inadequately secured or it has rotted. Depending on the size of the free-standing deck, it will require an additional beam and two or more posts situated next to the house. The posts should be mounted on footings that extend below the frost line. In order to be able to resist lateral and horizontal movement, diagonal bracing between the posts and the beams must be provided. (See FIG. 6-21.) As with attached decks, these decks should be checked for cracked, broken, loose, or rotting sections.

Composite decks

Most residential decks are constructed entirely of wood. However, the use of composite lumber for the visible portions of the deck, such as the deck planks, handrails, and trim has become very popular. Its popularity stems from the fact that it does not require waterproofing and resists damage from weather and insects. It is nevertheless not maintenance-free. The composites, which are generally made of recycled plastics and waste wood fibers, are

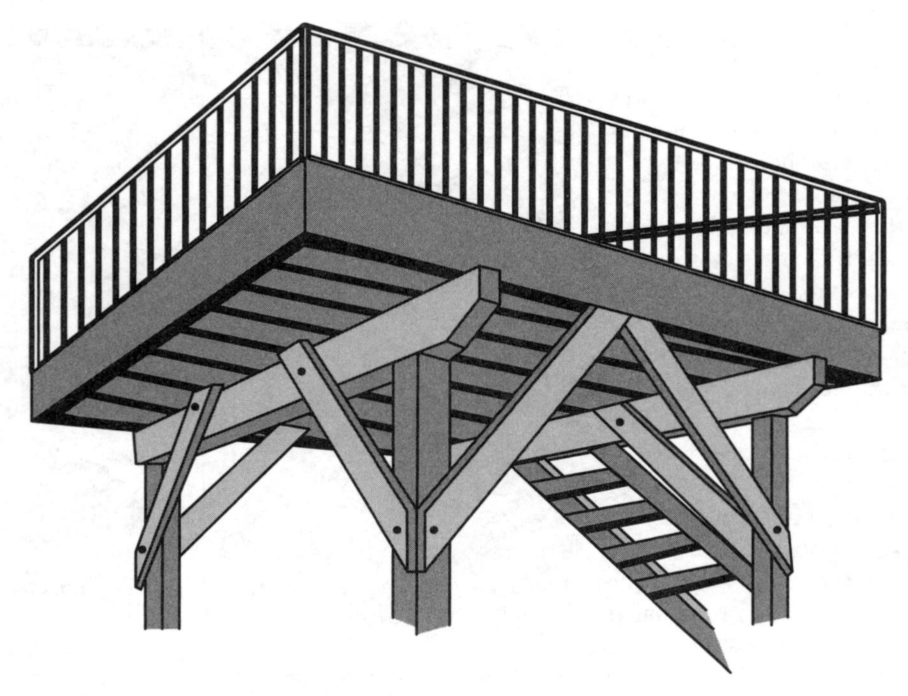

Fig. 6-21. *Diagonal bracing on free-standing deck.*

not intended for structural use. The structural support for those decks that use composite lumber (the joists, beams, and posts) will still be made of wood and should be inspected as previously discussed.

Trex and Timber Tech are the leading brands of composite lumber for decks, although there are a number of companies that also make composite lumber. Not all composite decks are made with the same quality ingredients. Because the material expands and contracts with temperature, proper spacing during installation is critical. Cupping deck planks and joints that don't line up have been noted. In addition, some of the other problems that have been noted, which should be recorded on your worksheet, are cracks, scratches, staining, mold (mildew) blotches, and fading.

One last point: If the deck was not built when the house was constructed or the deck is a complete replacement of a previously deteriorated one, check with the seller to see if a certificate of occupancy (CO) was issued by the local building department at the completion of construction. A CO is required by most municipalities. If one was not issued, record that fact on your worksheet for future discussion with your attorney.

Fences

If there is a fence on the property that you are inspecting, you should check its overall condition. The problems encountered with fences are normally not major and are usually not costly to correct. However, you might find a fence that has deteriorated to a point where it needs complete rehabilitation or replacement.

Wood fences should be inspected for cracked, broken, loose, and missing sections. In addition, they should be checked for deterioration from rot and termite infestation.

(Termites and rot are discussed in chapter 8.) The gates for the fence should also be checked for cracked, loose, or broken sections and ease of operation. Wooden gates often sag as they age and require periodic maintenance.

Metal fences should be inspected for rusting, loose, and deteriorated sections. Rusting sections should be scraped, primed, and painted. Chain link fences are generally constructed of galvanized steel. Galvanizing (zinc coating) protects the steel against rusting and is usually applied by hot dipping or electroplating. The hot-dipped process produces a heavy zinc coating that is very effective, in contrast to the thin coating produced by electroplating. Chain link fences that have been galvanized by electroplating have a tendency to rust and require periodic maintenance. Some chain link fences have a vinyl coating that protects against rust. The vinyl coating is quite effective and lasts for many years.

If there is an in-ground swimming pool on the property, there should be a fence around the pool area. Most communities have an ordinance requiring a fence of a specific height as a protective barrier. If you do not see a fence around a pool area, you should check the requirements with the municipal building department; otherwise, you might find after buying the house that you are legally obliged to install one.

Checkpoint summary

Drainage

- ❍ When approaching the house, take note of the overall topography.
- ❍ Is it level or inclined?
- ❍ Are there gently or steeply sloped areas?
- ❍ Is the house located near or at the bottom of an inclined street?
- ❍ Note whether there is a storm drain (catch basin) nearby.
- ❍ Are there nearby streams or brooks?
- ❍ Are you able to determine if the house is located in a flood plain or flood-prone area?
- ❍ Is the ground immediately adjacent to the house graded so that it slopes away on all sides of the structure?
- ❍ Are there natural drainageways to direct surface water away from the house?
- ❍ Are there low or level areas that are vulnerable to water ponding?
- ❍ Are there areas of ponded water on the lot?
- ❍ Are you able to determine whether the house has footing drains?
- ❍ Can you locate the outlet for these and any other drainage pipes?

Retaining walls

Timber, railroad tie, dry stone wall, gabion

- ❍ Inspect for missing, loose, and crumbling sections of stone.
- ❍ Check timber and railroad-tie walls for cracked, loose, rotting, and heaved sections.
- ❍ Are the wood-constructed walls properly anchored (tiebacks)?

Concrete, concrete block, wet stone wall

- ❍ Inspect for cracked and heaved sections.
- ❍ Check for loose, deteriorated, and missing mortar joints.
- ❍ Is the wall vertical, or does it lean?
- ❍ Are portions of the wall heavily covered with vines?
- ❍ Did you inspect these areas for cracked and heaved sections?
- ❍ Try to determine whether the area behind the retaining wall is adequately drained.
- ❍ Are there weep holes at the base of the wall?
- ❍ Are they blocked?
- ❍ Are the weep holes adequately sized and spaced?

Landscaping

Lawn

- ❍ Inspect for holes, sunken sections, bald spots, and eroding areas.
- ❍ Estimate areas that will require recultivation.
- ❍ Note soft sections or ridges (possibly due to moles).
- ❍ Inspect terrace steps for cracked, loose, rotting, or missing sections. Check steps for handrails, uneven treads, and variations in riser heights.

Shrubs

- ❍ Inspect shrubbery for overcrowding, dying sections, blocked walkways, steps.
- ❍ Note areas in need of pruning, transplanting, or removal.
- ❍ Note areas of the house that are covered with vines.

Trees

- ❍ Check for dead trees and limbs, especially those close to the house.
- ❍ Note tree limbs that are overhanging or resting on the roof.
- ❍ Note for future professional evaluation any trees that show evidence of rot, split sections, or insect infestation.

Decks

- ❍ Check and inspect the various deck components for safety rather than for appearance.
- ❍ Check concrete or brick piers for cracked, loose, and deteriorated sections.
- ❍ Inspect wood columns for rot and termite activity.
- ❍ Inspect metal columns for rust deterioration.
- ❍ Are columns supported on concrete pads, or are they in contact with the ground?
- ❍ Note any loose columns.
- ❍ Check for open and weakened joints between the deck and the house.
- ❍ Is the deck attached to the house with nails or lag bolts? (Lag bolts are preferred.)
- ❍ If the deck was not built at the time the house was constructed, does it have a certificate of occupancy (CO)?
- ❍ Inspect deck-joist supports at the portion of the deck attached to the house.
- ❍ Are deck joists supported by metal brackets (preferred), or are they toenailed into a header beam with a ledger board below the joist?
- ❍ Where the joists have been toenailed, check for missing ledger boards.
- ❍ Does deck contain diagonal bracings?
- ❍ Inspect the underside of deck (girders, joists, floor planks) for missing, cracked, and rotting members.
- ❍ Inspect wood step treads, stringers, and handrails for cracked, loose, missing, and rotting sections.
- ❍ Are the stringers supported on a concrete pad, or are they in contact with the ground?
- ❍ Check top portion of deck for cracked, loose, missing, and rotting sections of deck planks, railings, and railing posts.
- ❍ On free-standing decks, check for diagonal bracing, as well as for cracked, loose, and rotting members.
- ❍ On composite decks, inspect the wood structural support members. Check the composite decking, handrails, and trim for cracks, scratches, staining, mold (mildew) blotches, fading, and joints that don't line up.

Fences

- ❍ Inspect wood fences for cracked, broken, loose, and missing sections.
- ❍ Check for areas of rot and insect damage.

- ❍ Inspect metal fencing for loose, missing, and rusting sections.
- ❍ Check gates (metal and wood) for sag, missing hardware, and cracked, loose, broken, and missing sections.
- ❍ If property contains an in-ground pool, is the area around the pool adequately fenced off? (It might be a legal requirement.)

7

Garage

The garage should be inspected after the exterior inspection has been completed. There are two basic types of garages: attached and detached. An attached garage is a part of the main building. It might be located below a habitable portion of the structure or connected to the side of the building. A detached garage is a separate structure, not part of the main building. There might be a connecting breezeway or porch between the two structures.

Attached garage

Since the principal use of the garage is car storage, the possibility of dripping oil and gasoline presents a potential fire hazard. Because the attached garage is connected to the main structure, certain precautionary measures should be taken during construction to minimize the hazards. Look around the garage to see if there are any potential problems.

Fire and health hazards

Is there an interior door between the garage and the house? If there is, is there at least one step leading up to the door? There should be. (See FIG. 7-1.) It is surprising how often I find that the garage floor slab is at the same or a higher level than the adjacent living area. (See FIG. 7-2.) The living area should be above the level of the garage floor to prevent toxic exhaust gases and gasoline vapors, which are heavier than air, from entering the house whenever the interior door is opened. As a precautionary measure, the interior door should have a tight seal around the joints to prevent seepage. This door should be fire-resistant, such as metal-clad, solid wood, or hollow core, with a sheet-metal covering on the *garage side*. As a safety feature, the interior

Fig. 7-1. *Steps connecting the garage floor slab to the adjacent living area, which is at a higher level.*

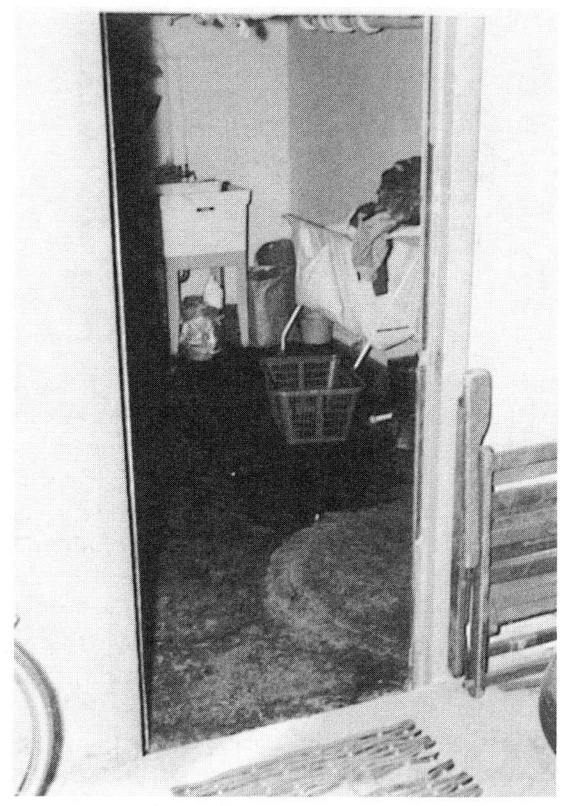

Fig. 7-2. *Garage floor slab at a higher level than the adjacent living area—a potential hazard.*

door should also be self-closing. In most homes it is not. It seems that homeowners have found self-closing doors inconvenient, especially when carrying in packages from the local supermarket. Nevertheless, safety should not be sacrificed for convenience.

Next, look at the walls that separate the garage from the living area. Are there exposed wood-frame members? There should not be. Exposed wood framing in this area is considered a fire hazard and should be covered with a fire-resistant material such as plaster or stucco on lath or ⅝-inch plasterboard. This wall should be insulated to reduce heat loss. If there is a living area above the garage, the ceiling should be insulated and have a fire-retardant covering.

In some garages an access hatch to the attic is located in the ceiling. Occasionally the hatch cover for this opening is missing or open. (See FIG. 7-3.) This is a fire hazard. If a fire should start in the garage, the open area in the ceiling could act as a flue and draw the flames up into the attic where they would quickly engulf the house. The attic hatch cover must be in place at all times.

Some homes have a garage in the basement. The garage is at the basement level with no partition walls separating the garage area from the basement area. This is a fire and health hazard in addition to being inefficient from an energy-conservation point of view. When the garage doors are opened, there will be a loss (from the basement) of warm air in the winter and cool air in the summer.

Occasionally the heating plant (furnace or boiler) is located in the garage. (See FIG. 7-4.) This is perhaps the least desirable location for a heating unit. There is always the possibility that a leak could develop in the gasoline tank or fuel line of an automobile. If the garage is inadequately ventilated, the resultant flammable vapors could be ignited by the flame in the heating system. Of course, this

Fig. 7-3. Open attic-access hatch in the garage ceiling.

Fig. 7-4. Furnace located in the garage. Note that the ceiling should have been covered with fire-rated gypsum board.

risk can be minimized by locating the heating unit on a platform or by building a low wall around the heating system (see FIG. 7-5) (since gasoline vapors are heavier than air and will accumulate at ground level) or by enclosing the heating unit in a room with a tight seal on the door. If the heating unit is enclosed, the area must be vented to provide outside air for combustion. Another problem with a water-heating system located in the garage is that the pipes are more vulnerable to freezing should the system malfunction or run out of fuel oil.

Plumbing check

While looking at the walls and ceiling of the garage, look for signs of plumbing leaks. Check the ceiling for water-leakage stains. If the garage has an overhead door, be sure to close the door and then look at the ceiling. In the open position, the overhead door will block about 25 percent of the ceiling; if there are leakage stains in that section, you might not see them. Ceiling stains are often caused by leakage from a bathroom above the area. (See FIG. 7-6.) When you do the interior inspection of the house, all the plumbing fixtures (sinks, bowls, and tub-shower) should be operated. After that portion of the inspection is completed, if the garage ceiling showed signs of past problems, you should reinspect it for indications of current leakage.

In some cases there are exposed drainpipes or water pipes in the garage. If your home is located in the northern part of the United States, the water pipes should be insulated as a precautionary measure against freezing.

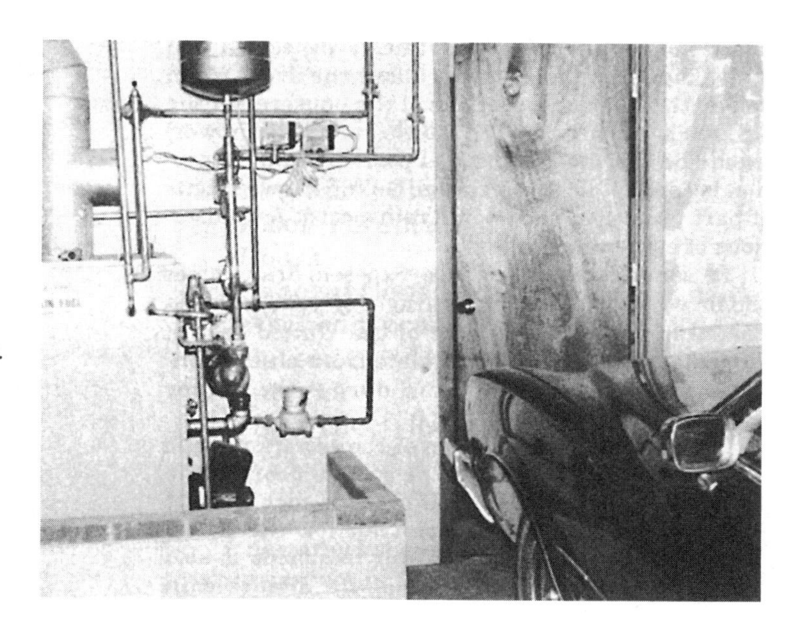

Fig. 7-5. *Heating system (oilfired, forced hot water) boiler located in the garage.*

Fig. 7-6. *Plumbing leak caused water-damaged ceiling in garage.*

Depending on the location of the sewer or septic tank, there might be a pit in the garage floor covered with a metal plate. It might contain a cleanout and trap for the house waste line. Sometimes the water inlet pipe is also located in this pit. (These items are discussed in detail in chapter 13.) Lift the cover and look inside the pit. Often the builder neglects to remove the wood framing around the sides of the pit (used as a form when constructing the open area). Because of the dampness in the pit, a wood liner will eventually rot and might be termite-infested. (This is an area where termites are often found; see chapter 8 to learn how to determine their presence.) If there is wood in the pit, it should be removed, regardless of its condition.

Flood potential

The bottom of the pit should be relatively dry in all but very wet weather. If the bottom contains water, it is an indication that the level of the subsurface water (water table) in the overall area of the home is high. When this condition exists, there is a possibility that during rainy periods the water level can rise and seep into the garage through the pit or through cracks in the floor slab. (Water seepage into this area is discussed in detail in chapter 11.)

Cracks in the floor slab can be caused by shrinkage or differential settlement and are usually not a concerning factor. They should, however, be sealed because they can allow water to seep into the garage. An extensively cracked or heaved floor slab is of concern because it may indicate a water problem. Heaving and extensive cracking are very often caused by water pressure being exerted on the underside of the floor slab. This condition should be evaluated by a professional.

Note whether there is a drain in the floor slab. The floor should be pitched toward that drain. If there is no drain, the floor should have a slight pitch toward the automobile entry door. This will allow water from melting snow to drain to the exterior rather than puddle on the floor. Also, the floor slab should be slightly above the level of the driveway to reduce the possibility of water entry. It should be noted that a garage at the base of an inclined driveway is *always* vulnerable to water penetration. (This condition is discussed in the driveway section of chapter 4.)

Doors

When you inspect the garage, you should always check the operation of the exterior door or doors. If the door(s) does not have an automatic control, open and close each door and note whether it operates relatively easily. The most common type of door for an attached garage is the sectional overhead type. This door has the advantage of not taking up usable space when open. Look for obvious deficiencies such as broken or missing springs or guide wheels, loose and misaligned tracks, and so on. Check the door's operation. If it is difficult to lift, stuck in a fixed position, out of plumb, or does not stay in the up position, some minor maintenance is needed. When closing the door, give it a start and let it come down by itself. If the door closes rapidly and heavily, it is a hazard, especially for small children. Adjustment is needed to the spring tension, or a new spring should be installed. Check to see if there is a restraining cable that runs through the center of the spring. (See FIG. 7-7.) The cable is a safety feature that prevents the spring from whipping around and injuring a person or damaging a car in the event that the spring breaks. When the door is closed, use the lock mechanism. You might find that the lock bars need to be reset.

Many overhead doors are opened by an automatic control. Operate the control. The doors should open and close smoothly without binding in the tracks. Check to see if the control unit has a reversing feature. This is very important from a safety point of view. When the door is closing, exert a force in the upward direction at the bottom of the door. The door should stop and then reverse its downward travel. If it doesn't, record that fact on your worksheet. Repair or replacement of the control unit is recommended because it is a potential safety hazard. The newer garage door opener assemblies include a photoelectric

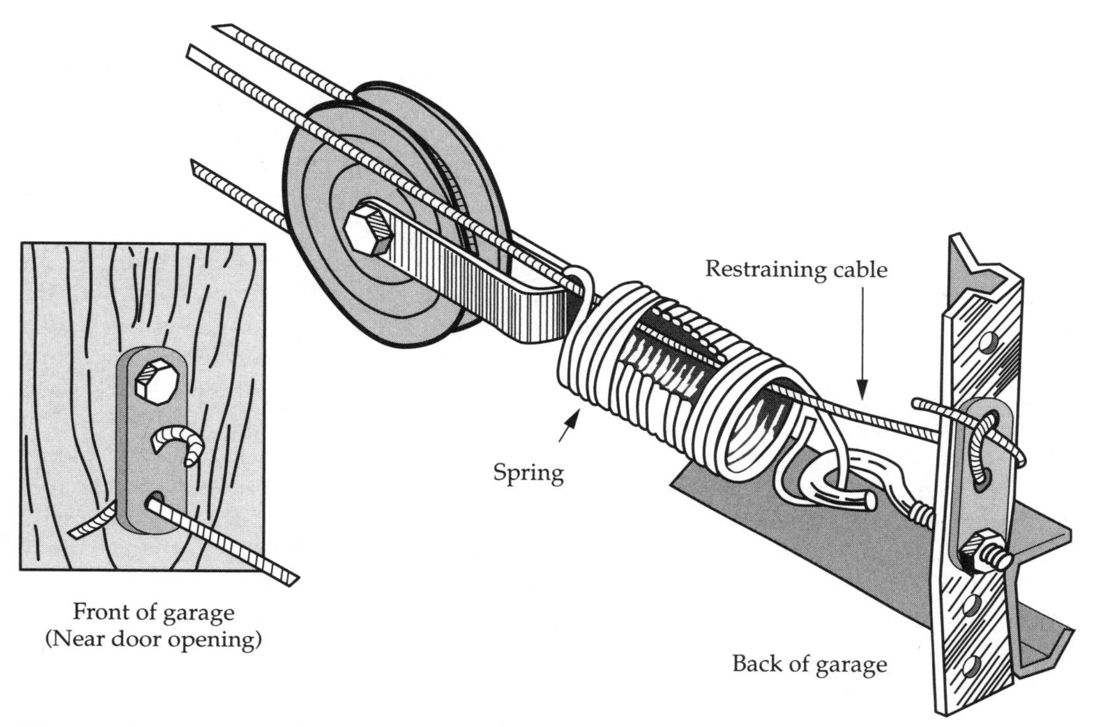

Fig. 7-7. *A restraining cable through the garage door spring prevents the spring from whipping around if it breaks.*

sensor that is mounted on both side tracks, about 12 inches above the floor. The sensor activates the controller, which will reverse the door when it is closing if a person or animal breaks the beam by passing through the opening. If the door does not reverse, it is very often because the sensor is misaligned causing the electronic beam to not contact the receiver's "bulls eye." This can be easily corrected by repositioning the sensor. Make certain that the radio controllers work. Ask the owner to demonstrate that they exist and work.

Some overhead doors are the one-piece, swing-up type rather than the sectional, roll-up type. These doors often require additional efforts to open, particularly during periods of snow and wind. Other doors found on a garage are the sliding and folding types. Sliding doors usually hang from overhead tracks. One disadvantage of such doors is that they take up valuable wall space when open. Also, small pieces of debris on the ground can interfere with their operation. Fold-out doors often sag, have loose hinges, and drag on the ground, making opening and closing quite difficult. In general, they require more frequent maintenance. These conditions can and should be corrected.

General considerations

Look around the garage for an electrical outlet. There should be at least one three-prong convenience outlet. Also, there should be overhead lights controlled by a switch near the interior door and the exterior door. Two desirable features, although not necessary, are windows that provide daylight and ventilation and a service door that can be used as access to the garage without opening the automobile entry doors.

Depending on the location of the house, the garage might be heated. Heat is usually provided by extending the central heating system (hot water, steam, or warm air) into the garage area. The radiators or heat registers providing the heat should be checked as part of the overall heating-system inspection. If there are warm air ducts in the garage, look for return grilles. There shouldn't be any because through them poisonous exhaust fumes could be brought back into the system and circulated throughout the house. When a garage is used solely for storage of automobiles, it is necessary (and not even always) to bring the temperature only above freezing. Heating the garage above that temperature is wasteful of energy.

Last, when inspecting an attached garage, look for termites in any exposed wood-frame members. The area most vulnerable to termite infestation (aside from wood found in a sewer cleanout pit) is the wood framing around the base of the exterior door.

Detached garage

If your house has a detached garage, you need not be as concerned with the fire and health hazards mentioned with the attached garage. True, the area is still considered a potential fire hazard; however, since the structure is physically apart from the main building, a fire would not usually result in the loss of life. The main concern with this type of garage is its structural integrity.

Exterior

The exterior of the detached garage is checked the same way you inspect the main house. Walk around the outside of the building twice. The first time, look at the roof and gutters. Do any of the roof beams appear to be sagging? If so, additional bracing might be needed. Have a professional make this determination. Do not assume that the roof over the garage and the roof over the house are in the same condition. Although the roof covering on the main house might be in good condition, the covering of the garage roof might be badly worn and require replacement. (Inspecting roofs is discussed in chapter 2.) Are there gutters all around the base of the roofs? If not, make sure you check all wood siding and trim for rot. The rain runoff from the roof can promote rot. A wood-frame garage with a pitched roof should have gutters. If there are long overhanging eaves or the garage is masonry-constructed, gutters are not a necessary feature, although they are often desirable. If there are gutters and downspouts, see if they need repair. (These items are discussed in chapter 3.)

After looking at the roof and gutters, walk around the building once more. This time look at the walls, windows, and doors. If the exterior walls are covered with wood siding, does the base of the siding extend to the ground? It should end about 8 inches above the ground. If the siding is in contact with the ground, it should be checked for termites and rot. Pay particular attention to the rear wall. You might see a wall that is bowed. This is usually caused by a car that did not stop in time. The wall stopped the car, and in the process, the supporting studs were broken. If such is the case, the wall is in need of rehabilitation. Also, you might sometimes see a wall that is offset; the bottom section of the wall extends about 3 feet beyond the upper section. This is done to accommodate longer cars than those for which the garage was constructed.

Finally, check the base of the wood framing and trim around the garage doors. This area is particularly vulnerable to rot and termite activity. (See FIG. 7-8.)

Interior

When entering the garage, check the doors first. Look for broken and cracked sections of wood framing and glass panes. Open and close the doors; they should operate smoothly and have all necessary hardware (see page 88).

Depending on the location of the garage, at times the entire roof is not visible from the outside. After you enter the garage, look up at the underside of the roof. If you can see daylight through a hole or crack, there is a problem with the roof. Look for signs of past water-leakage stains on the wood framing. These stains appear as dark-streaked discolorations on the wood. Leakage stains do not necessarily indicate a current leak—the problem might have been corrected. If you see stains on the wood framing, ask the homeowner whether repairs to the garage roof have been performed. Next, look at the walls. If the garage is located on an incline, look at those sections of walls that are below grade. These walls are usually constructed of brick, concrete, or rubble and also function as retaining walls. If proper drainage provisions have not been made, the walls will tend to crack and heave. (See FIG. 7-9.) If you see cracked and heaved walls, you should have a professional make a determination whether rehabilitation is required. Often the walls are covered with a stucco or plaster finish, and the wood-framing members that form the walls are not visible. However, any exposed studs and bottom plates should be checked for cracked and broken sections, rot, and termite activity.

To reduce the vulnerability of the bottom plate of a wood-frame wall to rot and termite infestation, the plate should be resting on a foundation wall that is at least 4 inches above the garage floor. In many older detached garages, this plate is found directly on the

Fig. 7-8. *Termite infestation and rot at base of garage door frame.*

Fig. 7-9. *Heaving garage wall. Section of wall was located below grade level.*

floor or in contact with the ground. If this is the case, look carefully at the plate and probe it with a screwdriver or ice pick. If it can be penetrated, there is probably rot, termite, or carpenter-ant activity.

Look at the condition of the floor. If there are cracked, broken, and settled sections, often found in older detached garages, rehabilitation is in order. This condition usually does not indicate an undermining of the structural integrity of the garage but a poor installation of the floor slab. In some garages, you will find a dirt floor rather than concrete or asphalt. This is not desirable because the dampness associated with this type of floor promotes rot in the wood-framing members and premature rusting of items stored in the garage.

Heat and electricity

Most detached garages are not heated. However, when they are, heat is usually provided by a space heater rather than by extending the central heating system. The heater should be checked to see if it is operational by turning up the thermostat. The thermostat will be wall-mounted or mounted directly on the unit. Most nonelectric heaters must be vented to the outside and should not have wood framing in contact with the exhaust stack, a fire hazard. If the heater is not vented to the outside, ask the owner to show you proof that the unit has been specifically approved for installation without a flue connection.

In new detached garages, the electrical service and wiring is usually not a problem. There should be an overhead light controlled by a wall switch and at least one three-prong outlet receptacle. If the garage is a distance from the house, a desirable feature would be to have either spotlights or row lights along the path between the two structures. The lights should be controlled by two three-way switches, one at the garage and one at the house. In many older garages, the electrical wiring and service is often makeshift and nonoperational. Look around. If you see loose and hanging wires, exposed junction boxes and wire splices, you are looking at electrical violations. In some cases, the service wire from the main house to the detached garage is interior wire, not exterior. This is a potential hazard and must be corrected. If electrical problems are found during the garage inspection, you should require that the seller provide you at closing with a certificate of approval for the electrical system. The approval should be made by the municipal electrical inspection agency. (See chapter 12.)

Checkpoint summary

Attached garage

Inspecting for fire and health hazards

- ❍ Are the garage and basement area combined into one open area?
- ❍ Is the interior garage door located at least one step above the garage floor?
- ❍ Does this door have a tight seal? Is it self-closing?
- ❍ Is this door fire-resistant, or does it have a sheet-metal covering on the garage side?
- ❍ Is the boiler/furnace unit located in the garage?
- ❍ Has it been placed on a raised slab?
- ❍ Inspect garage ceiling and walls for exposed wood-frame members.
- ❍ Check ceiling area for open or missing attic access hatch.
- ❍ Check for return grilles in warm-air heating systems.

General considerations

- ❍ Inspect ceiling area for signs of plumbing leaks, stains, and patched sections.
- ❍ If garage is unheated, are there uninsulated water pipes that are vulnerable to freezing?
- ❍ Inspect floor for extensively cracked, settled, and heaved sections.
- ❍ Check these areas for evidence of water seepage and silt deposits.
- ❍ Does driveway incline make garage vulnerable to flooding?
- ❍ Is there a drain protecting the garage entry? Is it adequate?
- ❍ Does garage floor contain a drain?
- ❍ Inspect exterior doors and trim for cracked, missing, rotting, and insect-damaged sections.
- ❍ Operate doors. Note broken and missing springs, guide wheels, locks, and misaligned tracks.
- ❍ Is there a restraining cable running through the spring?
- ❍ If the overhead door is electrically controlled, does it reverse its downward travel when an upward force is exerted on the door?
- ❍ Check overhead lights, wall switches, and convenience outlets.

Detached garage

Exterior

- ❍ Inspect walls/siding for bulging, cracked, loose, missing, and rotting sections.
- ❍ Note broken windows and patched sections.
- ❍ Check roof beams for cracked, rotting, and sagging members.
- ❍ Inspect roof shingles (as outlined in chapter 2 checkpoint summary).
- ❍ Check type and condition of gutters and downspouts. Note their absence.
- ❍ Inspect and probe wood framing and trim around doors (particularly doors that are in contact with, or in close proximity to, the ground).

Interior

- ❍ Check garage doors for broken, cracked, and rotting sections.

- ❍ Inspect doors for operation, sagging sections, missing hardware, and broken glass panes.
- ❍ Inspect underside of roof for damaged sheathing and signs of leakage.
- ❍ Inspect foundation/retaining walls for cracked, bowed, and heaved areas.
- ❍ Concrete, asphalt, or dirt floor?
- ❍ Check concrete or asphalt floor for cracked, broken, and heaved sections.
- ❍ Probe wood sills for insect-infestation damage (particularly if these members are in contact with the ground).
- ❍ Inspect for loose and hanging electrical wires, exposed junction boxes, wire splices, extension-cord wiring, and makeshift wiring.
- ❍ Is there a space heater? Check operation.
- ❍ Is unit properly vented?
- ❍ Are wood-frame members in contact with the exhaust stack?

8

Wood-destroying insects and rot

There are many types of wood-destroying insects—subterranean and dry-wood termites; carpenter ants; and powder-post beetles. The one that causes the most damage to residential structures in the United States is the subterranean termite.

Termites

Home buyers generally overreact after discovering a termite condition and on occasion lose interest in the house. Actually, the discovery of termite infestation should not be cause for alarm—concern, maybe, but certainly not alarm.

Termites work very slowly. It takes many years for termites to do serious damage to a house. A mature colony of 60,000 termites eats the equivalent of 2 to 4 feet of 2-by-4-inch board in one year. Some well-established termite colonies have been estimated to contain more than 2 million termites. A termite condition can be controlled through the application of chemical insecticides by constructing a chemical barrier in the soil around and beneath the house. Thus termites attempting to go through the termiticide-treated soil to reach the house are either killed or repelled.

Prior to the mid 1980s, the chemical most often used for termite treatment was *chlordane*. It had an effective life that often exceeded twenty-five years. However, because chlordane is considered a potential risk to human health, it was withdrawn from the U.S. market. Currently several different chemical termiticides are available for use by pest control operators for controlling a termite infestation. All of the chemicals are considered safe and have been found to be effective in the soil for periods of approximately 5 to 10 years.

Around 1995 the termite bait system came on the market for termite control. This system is an option for homeowners that don't want to use a chemical barrier treatment for termite control. Termite baits are considerably less toxic than most liquid termiticides, which are introduced into the soil by the hundreds of gallons in order to effectively control a termite condition in a house. Baits, on the other hand, deliver very small amounts of termiticides over a long period of time.

A bait system for termite control consists of installing plastic tubes or boxes in the ground at various locations around the house. Inside the tubes and boxes is a slow-acting poison combined with a termite food material such as paper or cardboard. Termite control depends on foraging termites finding the bait stations during their random search for a new food source, feeding on it, and carrying it back to the colony where the poisoned food is shared with other termites in the colony.

The length of time for termites to find the bait stations will vary considerably and depends on whether the stations are installed in the southern or northern states. It also seems to depend on whether the bait stations are installed in the spring or late summer. The time to locate the bait has been found to vary from as little as a day to as long as a year or more. Baiting to control a termite problem is a slow long-term solution, and it is not the recommended method to control a heavy infestation problem in a house.

In real estate transactions, if the house has a termite problem, a barrier treatment is the preferred control method rather than baiting. Successful control by baiting is a long-term commitment to frequent inspections and monitoring of the bait stations and rebaiting. In contrast, when necessary, a single application of a barrier treatment can be expected to last from 5 to 10 years after which retreatment may or may not be necessary.

A number of states have regulations requiring a termite inspection by a professional prior to, or as a condition of, the purchase agreement. The cost of this inspection is almost always paid by the seller. If your state has such a requirement, you should ask the seller or real estate agent to have the house inspected by a professional and have a report of the results sent to you. In many states (even in some that do not have a prepurchase termite inspection requirement), if a termite condition is found prior to the sale, the cost for correcting the condition (chemical treatment) is borne by the seller.

When termites are discovered, they should be exterminated professionally. However, because termites work slowly, termite-proofing the house need not be done immediately upon learning of an active infestation. Take your time and get two or three cost estimates from established termite-exterminating firms. After treating a house, most companies provide a one-year guarantee against reinfestation. The guarantee can often be extended annually for a fee, which covers inspection and retreatment if necessary. If the house had been treated previously for termites, find out if the owner has a guarantee and whether it can be transferred to you.

During an inspection, all exposed wood-framing members should be checked for structural deterioration from termite activity. There are very few houses on record that have been damaged by termites to a point where they are considered unsafe. Quite often the damage caused by termites (by the time termite activity is discovered) is minor, and repair or replacement of the infested wood members is not necessary. Even with a heavy infestation, usually only a portion of the house is affected. And even then, only a portion of the wood framing might be damaged to a point where it has lost its structural value. In this case, only the affected members require repair or replacement. If you are in doubt about the structural integrity of any of the affected members, you should consult a professional.

Termites play an important role in the natural ecological cycle. They feed on cellulose, the principal ingredient of wood, and help to break down dead trees in forests and other wooded areas, thus enriching the soil. Termites began attacking houses when the wooded areas were cleared for building construction and there was no other available source of food near their nest. Subterranean termites are found in every state except Alaska. Their overall distribution within the continental United States is shown in FIG. 8-1. As their name implies, subterranean termites live in a colony (nest) that is usually located in the ground below the frost line. Even when a house is infested with termites, they usually do not have a nest in the house. They are there only to gather food. The only condition under which a nest might exist in a house (a rare occurrence) is a constant source of moisture

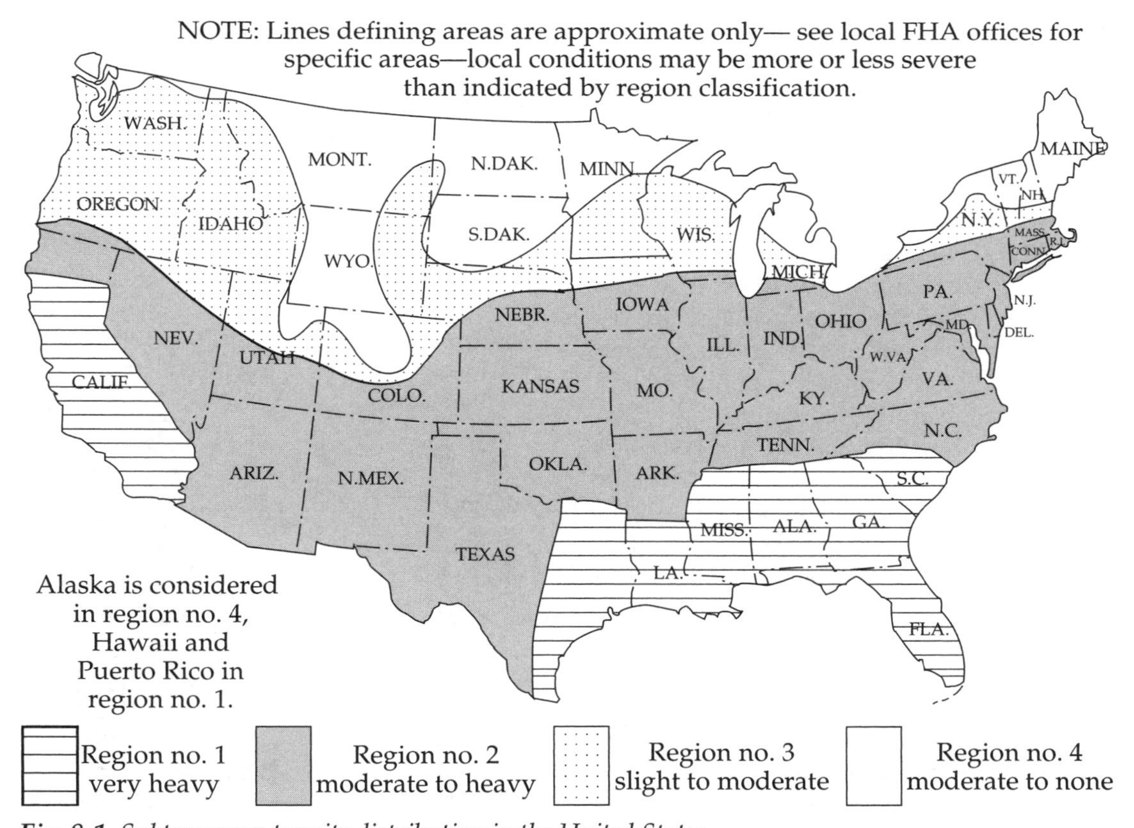

Fig. 8-1. *Subterranean termite distribution in the United States.*

such as a leaky waterpipe or drainpipe that wets the surrounding area.

Termites are social insects. Within each colony, there is a rigid caste system consisting of a queen and king, workers, soldiers, and reproductives. Each member of the colony instinctively performs its special task. The function of the queen and king is to propagate the colony. The fertilized queen lays the eggs and might live for as long as twenty-five years. The workers care for the eggs, feed the young and the queen, and generally maintain the colony. They also forage from the nest to the wood supply and return with food. The soldiers defend the colony against attack by other insects, mostly ants. The average worker and soldier live only two or three years. The function of the reproductives is to replace the queen and king in the event of their injury or death. They also lay eggs that rapidly increase the termite population.

When a colony matures, reproductives leave the nest (swarm) to set up a new colony. Although thousands of reproductives leave the nest, only a handful survive to establish a new colony. The remainder die because of adverse conditions in the soil or attacks by other insects. Reproductive termites sprout wings for the swarm. With their wings, they are only about ½ inch long. They are considered poor fliers and generally flutter around before falling to the ground. Some, however, might be picked up in the wind and carried great distances. Once the reproductives land, they shed their wings, pair off in couples, and return to the soil in search of a suitable place to build a nest.

In most parts of the country, swarming generally occurs in the spring, sometimes in the fall. However, swarming termites have been found in January in some heated houses. In the warm, humid parts of the country, swarming can occur at any time. Even if there are no other outward signs of termite activity, termite swarming in a house is an indication that there is a healthy established colony nearby from which worker termites are coming in their search for food.

Swarming termites do not attack wood. Their only function is to start a new colony. Even if a swarm is in your house, you might not see it. A swarm might last from fifteen minutes to one hour, and if you are not in the right place at the right time, it can be over by the time you enter the room. However, if there was a swarm, you can tell by the discarded wings. They are often found on windowsills and light fixtures, and beneath doors. Do not confuse swarming termites with swarming ants. To the untrained eye, they appear similar, but there are distinctive differences. (See FIG. 8-2.) The most obvious difference is that termites have a thick waist and ants have a pinched (hourglass) waist.

Subterranean termites

Subterranean termites require a dark, damp environment. In their search for food, worker termites build shelter tubes (tunnels) that help conserve moisture and shield the termites

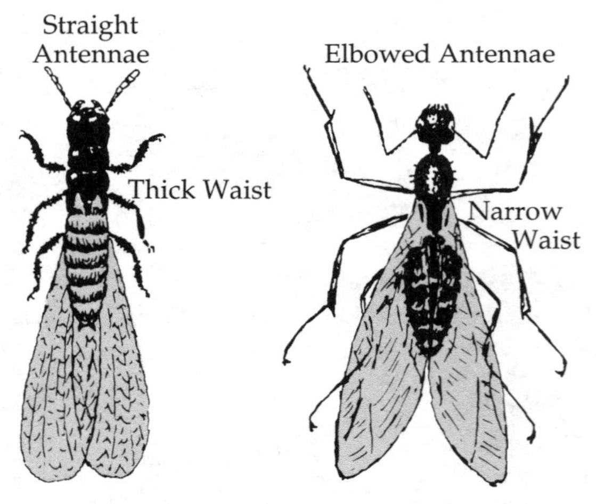

Fig. 8-2. *The difference between swarming ant (right) and swarming termite (actual size ½ inch).*

from the light. (See FIG. 8-3.) The tubes are about ¼- to ½-inch wide and provide a passageway between the ground and the food supply (wood member). They can be built at the rate of several inches per day and are mainly composed of soil, wood particles, and termite excreta. Shelter tubes, which might be noted on foundation walls, on the outside of wood framing, or even freestanding between the ground and an overhead pipe or beam, are visible evidence of termite infestation. If the tube is active, worker termites will be busy using it to go between the nest and the house. By breaking the tube, you can see the workers, who will try to repair the break. They are about ¼-inch long and have a whitish cream coloring.

Some tubes might be abandoned. If you find an inactive shelter tube, it does not mean that termites are no longer in the house. It might, if the house has been termite-proofed. However, if it has not, even an abandoned shelter tube, no matter how small, is sufficient evidence to consider termite treatment. Many shelter tubes emanate from a nest. There might be an active tube inside the voids of a concrete block wall that would not be visible during an inspection.

Whether a new building will be attacked by termites depends on the surrounding area and to a large extent the builder. Certain construction practices tend to increase the probability of termite attack. (See FIG. 8-4.) Some builders have been known to bury tree stumps and wood debris near the foundation or below the basement-floor slab. All stumps and debris should be removed from the building site. All form boards and scrap lumber should be removed before the excavated area around the foundation walls is backfilled. There should be no buried wood around the house. Otherwise, it can provide a source of food for a new termite colony that when it becomes large enough, will attack the house.

Fig. 8-3. *Termite shelter tubes: on foundation wall, header, and subflooring (top); hanging tubes in crawl space (bottom).*

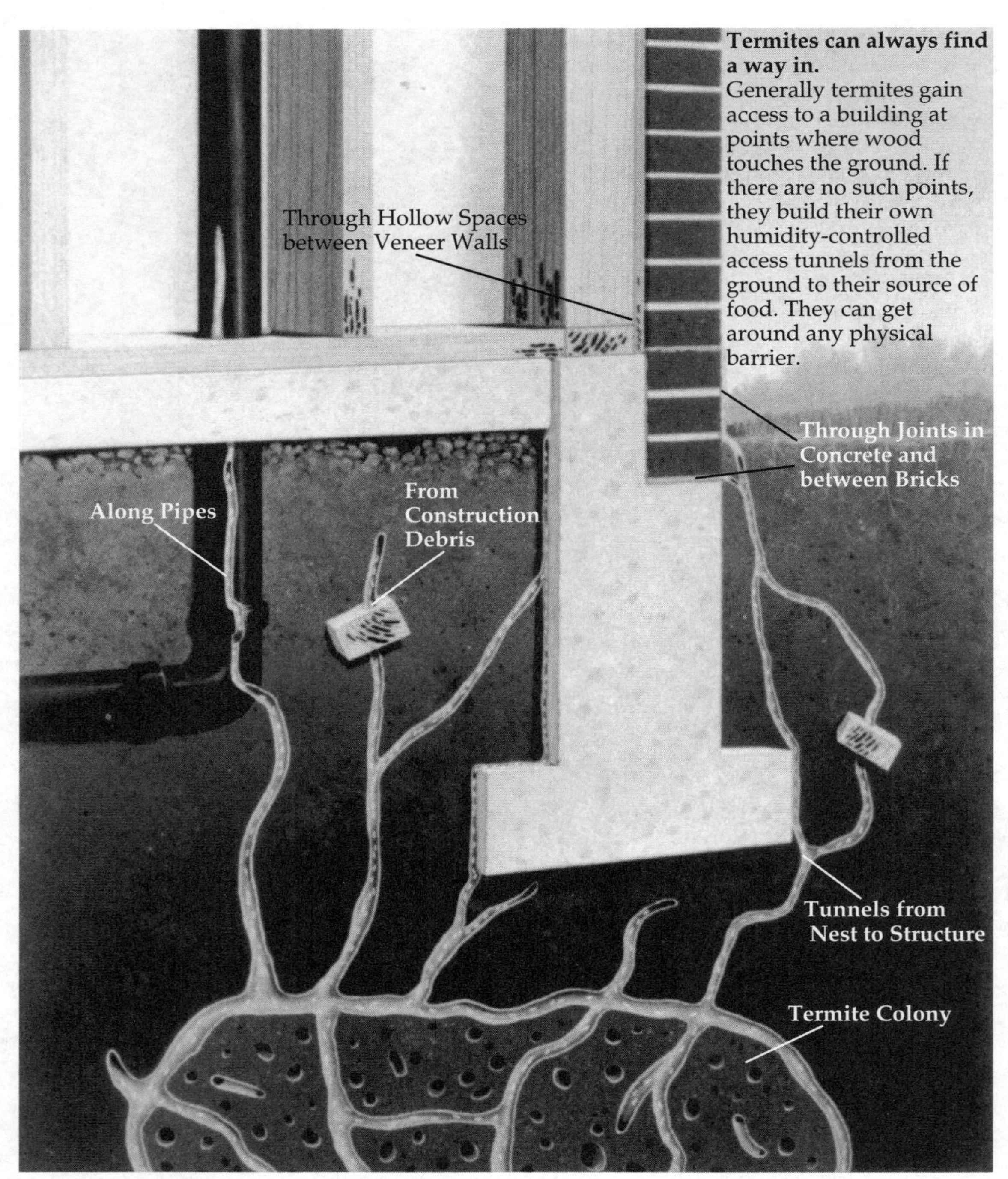

Fig. 8-4. *Termite entry points into a house.*

Most often termites enter a house by eating their way through untreated wood members that are in direct contact with the ground. Some of the more common points of entry are garage door frames, basement windowsills, wooden steps and supports, wood sills, and headers and studs on foundation walls that are located at or below grade. A particular area of attack is the wood framing adjacent to a concrete-covered, earth-filled porch, patio, or entrance slab. (See FIG. 8-5.) If there is no earth or wood contact, termites can build shelter tubes to provide passageways from the nest to the wood framing in the structure.

Some homes have a strip of metal (termite shield) between the foundation wall and the sill plate, that rests on top of the foundation wall. The purpose of the termite shield is to act as a barrier between the nest and the food supply. In most cases, the termite shield gives the homeowner a false sense of security. The shield does not prevent infestation. It only deters an attack. The problem is that the termite shield is rarely installed properly. An opening at a seam or a hole as small as $\frac{1}{32}$ inch is large enough for termites to pass through. All seams should be soldered, and any holes around bolts and pipes should be filled with coal-tar pitch. Even if you see a termite shield, you should look for termite infestation.

Inspection A complete subterranean termite inspection consists of an interior and exterior check of that portion of the house that is close to, or in contact with, the

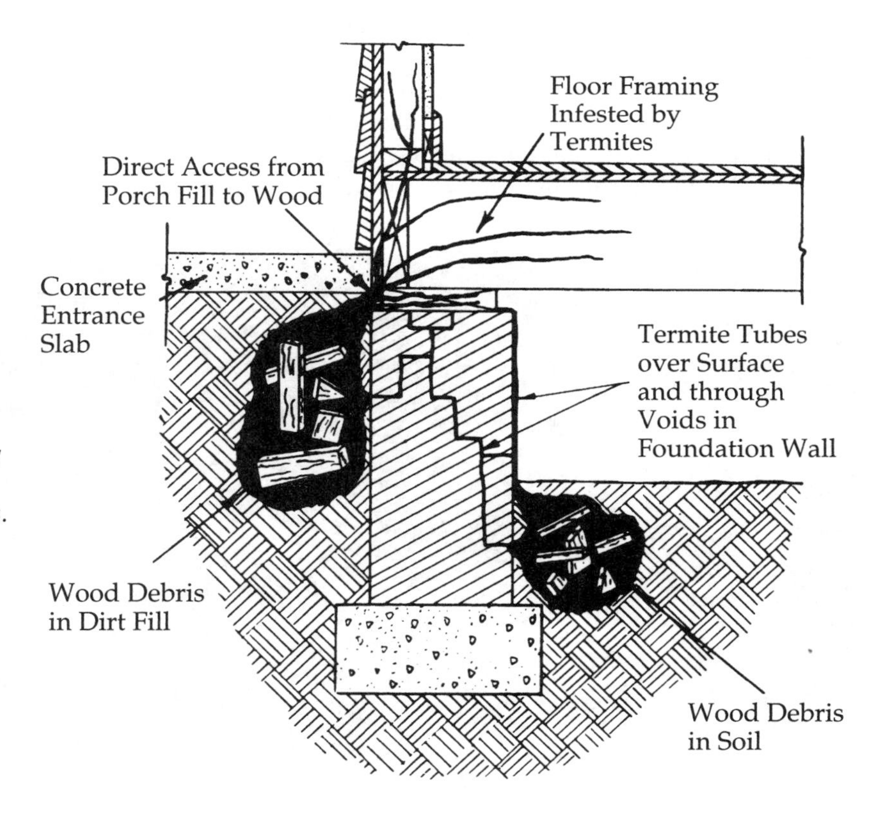

Fig. 8-5. Termite colonies can develop in buried wood debris and gain entrance into a building, particularly at earth-filled concrete entrance slabs or patios.

ground. The exterior termite inspection can be performed concurrently with the normal exterior inspection as described in chapter 1. As you walk around the outside of the house, look for termite shelter tubes along the outside foundation walls. (See FIG. 8-6.) In many homes, this area is covered or partially blocked by shrubbery. Part the shrubbery so that you can see the wall. This is especially important for areas just below a garden-hose spigot.

In some cases the base of the exterior wood siding is in contact with the ground, so that the foundation wall is not visible. This is a poor construction practice but unfortunately is fairly common and usually occurs during final grading and landscaping. The base of the wood siding should terminate at least 6 inches above the finished grade. Redwood or cedar exterior siding is often used. Both types of siding are resistant to termite attack and rot. However, it is important to understand that they are not immune to attack and might eventually succumb. If the base of the siding is in contact with the ground, probe it with an ice pick or a screwdriver. If the wood has not been attacked by rot, termites, or other wood-destroying insects, your probe will not penetrate much beyond the surface. If the probe penetrates the wood deeply, the wood has been attacked. To determine the cause for the deterioration, it is necessary to break open a section and look at the condition of the wood. Caution: Do not proceed beyond the probing without consent of the homeowner; in fact, it would be wise to obtain the homeowner's consent for the probing.

A section of termite-damaged wood reveals *galleries* (channels) that run parallel with the grain. (See FIG. 8-7.) Termites attack the softer portion of the wood grown during the spring and not the denser summer wood. The channels will not look polished, as they do with carpenter ants. Portions will be lined with grayish specks that consist of excrement and earth. Deteriorated wood could be the result of a combination of causes such as rot, termites, and carpenter ants. You should become familiar with the telltale signs. If you

Fig. 8-6. *Termite shelter tubes at base of exterior wall.*

Fig. 8-7. Termite-damaged wood. Note galleries that run parallel with the grain.

are not certain of the cause, have the wood evaluated by a professional. The appearance of wood damaged by carpenter ants, powder-post beetles, and rot will be discussed in their respective sections.

As you walk around the outside of the house, probe the attached wood trim, posts, and framing members that are on or close to the ground. Specifically probe garage door frames, basement or lower-level window frames, step stringers, deck posts, and the entry-door riser. Termite activity is of concern only when it is found in the house or in an attached structure such as a garage or deck. If you find termite damage or shelter tubes on a fence post in the yard (FIG. 8-8) or in a piece of wood debris on the ground in the yard, all that means is that those pieces of wood have had termite infestation. It does not mean that the house is infested with termites and should be treated.

After checking the outside of the house, the next place to look is in the crawl space beneath the house or porch. (Not all houses have a crawl space, in which case this step is omitted.) If the access to the crawl space is from the interior, it should be checked as part of the interior termite inspection. In the crawl space, probe the sills and headers for termite damage. Also check the first 15 inches of each

Fig. 8-8. Termite shelter tubes on a fence post.

joist that rests on the sill or foundation. As you move around in the crawl space, look for termite shelter tubes on the foundation walls and piers. Sometimes the shelter tubes can be spotted between *double joists* (two joists nailed together, used to provide additional support for a heavy load). While in the crawl space, note any items of wood storage, remembering that they are vulnerable to infestation.

The interior inspection for subterranean termites is generally conducted in the basement and crawl space. If the basement is finished so that there are no exposed sections of foundation wall or wood framing, a thorough inspection cannot be performed, even by a professional. There are, however, some sections more vulnerable to termite attack than others, such as sill plates, headers, and joists below grade or adjacent to a dirt-filled, cement-covered patio. If the ceiling is covered with suspended tiles, the tiles can be lifted or moved to expose the wood framing. If there are no accessible areas, termite activity will have to be determined by a swarm or exterior inspection. In areas where termite infestation is heavy, such as the South and Southwest, a termite inspection should be performed every year. In other areas, an inspection every two year is adequate.

When the basement is unfinished, the wood framing (sill plate) on top of the foundation wall should be inspected by probing as described for the exterior crawl space. (See FIG. 8-9.) Also look for termite shelter tubes. You might find a tube that appears to start in the center of the foundation wall. Actually, the tube is connected to the earth through a small crack in the wall at that point. As previously mentioned, termites can work their way through cracks as small as $\frac{1}{32}$ inch. While in the basement, pay particular attention to the area around the furnace/boiler and pipes through the foundation or floor slab. Often activity is discovered in these areas.

Homes built on a slab without a basement or crawl space are also vulnerable to attack by termites. Because there are generally no areas with exposed beams or foundation, detecting termites is quite difficult unless the infestation has advanced to the point where shelter tubes are

Fig. 8-9. Termite-damaged sill plate found by probing with a screwdriver.

visible in the finished rooms. When inspecting a slab house, look for soft spots in the baseboard trim along the exterior wall. Also check for shelter tubes around openings in the floor slab, such as around plumbing or heating pipes.

Dry-wood termites

Homes in the South and Southwest are vulnerable to attack by dry-wood termites as well as subterranean termites. Dry-wood termites cannot live outdoors in northern climates and have not established themselves in those areas. Isolated cases of dry-wood termite infestation have been found in homes located as far north as New York and Ohio, but those are rare.

Dry-wood termites are so called because they build their nest in perfectly good wood that is not decayed and not in contact with the ground. In fact, in many cases they establish a colony in the wood-framing members of the attic. As with subterranean termites, reproductive dry-wood termites swarm from the nest periodically in an attempt to establish a new colony. The swarming termites often do not fly more than a few feet before settling down. However, if aided by air currents, they can fly more than a mile. Once paired, the king and queen seek cracks or checks in nearby wood, whether a roof or lumber pile, and set up housekeeping. In homes, dry-wood termites can be found in rafters, studs, joists, sheathing, floorboards, window frames, door frames, and exterior trim.

Once a colony is established, it feeds on the wood around the nest. Dry-wood termites are general feeders and eat spring- and summer-grown wood. The galleries thus formed will cut *across* the grain. The cavities contain pellets of partially digested wood. These pellets are tiny, seedlike, and usually straw-colored. On occasion, some of the pellets are pushed through openings in the wood surface. If there is not much accumulation, the pellets can easily be overlooked. They are, however, often the first sign of infestation.

The maximum population of a fully established dry-wood termite colony is estimated to be about 3,000. This is considerably less than the typical number of subterranean termites in an established colony, which is estimated to be between 60,000 and 250,000. Consequently, it takes a longer period of time for serious damage to occur with dry-wood termites than with subterranean termites.

Control Since dry-wood termites do not nest in the ground, they must be chemically treated at the source of infestation. This is usually done by injecting insecticide into the galleries, a procedure that should be performed by a professional. To reach the galleries, holes are drilled in the infested wood members. The insecticide will be a liquid or a dust. When the wood is dry, dusting can be effective as far as 15 feet from the point of application. However, when the wood is wet, dusting usually is not effective because the dust tends to cake in the moist galleries. If dry-wood infestation is found in a fence or pole on the property, those wood members should be treated since the termites represent a potential source of infestation for the house.

For severe infestation, treatment is usually by fumigation. The entire house, including the roof, is wrapped with a plastic covering. After all the openings are sealed, a poisonous fumigant is introduced. The house should remain under fumigation for at least forty-eight hours. Since the insecticide is poisonous to humans, fumigation should be undertaken only by experienced fumigators.

Inspection Since dry-wood termites do not leave their nest in search of food, there are no telltale signs of infestation such as shelter tubes. They can be detected, however, by their pellets, which tend to accumulate in a small pile after being pushed from the wood by the termites or falling through a crack in the infested wood. An infestation in the house can also be detected by the presence of a swarm if you happen to be in the room during an occurrence. Since dry-wood termites can attack wood located anywhere in the house, from the attic to the crawl space, all the exposed wood should be checked for signs of infestation. The wood should be gently probed, so as not to break the surface. Infested wood has hollow sections and if heavily probed, can break open, spilling the seedlike pellets. Dry-wood termites often consume wood up to the paint itself, forming what appears to be a paint blister. If the slightest pressure is applied to the blister, it can break. Care should be exercised to maintain the integrity of the wood surface; a broken gallery is difficult to treat with insecticide.

Formosan termites

The Formosan termite is a subterranean termite native to the Far East. It was first discovered in the United States in a Houston, Texas, shipyard in 1965 and has since spread to a number of southern coastal and Gulf states. All indications are that Formosan termites might eventually establish nests in some northern states as long as the temperature and moisture conditions are satisfactory.

An established Formosan termite colony, which can have more than a million termites, is extremely destructive. They can destroy wood six times faster than our native species. Formosan termites have been known to penetrate lead, plastic, rubber, mortar, and plaster to get to their food—wood. They are able to penetrate the above materials by secreting an acid substance from their frontal glands. In at least one recorded case, they have caused short circuits by damaging electric cables.

Identifying the worker Formosan termite is difficult because there are no obvious characteristics distinguishing them from our native species. However, in the latter colony, only 2 percent of the population are soldiers; whereas with the Formosan's, 25 percent are soldiers. As a result, if an active Formosan termite shelter tube is broken or an infested section of framing is examined, many more soldiers would be visible than if it was from our native species.

Native subterranean termites ordinarily have a ground connection and live partly in ground and partly in wood. However, the Formosan termite can live without a ground connection if there is a suitably located constant source of moisture. This characteristic is what makes the control of Formosan termites so difficult. For example, because of a plumbing leak, the termites might have a nest in an exterior wall. Consequently, chemical treatment of the soil around the house, which is the control procedure for native subterranean termites, will not exterminate the termites in the wall. Fumigation is needed.

Since the location of a termite nest is normally not known—it can be in the ground or in a wall—to control an infestation, it would be necessary to treat the soil around the house chemically and fumigate. With a single detached building, this combination of control procedures is possible; however, in urban areas with row houses and attached structures, fumigation is not a viable procedure. Unfortunately for these buildings, the condition cannot be corrected.

Other wood-destroying insects

Carpenter ants

Carpenter ants are also social insects. They live in colonies with a rigid caste system consisting of a queen, workers, and reproductives (sexually mature males and females that periodically swarm and set up new colonies).

Although worker ants can live four to seven years and the queen for as long as fifteen, colonies have been known to last for thirty to forty years. When the queen dies or is accidentally killed, specially fed workers take over the egg-laying function.

Carpenter ants differ from termites in that they do not eat wood. They merely excavate it to build a nest. The nest consists of irregularly shaped galleries that generally follow the grain. The small fragments of shredded wood that are generated during the excavation are removed from the galleries and deposited on the outside. Consequently, the galleries do not have the earthy appearance of the termite galleries. Rather, they have a polished or sandpapered appearance.

A carpenter-ant colony can be located on the ground in a decaying log or tree trunk or in the roof framing of a house. The ants also nest high in trees and can fly from there to set up new colonies in a house. They build their nests in a variety of locations, preferring wood that is moist or softened by decay. However, they will also build their nests in wood that is perfectly dry and sound.

When inspecting for carpenter ants, look specifically at sections of wood that have begun to decay as a result of a past or current moisture condition. Even though the source of the moisture might have been eliminated (as by correcting a leak), an ant colony might have been established already. Typical locations to inspect are portions of the wood framing, siding, or trim that are in contact with the ground; wood that has been dampened by the overflow from defective roof gutters; the area around a damaged section of siding or flashing; the base of hollow porch posts and columns, and areas with large open joints as might occasionally be found around exterior windows and doors. These areas should be probed with a screwdriver or an ice pick. If the wood yields, breaks, or cracks and ants come crawling out, there is a good chance that you have located a nest.

Consider yourself lucky if you do, because a carpenter ant nest is usually quite difficult to locate; it is often established in an inaccessible location in the wall or roof assembly. One indication of the existence of a colony is unexplained piles of sawdust. (See FIG. 8-10.) Some people, however, think that piles of sawdust are an indication of termites. They are not. Subterranean termites completely devour the wood that they are attacking and leave absolutely no trace of wood particles. Drywood termites also eat the wood completely. However, they do drop tiny, well-formed seedlike pellets. If the pellets are observed

Fig. 8-10. *Sawdust is an indication of carpenter ant activity in window trim.*

closely, they can be differentiated easily from irregularly shaped particles of sawdust.

When a house is infested with carpenter ants, there is little likelihood that the people living there are unaware of the condition. Numerous worker ants will be seen walking around the rooms as if they live there—which indeed they do, with free room and board. These ants feed on sweets, crumbs, and other foodstuffs normally found or spilled on a kitchen counter or floor. Carpenter ants are easy to recognize. They are among the largest ants in the United States, worker ants varying in size between ¼ to ½ inch long. They are black or black with a reddish brown midsection.

While the first sign of infestation is usually the presence of carpenter ants in the house, the fact that they are there does not mean that the nest is inside the house. It might be outdoors, and the ants may have entered the house foraging for food. A carpenter-ant infestation can be controlled only by destroying the nest, either directly or indirectly. Nests can sometimes be located by watching the ant traffic. Ants continually entering and leaving an area are generally an indication of the nest location. If the nest is found, it can be treated directly with insecticide. If not, dusts or sprays can be used where the ants are commonly seen. The latter might not eliminate the infestation, but it should reduce it.

Powder-post beetles

There are many types of wood-boring beetles. The ones whose larvae or grubs feed on seasoned wood and break it down to a powdery residue are commonly called powder-post beetles. These beetles exist all over the United States, although the greatest concentration will be found in those states with a warm, humid climate. The two principal varieties of powder-post beetles are the lyctid and the anobiid beetles. The lyctid beetle attacks only hardwoods; the anobiid beetle attacks both soft and hardwood timbers.

For the most part, powder-post beetles are usually brought into the house via the wood that had been used in its construction. Building materials might become infested while being stockpiled in the lumberyard. The insects might also be brought into the house in finished wood products such as oak flooring, paneling, and furniture.

The beetles lay eggs in the open pores, cracks, and crevices in the surface of unfinished wood. After the eggs hatch, the larvae feed and tunnel their way through the wood, reducing it to a powder. Depending on the temperature and moisture content of the wood, the larval stage can be as short as a few months or as long as a few years. Just prior to emerging, the newly formed adults chew small round exit holes in the wood surface (1⁄32 to ⅛ inch). In the process of emerging, finely powdered wood called *frass* is usually pushed out in front of the body. This is often the first external sign of infestation. Shortly after emerging, the beetles mate and lay eggs. They occasionally deposit the eggs in the mouth of an old exit hole, thereby reinfesting the same piece of wood. Some wood members have shown signs of extensive damage as a result of infestation by several generations of beetles. (See FIG. 8-11.)

On occasion the homeowner might be the one responsible for the powder-post beetle infestation in the house. Under natural conditions, the beetles breed in dead branches and limbs of trees. When gathering wood for the fireplace, it is possible to pick up infested pieces and store them in the basement or under the stairs for later use. If wood is left in storage through the following spring and summer, the emerging beetles might attack unfinished lumber such as girders, joists, studs, sill plates, and subflooring.

Inspection Inspection for powder-post beetles should be performed along with the

Fig. 8-11. *Wood post with extensive deterioration caused by powder-post beetle infestation. Note beetle exit holes at the lower section of the post.*

inspection for termites. When probing the exposed wood-framing members, look for the small round emergence holes of the beetles. Since it is possible for the beetles to emerge without reinfesting the wood, the fact that there are emergence holes does not mean that the wood member is currently infested. Newly formed flight holes are light and clean in appearance, like a fresh saw cut; older holes are darker in color. If the infestation is well established, there will usually be more than thirty exit holes per square foot of surface. Even though the wood might no longer be infested, small amounts of larvae frass might continue to sift through the holes for many years as a result of normal vibrations of the wood. Look at the frass. If the infestation is no longer active, it will have a yellowish appearance or will be caked.

If there is any doubt about whether the infestation is active, call a professional pest-control operator for an evaluation. An infestation in a single wood-framing member can often be controlled by replacing that piece of wood or coating it with an appropriate insecticide. However, if the infestation is widespread, chemical treatment by a professional is necessary.

Rot

Wood products used in construction are susceptible to decay (rot). However, if properly maintained, they can easily last for hundreds of years. There are three basic types of fungi that attack wood: *stain*, *mold*, and *decay*. Wood rot is caused by an attack of the decay fungi. Stain and mold fungi mainly grow on the wood surface, causing discoloration. By themselves, they do not weaken the wood; but their presence does indicate a moisture problem and should serve as a warning of conditions favorable to the growth of the decay fungi. The decay fungi are microscopic threadlike plants that grow in the wood and attack its thick cell walls. They break down the walls and feed on the contents of the cells. With the destruction of the cells, the wood disintegrates, and decay becomes evident.

In the early stages, it is difficult to recognize that a section of wood has been attacked by the decay fungus. The wood might merely be discolored. However, the advanced stages of decay are easily recognizable because the wood undergoes changes in properties and appearance. The affected wood might be brownish and crumbly or white and spongy. In either case, the decay greatly reduces the strength and structural value of the wood member. The brown, crumbly rotted sections readily break into small cubes and in the final stage of deterioration are often quite dry. Most people when seeing this condition refer to it as "dry rot." This is really a misnomer. The actual decay occurred when the wood was wet since decay fungi cannot survive in dry wood.

Sometimes the physical changes in the wood are not apparent on the surface. They can be detected easily, however, by probing the wood with an ice pick or a screwdriver. If the wood is in good condition, the probe will not penetrate much beyond the surface. However, if the wood has deteriorated, the probe will easily penetrate into the wood. The conditions that promote the growth of the decay fungi also promote subterranean-termite and carpenter-ant activity. Consequently, when probing wood-framing members, you might find deteriorated sections that are caused by a combination of insect damage and decay. Under suitable conditions of temperature and humidity, the decay fungus gives rise to a fruiting body that contains enormous numbers of microscopic spores. The spores are the seeds of a new generation of decay fungi and are readily distributed by air currents. The spores are always present in the air and under normal conditions cannot be kept away from wood. The presence of decay-fungi spores on wood is of no concern unless the moisture content of the wood and the temperature are such that the spores will germinate and grow. Figure 8-12 shows the decay hazard zones in the United States.

Decay fungi will grow and develop only when the moisture content of the wood is in excess of 20 percent and the temperature is in a range from 40° F to 115° F. Temperatures above the upper limit kill the decay fungi; temperatures below the lower limit cause the fungi to become dormant. The latter is a condition that readily occurs in the northern states during the winter months. In the spring, when the temperature rises, the decay fungi in infested lumber resumes growth, assuming the moisture content of the wood has not changed. The moisture content of wood is defined as the weight of water in the wood expressed as a percentage of the weight of the wood when oven-dry. The decay fungi thrive when the moisture content is about 25 percent. However, when the wood is saturated with moisture, the decay fungi are inhibited from growing because of the lack of oxygen.

The moisture content of green lumber can be as high as 200 percent; after the lumber is kiln-dried, its moisture content may be as low as 7 to 10 percent. Wood in a house that has been properly constructed and maintained will seldom have a moisture content over 15 percent. However, once the moisture content exceeds 20 percent, the wood becomes vulnerable to deterioration by the decay fungi. If the building design is such that some of the wood must be subjected to damp or wet conditions, those sections should be treated with toxic chemicals to prevent decay or be made from the heartwood of certain species (cypress, cedar, or redwood) that are resistant to rot.

Inspection

As with the inspection for wood-destroying insects, the inspection for rot should be conducted on the exterior and interior of the

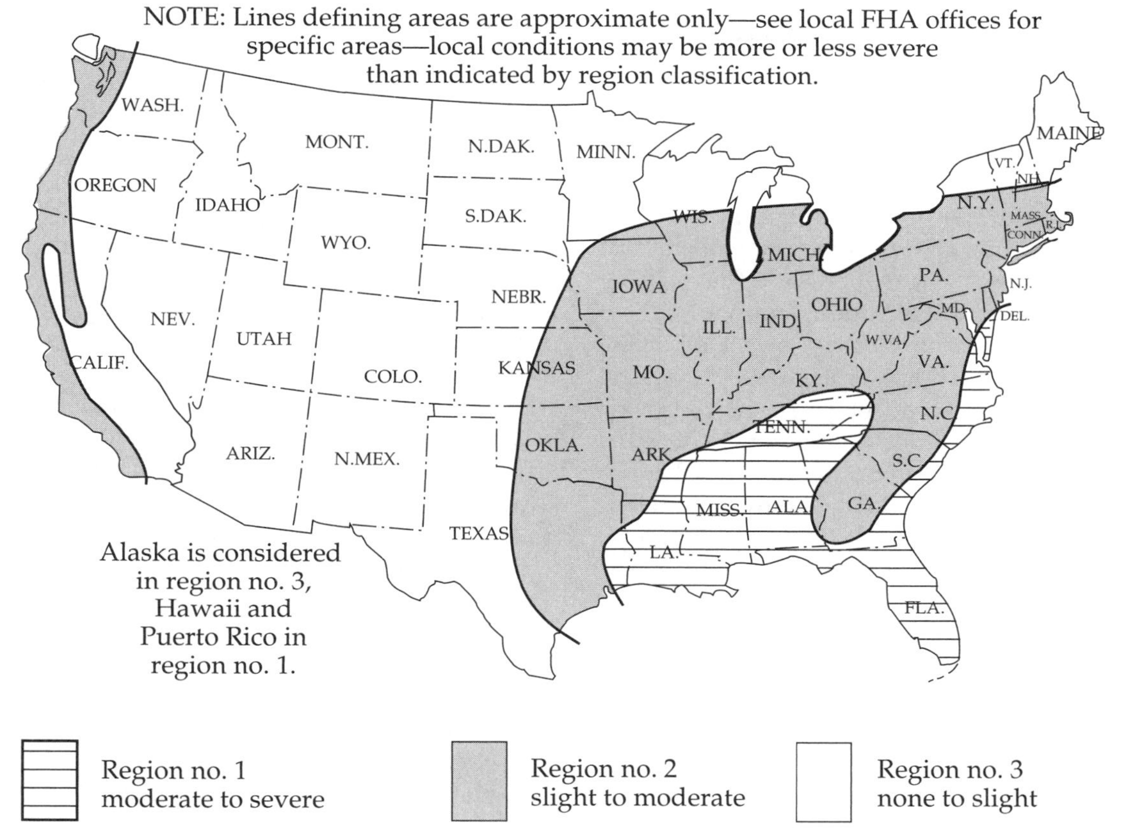

Fig. 8-12. *Decay hazard regions in the United States.*

structure. It can be performed concurrently with the inspection for termites. Areas that are vulnerable to attack by subterranean termites or carpenter ants are also conducive to the growth of the decay fungi. During the exterior inspection, in addition to probing the wood that is located near, or is in contact with, the ground, you should check wood members having cracks and open joints that are subjected to periodic wetting from rain. One area particularly vulnerable to decay is the end cut of exterior wood framing, trim, or siding. An exposed end cut (across the grain) absorbs water much more readily than a section that has been cut parallel with the grain.

During the interior inspection, you should check the unfinished attic and basement or crawl space for evidence of rot. In the attic the wood members can rot as a result of water intrusion through the roof because of a faulty roof covering or leakage around the joints of roof projections. Conditions conducive to rot can also result from condensation due to inadequate attic ventilation. In the basement or crawl space, the decay fungi thrives in the wood members if the humidity is constantly high. Check the overhead wood framing and subflooring for signs of rot. Probe the wood members above and around the top of the foundation wall and base of wood posts. Vulnerable locations for decay are the wood members through which plumbing pipes pass—because of the possibility of leakage or condensation. Probe the joists and subflooring below kitchen or bathroom fixtures.

When inspecting for rot, note on your worksheet all wood-framing members that have decayed to a point where they can no longer provide structural support. These sections should be replaced or rehabilitated. If in doubt, consult a professional. If natural decay-resistant wood or preservative-treated wood is not used for the replacement of rotted sections and the source of the moisture has not been eliminated, then it will only be a matter of time before the new sections begin to decay. Rotted wood trim that serves no structural function need not be replaced except for cosmetic reasons. Once the infected wood has been dried out and the source of the wetting has been eliminated, the decay is permanently arrested.

Checkpoint summary

Subterranean termites

Exterior inspection

- ❍ Check all exterior areas of the structure that have wood in contact with, or close proximity to, the ground.
- ❍ Note any termite shelter tubes on the foundation walls.
- ❍ Probe vulnerable areas such as garage door frames, basement windowsills and frames, deck posts, step stringers, and entry-door risers.
- ❍ Probe wood-frame members adjacent to concrete-covered, earth-filled porches.
- ❍ Inspect crawl areas under steps, porches, and so on.
- ❍ Probe sills and headers.
- ❍ Check wood fencing, dead tree stumps, wood debris, or stored firewood in close proximity to the house for infestation and rot.

Interior inspection

- ❍ Inspect for shelter tubes on foundation walls and piers and around all plumbing pipes that pass through the foundation.
- ❍ Pay particular attention to areas around the heating system.
- ❍ Probe exposed sill plates, headers, joists, and girders.
- ❍ Inspect wood support posts for infestation and rot.
- ❍ If the house is built on a slab, note any soft spots in the baseboard trim.

Dry-wood termites

- ❍ Inspect property fencing for infestation.
- ❍ Probe exposed wood framing throughout the house, from attic to crawl space.

Carpenter ants

- ❍ Look for small piles of sawdust below or around wood members.
- ❍ Did you see any ants walking around in the rooms, particularly the kitchen?
- ❍ Probe the sections of wood framing, siding, and trim that show evidence of decay or past wetting.

Powder-post beetles

- ❍ Inspect wood framing for clusters of small round holes.
- ❍ Newly formed holes are the color of a fresh saw cut and indicate an active infestation.
- ❍ Probe these wood sections for deterioration.

Rot

- ❍ Probe vulnerable areas such as wood members that are subject to periodic wetting from rain or garden sprinklers.
- ❍ Inspect roof sheathing from the attic for decaying sections around chimney, vents, and so on.
- ❍ Check subflooring and support joists below kitchen and bathroom fixtures and around plumbing pipes.
- ❍ Probe sill plates, headers, and the ends of joists and girders.

9

Attic

Once you are inside the house, if there is an accessible attic, it should be the first area inspected. There are basically two types of attic: full and crawl.

A *full* attic is one in which a person can easily walk around. Usually there is a floor in this type of attic, although the walls and ceilings are unfinished. There might be partition walls forming finished rooms with sloping or horizontal ceilings. Access to a full attic is usually through a finished staircase.

In a *crawl* attic, which is completely unfinished, the roof is sufficiently close to the floor so that to get around it is necessary to crawl or stoop over. The crawl attic usually does not have a structural floor. The ceiling joists from the level below are exposed. When getting around in this type of attic, be careful to walk only on the exposed joists. If you accidentally step between the joists, you will probably frighten the pants off of anyone in the room below because your foot will go right through the ceiling. Access to a crawl attic is usually through a ceiling hatch located in a closet or hallway or through hidden folding or sliding steps.

Inspecting an attic can reveal problems of which most homeowners are not aware, some of which might be potentially dangerous or costly to repair. It is not uncommon for a homeowner to say that he has lived in his house for over twenty years and has never gone into the attic. Incidentally, if the homeowner does say this, you can be sure that at the very least, the attic is inadequately insulated by current energy standards.

Insulation and roof leakage are probably the only items most people consider when thinking about the attic. However, there are other items of importance and concern, such as ventilation and its associated problems, fire hazards, electrical and plumbing violations, improperly discharging vents, and open duct joints.

Insulation

The attic area should be adequately insulated to minimize heat loss. The insulation needed in the attic will of course depend on the geographic location of the structure. (See chapter 19 to determine the proper insulation for your area and for a general description of the various types of insulation.) In both crawl and unfinished full attics, the insulation should be located in the floor (between the floor joists) and not between the roof rafters. (See FIG. 9-1.) Otherwise, heat from the rooms below will escape into the attic.

The insulation should be installed with a vapor barrier facing the heated portion of the structure and not the unfinished attic area. A *vapor barrier* is aluminum foil, a plastic sheet, or an asphalt-impregnated paper that prevents moisture movement from the heated portion of the house into the unfinished attic area. If the vapor barrier is incorrectly positioned (facing up into the unheated, unfinished attic), condensation problems can develop during cool weather. Moisture rising from the heated areas below condenses upon contacting the cool vapor barrier. Depending on the amount of vapor, the resulting condensation buildup can reduce the effectiveness of the insulation and cause peeling and flaking of the painted ceilings and walls in the rooms below.

Fig. 9-1. *The proper installation of insulation in the attic, between the floor joists and not the roof rafters. Note plumbing vent stack terminating in the attic—a violation of the plumbing code.*

If there are heating or air-conditioning ducts in the attic, check to see whether they are insulated. Metal ducts often have insulation in the inside, so tap the duct with your flashlight. If you hear a hollow sound, there is no insulation. If there is a dull thud, insulation is present. The more insulation on a duct, the less the heat loss during the winter and, in air-conditioning ducts, the less the heat gain during the summer. Generally, a minimum of 3 inches of insulation wrapped around the outside of the duct will substitute for missing insulation. If the duct is used for air-conditioning, the insulation should be covered with a vapor barrier to prevent condensation from forming.

In the northern sections of the country, when there is a furnace in the attic, insulation between the roof rafters as well as in the attic floor is recommended. To ensure proper ventilation and avoid condensation problems between the rafters, it is important to leave a ventilated air space between the top of the insulation and the underside of the roof deck. The insulation and the associated air space will help keep the roof deck cool. If the insulation is not between the rafters, heat from the furnace will warm the roof deck and melt the bottom layer of snow that has accumulated on the roof. This often results in an ice dam (see chapter 2).

Often, a homeowner adds additional insulation to the attic to bring the total insulation up to current energy standards. The insulation added should not have a vapor barrier if the existing insulation has one. All too often the homeowner adds insulation with a vapor barrier, which can then cause condensation problems. Look at the insulation in the floor. If there are two layers of insulation and both have vapor barriers, the upper barrier should be slit with a razor blade to allow moisture movement.

If there is a full attic with partition walls forming rooms, the insulation should be located on the unfinished sides of the partition walls and on the ceilings of the rooms. Occasionally, insulation is placed between the roof rafters and not between the floor joists and the partition walls. This installation is inefficient because the heat will escape into the unfinished areas.

Violations

When inspecting the attic, look for vent stacks that terminate in the attic area. (See FIG. 9-1.) This is a violation of the plumbing code, a condition that should be corrected. The vent stack should extend through the roof so that the sewer gases can discharge to the outside. In the past, according to code, every vent stack in the attic terminated above the roof. This requirement can now be modified if the local municipality approves it. With the advent of air admittance valves (AAVs), at least one vent stack must terminate above the roof, as long as all the other vent stacks in the attic are capped with an AAV. This valve is a pressure-actuated one-way mechanical vent that is designed to maintain a plumbing fixture's trap seal and also prevent sewer gases from discharging into the attic. The AAV (see FIG. 9-2) opens only under negative pressure, which is created when a toilet is flushed or when water is flowing down the drain. The valve closes by gravity

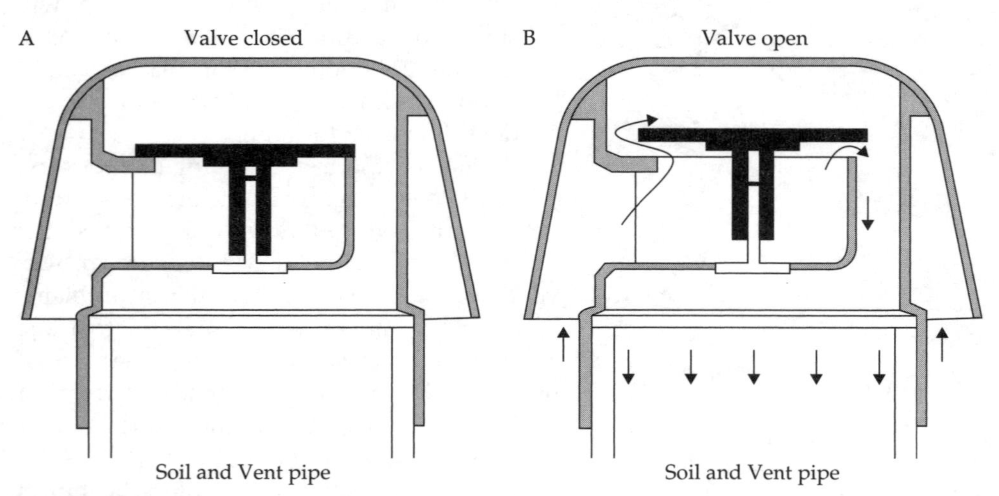

***Fig. 9-2.** Air admittance valve (AAV). A—During times of nonactivity in the drainage system, the diaphragm forms a tight seal to close the valve. B—When a toilet is flushed, a negative pressure develops in the pipe, which causes the diaphragm seal to open and air to enter the vent pipe.*

or slight spring pressure to form a tight seal when the water flow stops, or when there is a positive pressure, which would be the result of a blockage in the drain. AAVs come in various designs; however, they all operate under the same principle.

While in the attic, also look for ducts. Whether they are air-conditioning ducts or ducts from an exhaust fan, they should not have any open joints. (See FIG. 9-3.) All open joints should be resecured. Sometimes the exhaust fan from a bathroom discharges its moisture-laden air into the attic. This is undesirable because the moisture can cause condensation problems. The duct from such an exhaust fan should be extended above the roofline so that the exhaust is discharged into the atmosphere. Also look for open electrical junction boxes and makeshift electrical wiring such as extension-cord wiring and "pigtailed" hanging light fixtures. These are electrical violations that should be corrected.

Fig. 9-3. *Open joint in duct from kitchen exhaust fan.*

Leakage

It is important to check the underside of the roof for signs of past water leakage and, if the structure is inspected during a rain, current water leakage. Water stains can show up on the sheathing or roof rafters and appear as dark-streaked discolorations on the wood. Sometimes there is a separate masonry chimney for the fireplace and a prefabricated chimney for the heating system. Joints vulnerable to water leakage are those between the chimney and the roof and between vent stacks and the roof. These joints should be checked for leakage. Water leakage through joints is a relatively minor problem and can usually be corrected by sealing the joints with an asphalt cement.

Fire hazards

Of particular concern is the joint between a prefabricated chimney and the attic floor. (See FIG. 9-4.) According to building codes, there should be 2 inches minimum clearance between the chimney and adjacent wood framing. The clearance is a fire safety measure because wood, which normally burns at temperatures between 400 and 600°F, can ignite spontaneously at a reduced temperature of about 200°F if it has been exposed over the years to temperatures between 150 and 250°F.

If your house has a prefabricated chimney, check the clearance between it and adjacent wood framing. While the clearance space around the chimney prevents a problem, it also creates one because the open space around the chimney generally runs from the

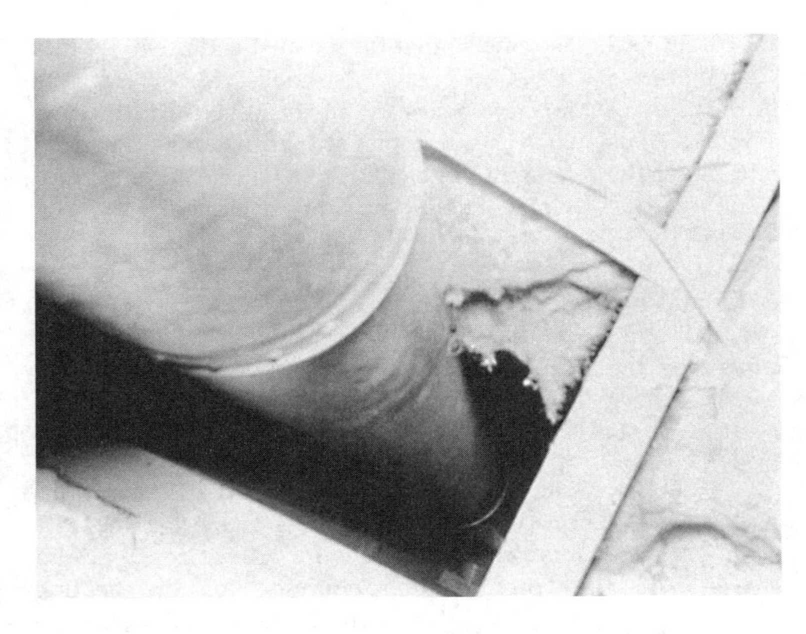

Fig. 9-4. *Open joint between chimney and attic floor—a potential fire hazard.*

boiler/furnace room to the attic. If a fire should develop in the boiler/furnace room, the open area around the chimney will act as a flue and draw the flames up into the attic where they can very rapidly consume the structure. Fortunately, the condition can easily and inexpensively be corrected by blocking (fire-stopping) the opening with a noncombustible material such as sheet metal.

If the house has a fireplace or wood-burning stove, check the section of the chimney that is exposed in the attic. Look for soot or a creosote buildup around the joints. This indicates cracked or open joints through which the exhaust smoke is seeping into the attic. This problem exists more often with chimneys that have unlined flues than those with lined flues. This condition is a potential fire hazard that must be corrected. Record this fact on your worksheet.

Ventilation

Ventilation in an attic is very important. It allows moisture that accumulates in the area to dissipate and also helps to reduce the heat buildup that normally develops during the summer months. Some of the moisture generated in the structure by bathing and cooking works its way up into the attic. If the area is inadequately ventilated, this moisture can eventually cause problems such as delaminating roof sheathing, water streaks on interior walls, peeling and flaking paint, and in severe cases some rotting of the wood framing. When walking around the attic, make sure that all vents are completely unblocked. Sometimes the homeowner blocks a vent opening (see FIG. 9-5) to cut down on the cold air entering the attic and reduce the heat loss through that opening. What he doesn't realize is that he is creating a problem. If the attic is properly insulated, the heat loss through the vent openings is minimal. Of greater importance is the need for vent openings so that moisture can escape. Look around; you'll be able to tell whether the attic is adequately ventilated.

Look at the roofing nails projecting through the sheathing. If you happen to be in the attic on a winter day when the temperature is around 20° F and the area is inadequately

Fig. 9-5. Partially blocked gable louvers. Louvers are used to ventilate the attic.

Fig. 9-6. *Frost on roofing nails due to inadequate attic ventilation.*

ventilated, you will find frost on the roofing nails. (See FIG. 9-6.) The frost that forms on the nails melts and drips onto the floor during the warmer periods of the day. Many droplets of water show up as circular stains on the floorboards and on top of the insulation. (See FIG. 9-7.) If you are in the attic during the warmer months, you'll find rust stains and mildew on the sheathing near the nails. (See FIG. 9-8.)

When inspecting the attic, pay particular attention to the north slope of the roof. If the area is inadequately ventilated and plywood is used for the roof sheathing, delamination of the sheathing might be a problem. The northerly slope begins to delaminate before the southerly slope. Figure 9-9 shows an advanced stage of delamination as a result of moisture buildup in the attic. At this stage the entire roof sheathing and shingles must be replaced. The condition could have been avoided if the attic area had been adequately ventilated. Correcting an inadequately ventilated attic is relatively easy. All that is required is to increase the size of the existing vent openings (if there are any) or to provide additional openings such as roof vents, ridge vents, soffit vents, or a power ventilator. These vent openings can be used individually or in combination. The exact number of vent openings needed should be determined by a professional.

Fig. 9-7. Water-droplet stains on catwalk in attic due to condensation on protruding roofing nails because of inadequate attic ventilation.

Attic fan

If there is a fan in the attic, it is considered a desirable feature. (See FIG. 9-10.) Because of trapped air, an attic area can reach 150° F on a hot summer day. An attic fan that is thermostatically controlled greatly reduces the heat load on the building. Usually the thermostat is set to activate the fan when the attic temperature is about 100° F. Along with a thermostatic control, some attic fans also have an additional humidistat control. The humidistat turns the fan on and off regardless of temperature whenever the relative humidity is too high in the attic. The fan provides two additional benefits regarding air conditioning. By reducing the heat load on the structure, less electrical energy is required for air-conditioning units. And for those homes with a central air-conditioning blower coil located in the attic, the unit operates more efficiently at lower surrounding temperatures.

The attic fan should normally be operating during the warm days of summer. If it isn't, either the unit is malfunctioning or the thermostat is improperly set. On many units the thermostat is exposed and can be manually adjusted. But if the thermostat is factory set and not accessible, the unit will require professional maintenance. On cool but not cold days, you can usually check the attic fan, assuming the thermostat is exposed, by lowering the temperature control setting until the fan is activated.

Whole house fans

Some structures have whole house fans, which draw outside air in through screened windows and doors and blow it out through the opening around the fan or movable louvered vents. Check the fan before you go into the attic. If the fan is turned on while you are in the attic, be careful when walking around. In some homes, the area of the vent openings for the fan is too small for the size of the fan; consequently, the air that is moved by the fan is partially blocked from leaving the structure. As a result, pressure is built up in the attic, which decreases the efficiency of the fan. If the fan is turned on before you go into the attic, put your hand over any electrical outlet box located on the level just below the attic. If you feel air rushing onto your hand, the attic vent openings are inadequate and should be increased. Some fans are controlled by manual snap switches. From a convenience point of view, this is not as desirable as a timer switch or a thermostat switch. If this is the case, consider its replacement.

Whole house fans are mounted either in the attic floor or on the exterior sidewall of the attic.

Fig. 9-8. *Mildew on underside of roofing deck.*

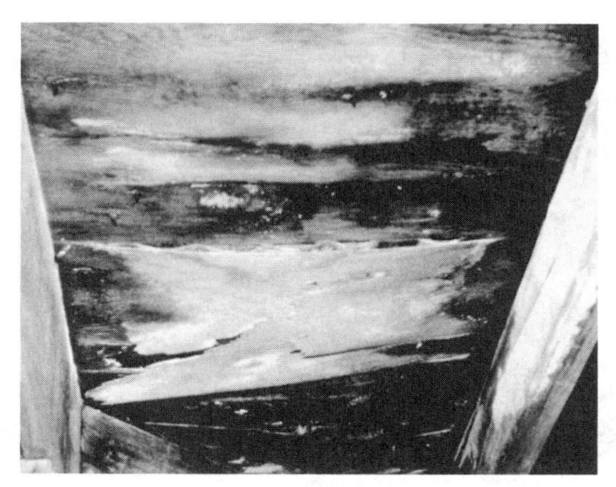

Fig. 9-9. *Delamination of roof sheathing resulting from inadequate ventilation of attic.*

They have movable louvers that open when the fan is activated and remain closed when the fan is off. This operation creates a potential problem. During the winter months and during portions of the summer months when the fan is not operating, the louvers will be in a closed position and will thereby completely close off the vent opening, causing the attic to be inadequately ventilated. This problem can be avoided by adjusting the louvers so that even when the fan is not operating, they remain in a partly open position.

Some houses have no access to the attic area. If such is the case in your home, the probability is very great that the area is inadequately insulated by current energy standards. The amount of ventilation in the area is also questionable. Installing an access hatch to the attic is recommended.

Structural

While in the attic, be sure to inspect the roof framing. Check the rafters and trusses for cracked, broken, and sagging sections. (See FIG. 9-11.) Are the rafters spreading apart near the ridge? Although these problems are uncommon, they indicate a structural defect that should be evaluated by a professional. Look at the floor

Fig. 9-10. Thermostatically controlled attic fan mounted between the roof rafters.

Fig. 9-11. Cracked and broken roof rafters.

joists. If they've been cut to accommodate an opening for an attic fan or pull-down stairs, have the cut ends been properly secured to a header? They should be. If not, record that fact on your worksheet for future correction.

When trusses have been used for roof framing, check to see if any of the webs or chords have been removed. Homeowners have been known to cut and remove them to accommodate storage items. This affects the structural integrity of the truss and should also be recorded on your worksheet for needed repairs.

Checkpoint summary

- ❍ Is attic insulated?
- ❍ Is additional insulation needed?
- ❍ Does existing insulation contain a vapor barrier?
- ❍ Is insulation properly installed?
- ❍ Is ventilation provided?
- ❍ Are vent openings blocked?
- ❍ Check operation of attic fan.
- ❍ If attic is poorly ventilated, look particularly at northerly slope for delaminating plywood or warped roofing boards.
- ❍ Look for signs of past or current roof leakage.
- ❍ Pay particular attention to area around the chimney and plumbing vent stacks.
- ❍ Are air-conditioning and heating ducts insulated?
- ❍ Look for open joints in the duct work.
- ❍ Do any plumbing vent stacks terminate in the attic? If so, are they capped with AAVs?
- ❍ Are kitchen or bathroom exhaust fans discharging into the attic?
- ❍ Are there open electrical junction boxes or makeshift electrical wiring?
- ❍ If house has a prefabricated chimney, is there a need for fire-stopping around the joint between the chimney and the attic floor?
- ❍ Are there any cracked, broken or sagging sections of rafters or truss members?

10

Interior rooms

After inspecting the attic, you should inspect every room and closet in the house. Don't pass a door without opening it and looking inside. If a door is open, close it to see if it operates properly. All *pocket doors* should be checked to see if they slide open and close easily (pocket doors are doors that when open are housed within the wall cavity). If there is a problem, it generally requires opening a section of wall to provide access to the tracks for repair.

Start your inspection of the rooms at the upper level and work your way down to the basement. As you walk from one floor to another, inspect the hallways and connecting staircase.

The problems normally encountered during an interior room inspection are usually of a cosmetic nature and are not costly to correct. Occasionally you might see cracks or uneven floors or walls that are symptomatic of structural problems, but that is rare. Usually cracks and uneven floors are caused by shrinkage, warpage, or slight movement of the house. Slight movement is considered quite normal. Few people realize that a house is constantly in motion. As the outdoor temperature and humidity vary with the season and the time of day, they cause differential expansion, swelling, and contraction of the various structural and nonstructural elements. This movement, although slight, is very often enough to cause cracks at points of stress concentration, such as over windows or doors.

When you walk into a room, try to look beyond the cosmetics. Don't dwell on the position of the furniture or the pictures on the walls. Look at the walls, the floor, ceiling, and trim, but do not be concerned about minor cosmetic problems. Hairline cracks, small holes, chipped sections, dirty and marked-up areas can all be corrected easily by spackle, wood filler, paint, or stain, depending on the finished surface.

Walls and ceilings

The materials most often used for covering walls and ceilings are plaster and plasterboard, also referred to as gypsum wallboard, drywall, or Sheetrock, which is a trademark of the U.S. Gypsum Company. Plaster—usually made from a mixture of gypsum (a common mineral), sand, and water—has qualities that are very desirable for a wall finish: structural rigidity, durability, resistance to sound transmissions, and high fire resistance. In addition, it can readily be applied to curved or irregularly shaped surfaces. A plaster wall is usually applied in two or three coats to a backing called lath. Depending on the type of lath, the thickness of the plaster usually varies between ½ and ¾ inch.

Plasterboard is a sheet material that basically consists of a gypsum mixture surfaced with a treated paper. Depending on the application, the thickness of the plasterboard can range from ¼ to ⅝ inch. The sheets are usually 4 feet wide and are available in lengths up to 12 feet. When constructing walls using this material, the plasterboard is fastened directly to the studs, and the joints are covered with tape. The edges of the plasterboard are recessed slightly so that when the tape is applied, it is level with the surface. This type of construction requires that the wood framing be perfectly straight and true. Otherwise, the wall surface will be uneven, and there will be visible joint lines. If the moisture content of the wood used in construction is not very nearly that which will be attained in service, there might be nail pops and joint cracks. Popping nails is a condition most often found with plasterboard construction. It can be corrected easily. It often does, however, indicate low-quality installation.

In houses where the roof framing consists of trusses rather than rafters, you might see cracks at the wall-ceiling intersection in the interior rooms below the attic. The cracks are usually the result of truss uplift, which causes the ceiling to move up and down in the course of a yearly weather cycle. The size of the cracks can vary from a hairline to an actual opening of an inch or more. This condition is a cosmetic problem rather than structural. The easiest cosmetic repair is to cover the crack with decorative molding that is nailed to the ceiling and not to the wall. As long as the molding is wide enough, it will cover the crack as the ceiling moves up.

When you inspect a room, look specifically for areas that need major cosmetic rehabilitation, such as broken walls or ceiling, loose and bulging sections of plaster, and disintegrating plaster. Sagging or bulging sections in a plaster ceiling are a safety concern because of the possibility of collapse. This problem, if it exists, is usually found in older homes. Plaster is applied over a wood, metal, or gypsum lath. The sagging condition results from broken plaster keys, which no longer lock the plaster layers to the supporting lath. (See FIG. 10-1.) The condition occurs over a period of time and is usually caused by vibration and wetting from roof and plumbing leaks, which weaken and crack the keys. If there are sagging sections of the ceiling, record this problem on your worksheet for future repair.

Very often, major cosmetic damage is the result of a water condition. As you look at these areas, also look for water stains. Stains

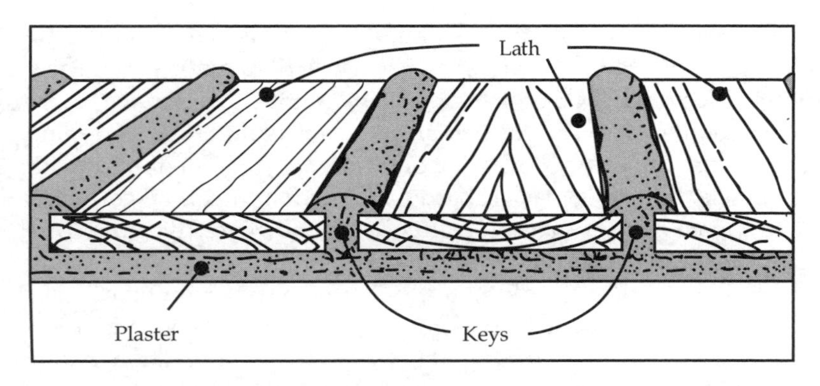

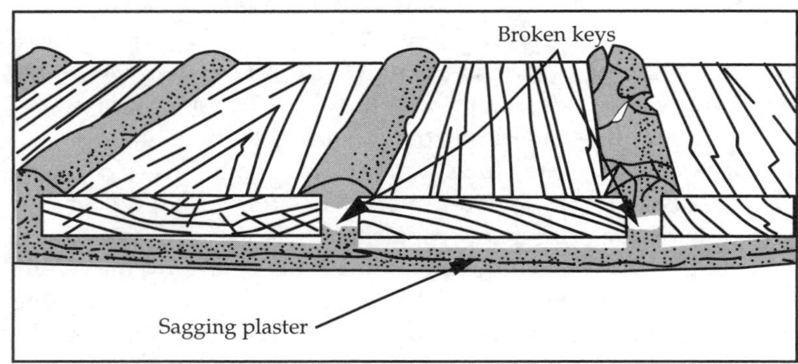

Fig. 10-1. Ceiling plaster can sag down from lath when keys securing base coat give way.

often show up as discolorations circled by a light brownish ring. If you see water stains, try to determine the source of the water. All stains do not indicate a current problem condition. You might be looking at the results of a problem that has been corrected. If you have a moisture meter, you should check all water stains to determine whether or not the leakage condition causing the stain is active. If it is, record it on your worksheet.

Mold-resistant drywall

Because of the concern in recent years about mold developing on wet or damp drywall, manufacturers have developed and are now supplying a mold-resistant gypsum wallboard. Mold growth requires moisture and a food source. When traditional drywall gets wet, the paper becomes the food source. With mold-resistant wallboard, either the paper has been chemically treated to resist or minimize the absorption of water or the paper has been replaced with a fiberglass mat facing. Mold-resistant wallboard will reduce the chances of mold developing. However, it does not mean that mold cannot grow. The installation of mold-resistant wallboard is exactly the same as for traditional wallboard.

Water stains on a ceiling can be caused by several types of problems, such as water leaking between the tile joints around a shower or tub located in a bathroom above the ceiling, an ice dam on the roof, an open joint between a plumbing vent stack or chimney and the roof, leaky plumbing, or condensation on cold-water pipes located above the ceiling. Sometimes the cause is obvious, such as a faulty roof. However, you should be aware that leaks that originate in the roof can stain ceilings two or more levels below. The stains are not limited to ceilings just beneath the roof.

Sometimes the source of a stain cannot be explained by any of the above. On one inspection, I noticed that the ceiling of one room had several stains that were still damp. I knew that there were no plumbing pipes above the ceiling and that the roof was in good shape. Frankly, I couldn't figure out why the ceiling was damp. All of a sudden, I noticed that water was dripping from one section of the ceiling. I then went upstairs to the room above and found what was causing the leak. Apparently, the owner kept a puppy that was not housebroken in the room. Whenever the puppy missed the newspapers that were placed there for training purposes, the urine seeped through the floor joints, wetting the ceiling below.

Ceilings can also be covered with composition tiles. These tiles are generally made from asbestos, glass fiber, or fiberboard. They can be applied directly over plaster or plasterboard ceilings or over furring strips and are usually interconnected with tongue-and-groove joints. If you are inspecting an older home with plaster walls and ceilings and find yourself in a room that has a tile ceiling, the tiles were probably installed to cover broken and cracked sections in the original ceiling. It is becoming increasingly difficult to find a good plasterer. Consequently, when renovating an older home, many people cover a ceiling with tiles or plasterboard rather than have it replastered. Look specifically for loose and sagging sections. These sections must be resecured; otherwise, they can cause adjacent tiles to loosen, which can eventually result in a "domino" effect, causing most of the tiles to come down. (See FIG. 10-2.) I inspected a house the day after two-thirds of the ceiling tiles in one room came thundering down. The owner told me that there always had been a few loose tiles.

Water stains on walls facing the exterior might be an indication of water seepage through open joints in the exterior siding. In particular, this condition has been found in older homes with an exterior brick or stone siding and is usually caused by wind-driven rain penetrating cracked and deteriorating mortar joints. When water stains and peeling and flaking paint are found on walls just below windows, it is usually an indication of

Fig. 10-2. *Ceiling that had been covered with tiles, which came loose and fell down. Note that the tiles had been installed to cover a cracked, broken, and peeling ceiling.*

leakage through open exterior joints around the window.

Cracks in walls are normally merely of cosmetic concern. However, when walls are cracked and the windows and door frames in that room are not level, the condition might be indicative of a structural problem and should be evaluated by a professional.

When the walls are covered with wood or hardboard panels, cracks will not be visible. In quality homes, panels are normally installed over a plasterboard wall. This type of backing provides added rigidity and, when applied on exterior walls, added insulation. If the room you are inspecting has a panel wall, push on the panel at a point between the studs (the studs are usually located 16 inches apart, in some houses 24 inches apart). If the wall panel yields under your push, there is no plasterboard backing. I inspected a new house just after construction and prior to occupancy. The buyer specifically asked me to check the plasterboard because he was paying extra for ⅝-inch rather than ½-inch plasterboard. The builder did install the thicker plasterboard, but in those rooms that were paneled, the builder completely omitted the plasterboard behind the panels because it was not visible to the buyer.

The trim around the various joints in the room should be inspected and checked for missing, loose, cracked, and broken sections. Although a minor element, trim in this condition is an indication of shoddy workmanship in new construction and neglect in a resale.

Chinese drywall

Between 2004 and 2006, because of the housing boom and the need to rebuild the homes destroyed by Hurricane Katrina, many millions of pounds of drywall was imported from China. It is estimated that enough material was imported to build about 100,000 homes. Shortly thereafter, problems began to develop in some of the houses in the southern states where the interior walls were built using drywall made in China. Chinese drywall emits higher levels of sulfuric and organic compounds than drywall made in the United States. The sulfur fumes emitted caused air-conditioner coils, copper plumbing, and electrical wiring to corrode, as well as producing a foul odor such as that of a burning match or rotten eggs. The fumes have also been associated with respiratory and sinus problems in some residents. So far, most of the drywall problems have been limited to the southern states, where a warm, humid climate encourages the emission of sulfur fumes. If the drywall was also used in other parts of the country where it is dryer and cooler, it could be years before homeowners begin seeing the problems associated with the material. Correcting the problem would require stripping and removing all the drywall. There are also indications that it may be necessary to replace some of the material in the house such as carpeting and furniture.

Floors

When walking around the house, you might notice that the floor squeaks on one or more sections. This condition is found in both new and older homes and indicates slightly loose floorboards. It is usually difficult to eliminate. One corrective procedure is to wedge the floor from below. However, the underside is usually not accessible. If a hardwood floor is nailed from above, the nail holes can ruin the finish. With regard to the structural integrity of the house, squeaking floors are not a concern, although if excessive, they can be annoying.

If the floors in a portion of the house are finished hardwood and the remaining floors (other than those in the bathrooms and kitchen) are covered with wall-to-wall carpet, do not assume that there is a hardwood floor below the carpeting. There might be, but quite often there isn't. It is not uncommon for a builder to give a buyer the option of hardwood floors or carpeting. When the buyer selects carpeting, it is laid over a plywood floor. If you are inspecting a room and see a hardwood floor in a closet, though the floor in the main portion of the room is carpeted, do not assume that there is a hardwood floor beneath the carpeting. If you cannot see it, you really cannot be sure. If a hardwood floor is important to you, get a representation in writing from the owner that the floors beneath the carpets are hardwood.

Just because the floor looks like a hardwood floor, it may not be, in the classic sense; that is, solid wood from top to bottom. It may be an engineered hardwood floor or a laminate floor. An engineered hardwood floor is made up of three to 10 thin layers of wood with a core of hardwood, plywood, or high-density fiber. The top layer is a hardwood veneer that is glued on the top surface of the core and is available in almost any hardwood species. It is less expensive than hardwood, and due to its multiply structure, engineered wood is much more stable than solid wood and is less susceptible to shrinking and expanding with changes in temperatures and humidity. It can be installed on any grade level, directly over concrete and can also be installed below ground level, whereas solid hardwood should only be installed above grade. Unlike engineered wood, moisture and extreme temperature changes can cause solid wood to shrink and expand, potentially causing gaps between boards during colder or drier seasons. If there is excessive moisture on the underside of the solid hardwood floor, it will cause the top surface to cup, that is, to form a concave surface with the edges raised.

Laminate flooring is the least expensive between solid hardwood and engineered flooring. Although it looks like wood flooring, there is actually no solid wood used in its construction. Laminate floors are made up of several materials bonded together under high pressure. Most laminate flooring consists of a moisture-resistant layer under a layer of HDF (high-density fiberboard). This is topped with a high resolution photographic image of natural hardwood flooring, which accurately reproduces the grain and color of natural hardwood. To protect the laminate, it is then finished with an extremely hard, durable clear coating made from special resin. Laminate flooring can be installed above or below grade and over virtually any other flooring surface. It is usually laid as a floating floor. It is not nailed or glued to the subfloor.

If the wall-to-wall carpet covering the hardwood floors is brand new, ask the seller for a representation in writing that the surfaces of the hardwood floors are in good condition. Several years ago I inspected a split-level house with hardwood floors. All the floors were covered with new wall-to-wall carpet. Lifting a corner of the carpet in each room did not reveal a problem. However, after the buyer moved in, he removed the carpet from all the hardwood floors and was dismayed by what he saw. Every one of the floors was very badly stained by cat or dog urine and had to be refinished.

Another type of hardwood floor is a bamboo floor. The flooring is typically made by slicing bamboo poles into strips. To prevent potential mold, fungus, or insect problems, these strips are boiled and then sent to a drying kiln to reduce moisture levels. The strips are then glued and assembled under extremely high pressure and heat either face up for horizontal flooring or side by side for vertical flooring. Although vertical flooring results in a smooth uniform appearance, horizontal flooring has a more natural look. The pressure and heat applied assure a strong, stable bond. After milling, the bamboo is given an aluminum oxide coating to increase its strength and endurance, and to protect it from stains. Direct sunlight can cause some types of bamboo floor to become discolored. It may happen if the floor is exposed to highly concentrated sun rays for long periods. Bamboo flooring needs to be kept free of dust, dirt, and grit. Even the tiniest particles of dirt can cause minute scratches on its surface. It is like any wood floor—it is damaged by dents, scratches, and shoes with high heels.

If you walk into a room and notice that the floor is not level, do not be alarmed. In all probability, it is a condition caused by past shrinkage, warpage, and settlement of the wood framing and is not a concern. If the floor is sagging in one section, it might indicate that that portion of the floor is or was improperly supported. Go down to the level below to check the ceiling to see if it too is sagging. If it is, have a professional evaluate the condition. Occasionally this condition occurs in a kitchen floor when the refrigerator is placed in a location other than that intended by the builder.

Depending on the moisture content in the wood used for framing, you might see a large open joint between the floor and the partition walls. The joint might be open as much as 1 inch and might run the entire length of the partition wall. When you see this condition for the first time, it is quite unnerving. The trim that normally covers the joint between the floor and the wall is about an inch above the floor, and the joint is wide open. The condition is caused by excessive shrinkage of the wood framing and is only a cosmetic problem. By covering the open joint with trim or lowering the existing trim, the room will look almost as good as new.

The floor on the lower level of bilevels is a concrete slab. Depending on the quality of construction, the concrete floor slab could settle to a point where there is also a large open joint between the floor and the walls. (See FIG. 10-3.) During construction, if the ground below the floor slab is not properly compacted, the slab might eventually settle, resulting in open joints. The foundation walls all have support footings and are normally independent of the floor slab. Consequently, settlement of the floor slab usually does not indicate a problem with the main structure. Nevertheless, the wall above the open joint should be shimmed; otherwise, the floor above might sag. If the settlement of the concrete floor slab is accompanied by cracked and settled sections of the foundation wall, a severe problem does exist and should be evaluated by a professional.

Sometimes the concrete floor slab has a raised wood floor that might be covered with carpeting or resilient floor tiling. When walking on this floor, test for soft or spongy

Fig. 10-3. *Settled concrete floor slab. Note open joint between floor and partition walls.*

sections. This condition can be caused by inadequate spacing of the wood framing. However, more often than not, it is caused by rotting wood. If water seeps into the area between the wood floor and the concrete slab, it eventually causes the framing to rot and the subflooring to delaminate. (See FIG. 10-4.) See chapter 11 for a discussion of water seepage into the basement level.

Heat

With the exception of homes in the sunbelt, all the finished rooms in the house should have provisions for heating. Depending on the type of heating (see chapter 14), look for a radiator or heat register. If you don't see any, ask the owner how the room is heated. It might be heated by radiant panels in the floor, walls, or ceiling that would not be visible. If there is no source of heat, record the fact on your worksheet. Adding heat to a room by extending the existing heating system can be costly. If there is a means of heat supply, is it properly located? For maximum efficiency, radiators and heat-supply registers should be located on an exterior wall, preferably below a window. This allows the heat to mix with the cooler outside air that normally infiltrates the interior from around the windows.

Windows

Although the condition of most of the windows is checked during the exterior inspection (see chapter 5), the condition of the windows on the upper levels and the operation of the windows should be checked during an interior inspection. Also, if the windows are the casement or awning type, the storms and screens are often not visible from the exterior and should be looked for during the interior room inspection.

The cost of repairing a broken window frame or replacing a broken pane is usually not high on an individual basis unless the window is one of unusual design or a large thermal pane (double or triple glazing). Keep track of the number of windows that need repair or rehabilitation. It is surprising how fast the dollars add up when you multiply the

Fig. 10-4. *Rotting and delaminating section of raised wood floor in finished basement. Condition is caused by constant wetting from water seepage into the basement.*

cost for one repair by the number of windows that are faulty.

Specifically, check the windows for cracked and broken panes that require replacement. BB holes or small cracks in the corner of a pane can be overlooked, providing that the window is covered by a storm pane. Is the putty around the pane cracked, dry, chipped, or missing? If so, the window joints must be reputtied. Look at wood window frames and exterior sills for cracked and rotting sections. Some windows have metal frames. Metal frames get quite cold during the winter in the northern part of the United States and have a tendency toward excessive condensation. This can cause peeling and flaking paint or disintegrating sections of plaster near the window frame.

Pay particular attention to steel casement windows. These windows are usually a problem. They rust easily and must be painted every few years to prevent deterioration. In many cases, they do not close properly, a condition often caused by a sprung frame or excessive layers of paint around the joints. Also, almost invariably you will find cracked panes. Usually casement windows are crank-operated. Check the hardware. Sometimes the cranking mechanism is not operational.

Open and close the windows. They should operate relatively easily without sticking or binding. Double-hung windows should not rattle in the channel and should stay in a fixed position when opened fully or partially. Many older double-hung windows use a counterweight to hold the sash in a fixed open position. The weight is usually tied to a sash by a cord or a chain. If a cord is used, check to see if it is broken. Broken cords are not uncommon and should be replaced. If the cord is frayed, you can anticipate its early replacement.

When looking at the windows, check to see if the glazing is a single pane or a thermal pane, that is, two pieces of glass separated by a sealed air space. You can tell by looking at the thickness of the joint between the pane and the frame. A single pane is usually no more than 3⁄16 inch thick; a thermal pane is about 3⁄8 inch to 1 inch thick. You can also tell by looking very closely at the pane. Usually, you can see dust or dirty spots on the opposite side that reveal the thickness of the pane. Sometimes fixed-pane windows and sliding glass doors have thermal panes, even though the openable windows throughout the house have only a single pane. One of the well-known names for thermal glass is Thermopane. It is also referred to as insulated glazing. Each manufacturer has its own term for it. You can usually find the manufacturer's name in the corner of the window pane.

Thermal-pane windows are fabricated by hermetically sealing dry air or an inert gas between the panes. This is done to eliminate the possibility of future condensation problems between the panes. (See FIG. 10-5.) When the seal breaks, accidentally or otherwise, water vapor can enter the space between the panes. Look at the windows for signs of a faulty seal. If a portion of the window appears cloudy (FIG. 10-6) or there are water droplets between the panes, the seal has been broken. Although the insulating characteristics of a thermal-pane window with a faulty seal are at least as good as a storm window, the pane is not desirable from a visibility and a cosmetic point of view. If you see a thermal-pane window with a faulty seal, record it on your worksheet. The window requires replacement. The better window manufacturers offer a 20-year warranty on the seal of these windows.

Electrical outlets

While inspecting the interior rooms, look for electrical hazards and violations and whether there are an adequate number of electrical outlets on the walls. These items are discussed in chapter 12. According to the electrical code,

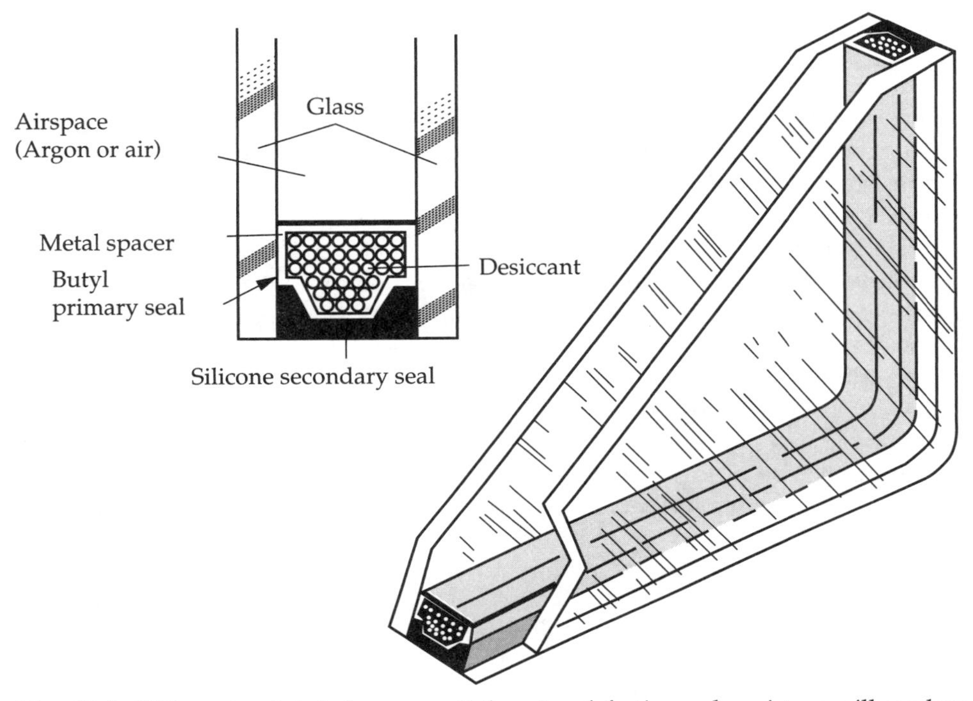

Fig. 10-5. With an insulated glass pane, if there is a defective seal, moisture will condense between the panes.

Fig. 10-6. Faulty seal in thermal-pane windows adjacent to a sliding glass door. Restricted visibility is caused by condensation between the glass panes.

the outlets in a new house must be located such that no point on the wall is more than 6 feet horizontally from an outlet. However, the actual number of outlets needed for a specific room depends on the room's usage and the position of the furniture.

Look at the electrical outlets. In new homes and in recently rewired older homes, the receptacles have three slots rather than two. The third slot is a ground connection that is used in conjunction with the three-prong plugs found on most modern appliances. It is a safety feature and is used for grounding the casing of electrical appliances and equipment. Should an internal short develop between the wiring and the equipment or appliance casing, the ground connection directs the leakage current harmlessly to the ground rather than through the user.

In homes built after 2008, check to see if the outlet receptacles are tamper resistant. If they are, the letters "TR" will be marked on the face of the receptacle. In 2008, the National Electrical Code (NEC) mandated the use of tamper-resistant receptacles in new or renovated construction. Tamper-resistant receptacles have a shutter mechanism that in most instances will block access to the electrical contacts from anything but an electrical plug. Although a worthwhile safety feature, not all municipalities have adopted the new code. Check with your local Building Department.

Not all electrical appliances and equipment have a grounding wire (three-prong plug). Some are made with double-insulated plastic cases that prevent the user from touching anything electrically charged, even if an internal short should develop. This type of equipment does not require a ground connection and can therefore be used with the older two-slot receptacles. As a safety precaution, you should use only electrical appliances and equipment that have been approved by Underwriters Laboratories, Inc., or some other nationally recognized testing agency.

The two-slot outlet receptacles can be used for the grounding of appliances with three-prong plugs. Adapters are available at hardware and electrical supply stores to enable two-slot receptacles to accept the three-prong plugs. To complete the ground connection when using the adapter, remember that the small wire (pigtail) must be secured to the center screw on the receptacle cover plate.

Check the receptacles (two- and three-slot) to determine if they are electrically hot, whether the receptacles have reverse polarity, and whether they are properly grounded. This can be done using a simple plug-in tester available at all electrical supply stores.

Reverse polarity is a problem because it indicates that the hot and neutral wires are connected to the wrong terminals in the receptacle. This is important because the switch on an appliance should disconnect the hot wire. This keeps everything beyond the switch inside the appliance from being electrically hot when the switch is off. With reverse polarity the switch disconnects the neutral wire so that everything beyond the switch will be hot, which could result in an electrical shock. The diagrams in FIG. 10-7 show how a polarized circuit can protect you from a shock hazard.

It is particularly important to check the receptacles in the bathroom and kitchen to see if they have ground-fault circuit protection. Receptacles that are not properly protected for ground faults are potential hazards. See chapter 12 for discussion of ground-fault circuit interrupters (GFCIs).

Fireplace

If a house has a fireplace, it is most often located in the living room or family room, although

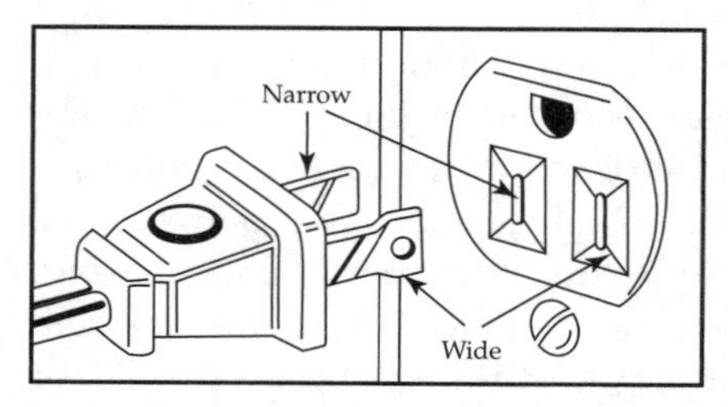

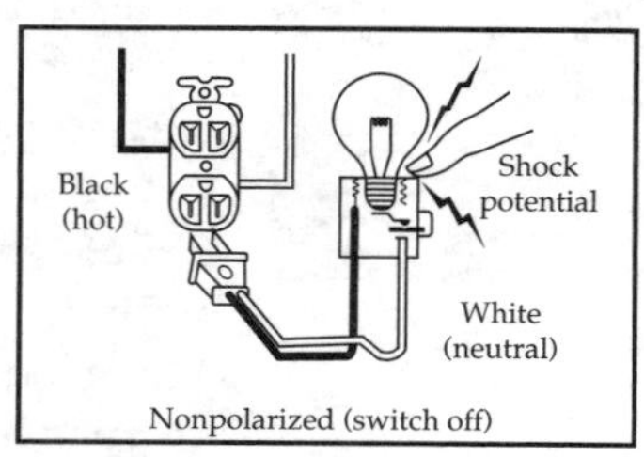

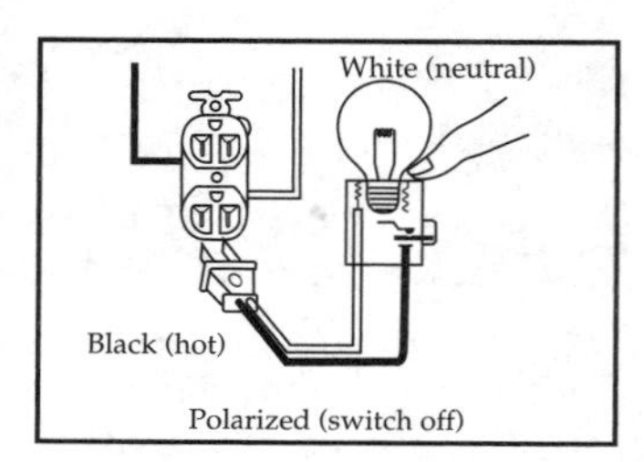

Fig. 10-7. *How a polarized circuit can protect you from a shock hazard.*

you might find a fireplace in a bedroom or kitchen. When inspecting the fireplace, look at the front face just above the firebox. If this area has a blackish tint or color, it is usually an indication of a smoky fireplace and is the result of a buildup over the years of layers of soot and creosote. This problem can usually be corrected.

A smoky fireplace might be the result of too small a flue for the size of the firebox opening. If this is the case, reducing the size of the opening by raising the hearth or installing a canopy on the top portion of the opening very often corrects the problem. However, determining the amount to raise the hearth or the size and shape of canopy to use requires experimentation. Sometimes the soot buildup is caused by backsmoking as a result of downdrafts. This is the result of wind currents bouncing off the side of the building or tall trees and then down the chimney. If the smoking is caused by downdrafts, the problem can usually be corrected by installing a concrete, stone, or metal cap on top of the chimney. Another possible reason for poor draft or backsmoking is that the chimney is undersized. A rule of thumb for the minimum height of the chimney states that the total height from the firebox floor to the top of the chimney should never be less than 15 feet.

Backsmoking can also be caused by a negative pressure condition within the house. This is a fairly common phenomenon in newer houses. To conserve energy, newer homes are better caulked, weatherstripped, and tighter than older homes. The negative pressure results when more air in the house exhausts to the outside from fans, heating-system chimney, and so on than flows in from infiltration. Because of the negative pressure, when a fire is first lit in the fireplace and the damper is opened, there is an onrush of incoming air down the flue. This condition can be eliminated by slightly opening a window or in newer fireplaces opening the fresh-air vent associated with the fireplace.

Look inside the firebox for a damper. The damper is used to close the flue when the fireplace is not in use. It prevents heat loss through the flue in the winter and also prevents small animals such as squirrels and racoons from entering the room through the flue. Check the operation of the damper. On many older fireplaces, dampers were omitted. If the fireplace does not have a damper, or if the damper is defective, record that fact on your worksheet. A damper or equivalent is considered necessary, and one should be installed. If there is a chimney-top damper, check its operation. See chapter 3, page 25.

The bricks or stones lining the firebox should be checked for cracked, chipped, broken, and disintegrating sections. Are the mortar joints intact, or are they in need of repointing? Cracked or open sections inside the firebox are a potential fire hazard and must be repaired.

Look up at the flue from inside the firebox. Normally this area will be coated with a layer of soot and creosote. If the layer is thick, have the chimney cleaned to minimize the possibility of a chimney fire. If the creosote layer is powdery, the flue can be cleaned by a chimney sweep with a brush. However, if the creosote buildup is of a tar consistency or a hard glaze, it cannot be brushed out by conventional means. It must be removed by chemical or mechanical means. Is there an obstruction in the flue, such as a bird's nest? If the flue has a slight offset and you can see daylight, there is no obstruction. When the flue is offset so that you cannot see straight up, determining whether there is an obstruction is difficult without lighting a fire. One trick is to blow up at the flue to dislodge fine particles of soot. If the flue is not obstructed, they will float up the chimney.

In some homes you might find a prefabricated wood-burning fireplace. These units are usually available with chimneys and have a specially insulated firebox shell. They are light in weight, do not require a special foundation, and can be wall-mounted or freestanding. The fireplace can be located in practically any part of a house, depending on local codes. If you find such a fireplace, check to see if it has been approved by Underwriters Laboratories, or some other nationally recognized testing agency.

Gas-fired fireplace

There are two basic types of gas fireplaces. They are either vent-free or have a direct vent. With direct vent fireplaces, the combustion gases are vented to the outside by a two-layer pipe that runs through a hole in the wall behind the unit. The outer pipe draws air in from the outside, which is needed for proper combustion of the gas. The inner pipe channels the gaseous products of combustion to the outside, or up a chimney. If a chimney is used, the local building codes may require the damper in your chimney to be permanently set to an open position because of the gas pilot light.

With a vent-free fireplace there is no need for a damper. These appliances have a thermal sensor located near the top of the fireplace opening. This is a safety device that in a properly operating fireplace would never be triggered. The sensor cannot be reset. It comes with the unit and is set in the factory. Vent-free fireplaces also have an oxygen depletion sensor. If the fire takes too much oxygen out of the room's air and is not replaced, the sensor is designed to turn off the gas before carbon monoxide reaches a dangerous level in the room.

Bedrooms

Every habitable room in the house must have at least one openable window. Sometimes do-it-yourself homeowners finish off an area in the basement or attic for use as an extra bedroom even though there is no window in the room. If you find such a room, be advised that it is a potential fire hazard.

When a room is used as a bedroom, certain items are necessary as fire-safety measures. As a means of escaping in the event of a fire, each bedroom should have at least one outside window whose sill height is not more than 44 inches above the floor. The window should also have a minimum net clear openable area of 5.7 square feet with no dimension less than 20 inches in width and 24 inches in height. If a bedroom is located two stories above the ground, a rope ladder or equivalent device should be provided in the room to enable the inhabitants to escape once they climb through the window. Some municipalities have local ordinances against converting a third-level attic into bedrooms without adequate fire protection. If you find bedrooms in a converted attic, you should check with the local building department to see if a certificate of occupancy had been issued for those rooms.

Last, from a fire-safety point of view, every bedroom must have an entry door. This door should be closed when the occupant goes to sleep. A closed door reduces drafts, thus reducing flame-spread time, and drastically reduces smoke infiltration. Check to see if there is a smoke detector on the wall or ceiling. There should be. In new construction most states require that all of the smoke detectors in the house be hard wired and interconnected. Check the operation of the detector. If it's interconnected with the others, hold your ears. The alarm noise is quite loud.

Don't forget to look for a closet. Is there a ceiling light fixture in the closet? More often than not the light fixture is a porcelain base with an exposed incandescent bulb. Check to see if the bulb is relatively close to the closet shelf. If it is, it's a potential fire hazard. If clothing or other fabrics come in contact with a bulb that has been inadvertently left on for an extended period of time, they could ignite and start a fire. In some homes where bedrooms have been added, closets are omitted in the renovation. There should be a closet in every bedroom. If there isn't one, record that fact on your worksheet.

Bathrooms

Because of the nature of the room, the bathroom must be well ventilated. Ventilation can be provided naturally through an openable window or mechanically by an exhaust fan. If there is an exhaust fan, try to determine where it discharges. The air drawn through the fan should discharge into the atmosphere. This can be accomplished through the use of a duct that terminates on the side of the building or one that extends up through the attic and terminates on the roof. When the bathroom is located on the level just beneath the attic, the exhaust fan very often discharges directly into the attic. This is undesirable because the moisture-laden air can cause condensation problems in the attic. From a convenience point of view, many builders often connect the exhaust fan to the lighting circuit so that both are controlled by one switch. This is not an energy-efficient installation, since the exhaust fan is not always needed. If you see such a setup, you might want to consider rewiring the fan so that it can be controlled by a separate switch.

A recurring problem in bathrooms is water leakage through cracked and open tile joints around the tub or shower. This condition requires periodic maintenance and if neglected can cause considerable cosmetic damage. Check the tiles for cracked and open joints. Lean over the tub or go into the shower and press on the tiles, particularly at the lower portion of the walls. Loose tiles will move slightly or might even come out. If the tiles yield, it is usually an indication that the plaster or plasterboard backing has suffered some deterioration from water seepage. Depending on the severity of the deterioration, it might be necessary to rehabilitate that portion of the wall. Sometimes the cracks are not readily visible but will nevertheless allow water to seep through and wet the wall behind the tiles. You can check for leakage by using your moisture meter on the tile wall.

If the walls around the tub and shower are not kept watertight, water can leak around open areas, wet the ceiling below, and eventually rot the wood framing and cause the ceiling to deteriorate. Missing tiles must be replaced; loose tiles must be resecured; and cracked and open joints must be regrouted or caulked. Caulking is the procedure most often used for making repairs. A tube of caulking compound is available at any hardware store.

If the walls around the tub or shower have a panel finish rather than tile, check the joints for cracked or open sections. Sometimes the tub or shower and its associated walls are an integral unit made of molded plastic. In this case, leaks through open joints are not a problem. Many homes have showers and tubs with doors to prevent water from splashing onto the floor. The doors must be made with safety glass or plastic and should be checked for cracked panes and ease of operation.

When the shower base is covered with ceramic tiles rather than molded plastic or is terrazzo-constructed, there is the potential for a problem because of a faulty shower pan. A large lead or plastic sheet is normally installed below the tiles at the base of the shower to collect water that seeps through cracked and open tile joints. If the shower pan is intact, it will direct the water down the drain without incident. However, as the shower pan ages, the joints very often deteriorate, resulting in water leakage through those joints. If you notice large water stains on or damage to the ceiling of an area below the shower, you should suspect a shower-pan problem. (See FIG. 10-8.)

If you have the seller's permission, you can test for a faulty shower pan by covering the shower drain and filling the shower base with about an inch of water. Let the water stand in the base for about forty-five minutes. If the shower pan is faulty and there are cracks in the base tile joints, water will seep through and wet the ceiling below. When the tile joints at the base of the shower are all sealed, even if the lead pan is faulty, there will be no leakage. Instead of replacing a faulty shower pan, many homeowners simply recaulk the tile joints at the base of the shower. This is considered a makeshift fix. If you find a heavy layer of caulking in the shower base, even if you do not find water stains on the ceiling below (the ceiling could have been repainted after the makeshift fix), you should suspect shower-pan problems.

Water pressure and flow

The water pressure in a house is an item of concern to most home buyers, and rightly so. However, water pressure per se is not usually the problem. If you see water trickling out of a faucet, it does not necessarily mean that the pressure is low. It means that the water flow is low. Low water flow can be caused by low pressure. However, more often than not, low flow is caused by a constriction in the supply pipes. The constriction can be the result of mineral or corrosive deposits on the inside diameter of the pipes, a kink in the pipe, or small-diameter piping. Quite often low flow at a fixture is caused by a partially clogged faucet aerator, which requires only cleaning. Water flow is measured in gallons per minute, and a sink faucet should be able to deliver 4 to 5 gallons per minute.

The water pressure available to a house at the meter varies, but it is usually in a range of 20 to 60 pounds per square inch (psi). I inspected

Fig. 10-8. Water stains on ceilings caused by leaks from radiator in room above.

a house that had a pressure gauge on the house side of the water meter. (A pressure gauge in the plumbing system is desirable but is not very common.) The gauge indicated that the available water pressure was 110 psi. Yet when three faucets were turned on at the same time, the water flow from each faucet was just a trickle. Regardless of how high the pressure is, if a constriction in the water pipes limits the water flow, the amount of water discharging from the faucet will be noticeably low. In this case, the pipes were made of galvanized iron. Over the years, corrosion and deposits had coated the inside of the distribution piping to a point where the pipes required replacement.

When you inspect the bathroom, check the water flow by simultaneously turning on the faucets in the sink and tub and flushing the bowl. There will usually be a drop in flow when the second faucet is turned on. If it is not very noticeable, do not worry about it—it is normal. You can trust your eye. If the water flow from the faucets looks good, then for all practical purposes, it is good. If the flow is noticeably low, record it on your worksheet. The condition requires correction. If there is a shower in the tub, check the operation of the shower diverter valve. When the shower is on, there should be very little or no water leaking out of the tub spout. If there is, the diverter valve is defective and will require replacement. Record this on your worksheet.

Check the flow for both the hot water and the cold water, but not at the same time. It is possible for the cold-water flow to be good and the hot-water flow fair or poor. Occasionally, when the domestic hot water is generated through the heating system, deposits form in the tankless coil, restricting the flow. If the hot-water flow is poor, you probably need to replace the tankless coil. (See chapter 16.)

Open and close the faucets rapidly. Do you hear a hammering and vibrating noise? You should not, but if you do, you are hearing *water hammer*, which occurs when the water flowing in the pipe comes to an abrupt stop. It introduces hydraulic shock and vibrations that can in extreme cases damage the pipe or fittings. The condition can be corrected easily by installing an air chamber or antiknock coil. These units provide a cushion of air to absorb the shock when the water flow stops.

While you are checking the water flow, look for water leakage around the faucets, and the drainage trap below the sink. If you notice that a portion of the trap is taped over, it is usually a makeshift correction for a leak. The drainage trap is not installed below the sink to catch rings or other valuables that fall down the drain, although very often this happens. The purpose of the trap is to provide a water seal that blocks noxious gases from seeping back into the room. If a sink trap is not properly vented, it can lose its seal from a siphoning action when the water is draining. (See FIG. 10-9.)

Usually the P-type sink trap is vented properly. However, the S-type trap is often not vented properly. You can tell by making the following test, although it is not foolproof. Fill the sink until the water is almost at the rim; then let the water drain. When most of the water has drained, if you hear a sucking and gurgling noise, the trap is not vented, and the water seal has been lost. When all the water has drained, if you let the water run freely for about five seconds, you will reestablish the water seal. If the main plumbing waste line has a house trap, the problem is minimized.

Check the drainage in the sink, tub, and bowl. Wrinkle up some toilet paper into a ball and throw it into the bowl. When the bowl is flushed, the paper should be carried down the drain. If the water in the bowl starts to rise to the top and then settles down without flushing, there is a blockage in the drain line that must be checked further and corrected. If the house has a septic system, this condition might indicate a problem with the septic

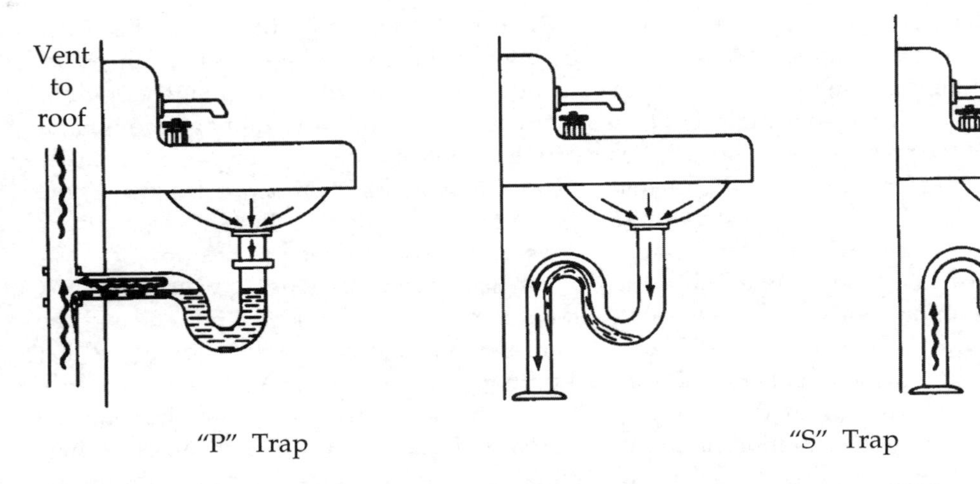

When the sink trap is properly vented, the trap holds enough water to form a seal against the entry of sewer gases. The gases vent harmlessly through the vent stack to the outside.

When the sink trap is not properly vented, the rush of the waste water can siphon the water seal out of the trap.

With no water seal, sewer gases can escape into the house.

A

B

Fig. 10-9. *Sink traps: A—Properly vented P-type trap; B—S-type trap, usually not properly vented.*

system. If the water in the sink and tub does not readily drain away or drains sluggishly, there is a blockage in those drains that must also be corrected. In this case, the correction might be as simple as removing hair that has accumulated around the pop-up drain plug.

Look at the sink and tub faucets. When the tub and sink are filled to the flood rim, there should be an air gap between the top of the water and the bottom of the faucet spout. (See FIG. 10-10.) If there isn't, the faucet is vulnerable to back siphonage (the flowing back of dirty water from the sink or tub into the potable water supply because of a negative pressure in the line). This condition is a violation of the plumbing code and is found periodically in old homes. However, it is not limited to old homes. The "do-it-yourself" homeowner could have installed the faucet without being aware that it is a potential problem. Are there shutoff valves below the sink and bowl for the supply pipes? Although not a necessary feature, they are very convenient when making minor repairs.

If the tub is equipped with a whirlpool bath, check its operation. Fill the tub to a level above the jet ports before turning on the bath's pump. Record any problems on your worksheet. For health reasons, before the whirlpool bath is used, it should be professionally cleaned to remove bath residue and scale deposits because no inspection method is available to ensure that proper maintenance procedures were previously followed. Also check to see if there is an access hatch for the pump motor. This is occasionally overlooked during the installation, and it is helpful in the event that maintenance is required.

After you flush the bowl, if the toilet is the tank type, lift the tank lid and look inside. If the lid has a cloth cover, be careful when lifting

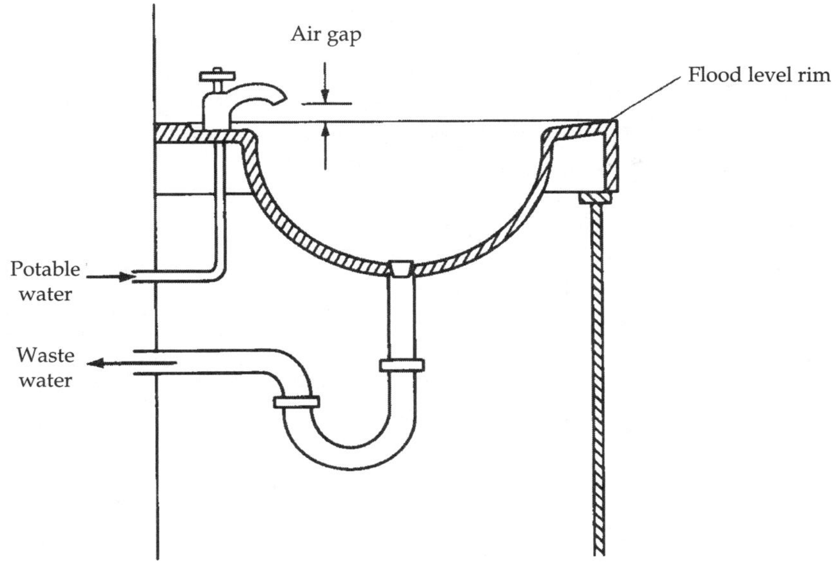

Fig. 10-10. *Air gap between bottom of faucet spout and flood rim of sink prevents the backflow of nonpotable water into the potable water supply.*

it. It might be cracked or chipped. If water is spurting out of the fill valve, maintenance is needed. You might even find the trip lever connected to the flush valve by a string rather than a metal chain or strip. Repair or replacement of the flushing mechanism is not costly.

Occasionally the seller, not being the original owner, inadvertently misrepresents the age of the house. Depending on the manufacturer of the toilet bowl, you might be able to verify the house's age by looking on the underside of the tank lid. There might be a date stamped on the underside that can be a clue, assuming the toilet is the original and was not installed during a later renovation. The date is the date of manufacture, usually within a few months of installation.

Kitchen

When inspecting the kitchen, in addition to the plumbing items mentioned above, the condition of the cabinets and counters should be checked. The appliances, although important, need not be inspected at this time. Because appliances can break down at any time, it is recommended that on the day of, but prior to, the contract closing, you come back to the house and operationally check every appliance included in the purchase. If an appliance is not operational, you can have your attorney request an adjustment at the closing for the cost of repairs.

Cabinets should be inspected for missing, cracked, and loose-fitting doors and drawers. Missing hardware for doors and drawers should be noted. The shelves should be checked to see if they are adequately supported. The counters should be inspected for cracked, burned, blistered, and loose sections. If there is a cutting board or hotplate on the counter, lift it up or move it aside. You might find that it is concealing a damaged section of the counter.

When inspecting the sink, in addition to checking water flow and drainage, look for a sprayer. If there is one, see if it is operational. I have seen many sprayers with a disconnected hose mounted in the sink fitting. The sprayers

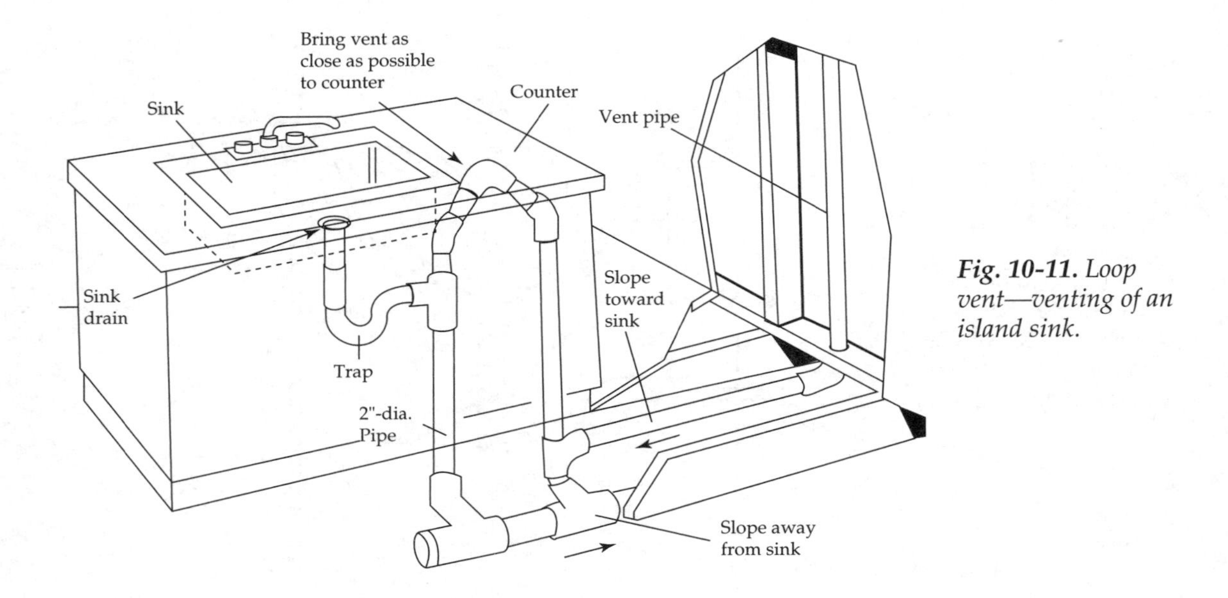

Fig. 10-11. *Loop vent—venting of an island sink.*

were not functional and served only as a decoration to cover the opening in the sink. Is there a sink in a central island cabinet? If there is, check the drain piping. Island sinks are notorious for being improperly vented. If the sink drain has an S-type trap rather than a P-type trap, it is not properly vented. Record this item on your worksheet. If approved by local codes, this condition can be corrected with a minor plumbing change by connecting an air admittance valve (AAV) to the drain. Another way of venting an island sink is with a "loop vent" (see FIG. 10-11), which is usually, but not always, installed during the construction phase of the house.

Is there a garbage-disposal unit connected to the sink drain? If there is and the house has a septic tank, there might be a problem. A garbage-disposal unit introduces solid wastes into the septic system at a greater rate than normal. To avoid overloading the system, some states have a design criterion calling for a larger-capacity septic tank when there is a garbage-disposal unit. Other states recommend that the tank be cleaned at more frequent intervals. (See chapter 13 for a discussion of septic systems.)

If the disposal unit was added after the house was constructed and provisions were not made for a larger septic tank or more frequent cleaning, the disposal unit might have been overloading the septic system. This can result in premature failure of the system. If the house has a garbage-disposal unit and a septic system, check with the local municipal building department to determine if the system was designed to accommodate the wastes from the disposal unit. Also, check with the owner to find out when the septic tank was last cleaned. If it was not cleaned or at least inspected for sludge buildup within the last three years, record the fact on your worksheet. The tank should be cleaned after you move in.

Hallway and staircase

The remaining rooms in the house should be checked as described previously. In addition, the connecting hallway should be inspected as you walk from one part of the house to another. The hallway should be treated as an interior room, and its walls, floor, ceiling, and trim should be inspected. Look for an overhead light in the hall. Is it controlled by three-way switches located at both ends of the hall? It should be. As a fire-safety measure, a smoke detector or equivalent should be mounted on the ceiling of the hallway leading to the bedrooms. If there is a smoke detector, is it properly located? The corners of a hallway where the walls and ceiling meet is considered dead-air space. This means that even though smoke will circulate and accumulate near the ceiling, it will not penetrate into those corners until the hall is completely filled with smoke. According to the National Fire Code: "Spot-type smoke detectors shall be located on the ceiling not less than six inches from a side wall, or if on the side wall, between six to twelve inches from the ceiling." Neither wall- nor ceiling-mounted detectors should be placed near a light fixture or a ventilation grille that could block smoke from reaching the detector. If you see an incorrectly located smoke detector, you should alert the seller and record it on your worksheet.

As you proceed from one level to another, check the connecting staircase. Squeaky treads, as with squeaky floors, indicate loose sections. They are not a concerning factor, but if excessive, they are annoying. Correcting the problem is often difficult because the underside is usually concealed. The steps should be uniformly spaced without any dimensional variation. It is not uncommon to find basement steps with uneven risers for the bottom and top steps. This condition is a potential hazard, since someone can easily trip.

As a safety precaution, all steps should have handrails. Check to see if both ends of the handrail have a return to the wall. Rails with returned ends will prevent clothing from catching on the end, which could result in a fall. Loose handrails must be resecured. Sometimes the handrail is secured in such a manner that allows little finger room. This is a hazardous condition, especially for children. Is there a window at the base of the staircase or on the landing? If there is, its sill should be at least 36 inches above the floor. This will prevent someone from falling through the window in the event of a fall down the stairs. If the sill is less than 36 inches high, a window guard should be installed as a precautionary measure.

Lighting the stairway is essential, especially for areas that are often neglected, such as basement steps. The location of the light fixture is unimportant as long as the light provides sufficient illumination for the entire staircase. The light should be controlled by three-way switches, one at the top and the other at the bottom of the staircase.

Checkpoint summary

Walls and ceilings

- ❍ Do not be concerned with minor cosmetic problems. Specifically check for:
  - — broken walls and ceilings;
  - — loose, missing, and bulging areas of plaster or plasterboard;
  - — missing, loose, and sagging sections of ceiling tile;
  - — sagging sections of plaster;
  - — truss uplift cracks (at wall-ceiling intersection);
  - — water stains on ceilings, particularly below roofs, bathrooms, and kitchens;
  - — disintegrating plaster, peeling and flaking paint, and water stains on walls facing the exterior (usually caused by open or exposed exterior joints).
- ❍ If there is a sulfur-based odor in the house, suspect Chinese drywall, especially if the copper plumbing and electrical wiring show signs of corrosion.

- ❍ Note rooms that have cracked walls in which the windows and door frames are not level (professional evaluation is recommended).
- ❍ Check paneled walls for plasterboard backing (often omitted).
- ❍ Check trim for missing, loose, cracked, or broken sections.

Floors

- ❍ Inspect for floors that are not level, have loose floorboards, or have sagging areas. Sagging floors that are also noted in the ceilings below should be evaluated by a professional.
- ❍ Inspect for large open joints between floor and partition walls (usually caused by excessive shrinkage).
- ❍ Note floors that have wall-to-wall carpeting. Do not assume that there is a hardwood floor beneath the carpeting. Request that owner make this representation in writing.
- ❍ Inspect concrete floor slabs for cracked and settled areas. Note areas that have large open joints between the floor slab and the walls.
- ❍ Check raised wood floors over concrete slabs for soft, spongy, and delaminated sections. Often these conditions are a result of water seepage with associated rot in wood framing.

Heat

- ❍ Check interior rooms for missing radiators or heat registers.
- ❍ If rooms are heated by other means (radiant panels), verify this with the owner.
- ❍ Are radiators and heat-supply registers efficiently located (preferably on an exterior wall and below a window)?

Windows

- ❍ Check windows for ease of operation.
- ❍ Inspect for cracked and broken panes; chipped, cracked, and missing putty.
- ❍ Check exterior sills for cracked and rotting sections.
- ❍ Inspect double-hung windows for broken or missing sash cords, loose or binding sashes, and missing hardware.
- ❍ Inspect steel-casement windows for cracked panes, rusting and sprung frames, loose, missing, or inoperative hardware.
- ❍ Check thermal-pane windows for faulty seals (water droplets or cloudy areas between the glass panes).
- ❍ Check casement and awning windows for interior storm windows and screens.

Electrical outlets

- ❍ Inspect rooms, stairways, and hallways for electrical hazards and violations (see chapter 12 summary).
- ❍ Note all rooms in which there are insufficient outlets and outlets that are loose or have missing cover plates.
- ❍ Check whether outlets are electrically "hot" and properly grounded. Do any have reverse polarity?
- ❍ Are the kitchen and bathroom receptacles GFI protected?

Fireplace

- ❍ Inspect brick or stone firebox lining for cracked, chipped, broken, or deteriorating sections.
- ❍ Check for cracked, loose, or disintegrating mortar joints.
- ❍ Check top of firebox for an operational damper.
- ❍ Is there a chimney top damper?
- ❍ Inspect area for obstructions.
- ❍ Is the chimney flue lined?
- ❍ Check for heavy layers of soot and creosote (heavy layers indicate the need for chimney cleaning).

Bedrooms

❍ Check that all bedrooms have at least one openable window with the following criteria:
  — sill height not *more* than 44 inches above the floor;
  — minimum openable area of 5.7 square feet with no dimension less than 20 inches.

❍ Do all bedrooms have entry doors and closets?

❍ Is there a smoke detector on the wall or ceiling?

❍ If a portion of the attic or basement has been converted to a bedroom, is there a certificate of occupancy for the room?

Bathrooms

❍ Check bathrooms for adequate ventilation.

❍ If there is an exhaust fan
  — is it operational?
  — does it have a separate on-off switch?
  — can you determine where the fan exhaust discharges?

❍ Inspect tiled areas, particularly around the tub or shower, for open joints, cracked, loose, and missing tiles.

❍ Note wall areas in the tub or shower that show evidence of deterioration (spongy or loose sections).

❍ Check shower doors for cracked panes (should be safety glass) and ease of operation.

❍ Check sinks, bowl, and tub or shower for cracked, chipped, and stained areas. Check that sinks and bowls are properly secured.

❍ Inspect sink and tub faucets for proper air gap (potential back siphonage).

❍ Check fixtures for individual shutoff valves.

❍ Inspect fixture plumbing for leaks, kinked lines, patched and makeshift corrections (taped joints and rubber-hose connections).

❍ Inspect sink drain lines for improper venting (S-type traps).

❍ Does shower diverter valve in the tub leak?

❍ Check operation of whirlpool bath.

❍ Is there an access hatch for the pump motor for the whirlpool bath?

Water pressure, flow

❍ Check cold-water flow by simultaneously turning on the faucets in the sink and tub or shower and flushing the bowl.

❍ Perform a similar check for hot-water flow.

❍ Check for water hammer when faucets are opened and closed rapidly.

Kitchen

❍ Check sink for low water flow and proper drainage.

❍ If there is an island sink, is the drain vented properly?

❍ If sink contains a sprayer, check operation (often this unit is disconnected).

❍ If sink drain contains a garbage-disposal unit and house has a septic tank, you should determine the following:
  — Was disposal unit added after the house was constructed? (Septic tank might be undersized.)
  — When was septic tank last cleaned? (If over three years, the tank should be cleaned.)

❍ Inspect cabinets for missing, cracked, or loose-fitting doors and drawers.

❍ Check for missing hardware on cabinet doors and drawers.

❍ Check shelving for adequate support, cracked, warped, or missing sections.

❍ Inspect counter and countertops for cracked, burned, blistered, or loose sections.

❍ Check all appliances for operational integrity on the day of, but prior to, contract closing.

Hallway and staircase

- ❍ Check for properly located smoke detectors in hallway areas leading to the bedrooms.
- ❍ Check hallways and staircases for adequate lighting. Are three-way switches located at both ends of the hallway?
- ❍ Check walls, floor, ceiling, trim, and so on as you would for interior rooms.
- ❍ Inspect stairways for uneven risers, loose treads, missing handrails, and handrails with tight finger room. Do the ends of the handrail have a return to the wall?
- ❍ If there is a window at the base of the stairway or landing, is the sill less than 36 inches above the floor? (If this condition exists, you should install a window guard.)

11

Basement and crawl space

Problems in a basement or crawl space are often among the most costly to correct. Look specifically for signs of water penetration, structural deterioration of the wood support members, and structural deficiencies of the foundation wall.

Foundation

The purpose of the foundation is to support the main portion of the house and transmit its load to the ground. The foundation of most residential structures consists of walls that rest on an enlarged base called a *footing*. The footing spreads the transmitted load directly to the supporting soil and is usually resting on undisturbed earth. Where there are freezing temperatures, the footings must be located below the frost line. Otherwise, the footings become vulnerable to frost heaving resulting from the freezing of soil moisture. Pilasters, piers, and columns are also used to support the main structure. (See FIG. 11-1.)

Foundation walls are normally designed to support the vertical loads from the house and to resist the horizontal forces resulting

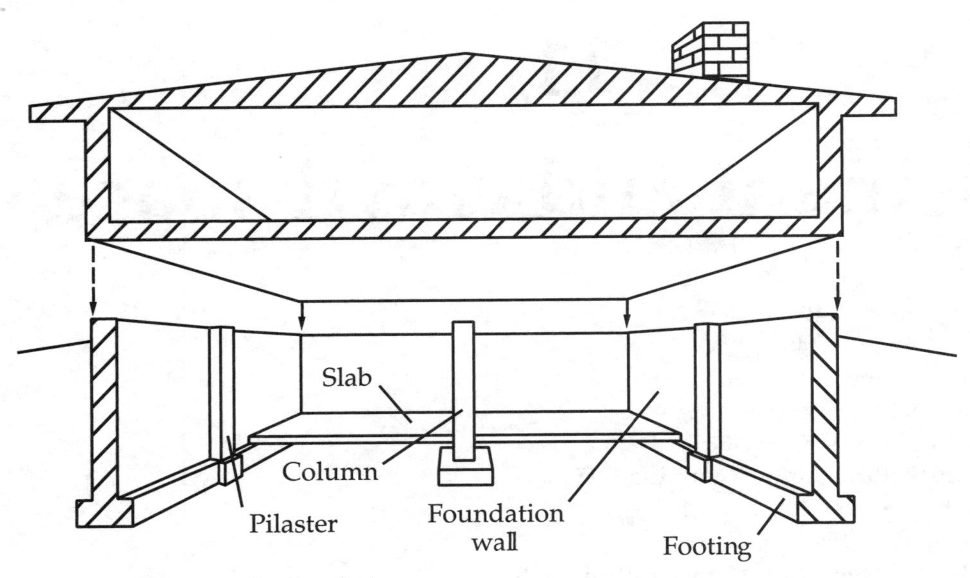

Fig. 11-1. Elements of a foundation.

from the earth's pressure. In most parts of the country, the house must be anchored to the foundation to provide resistance to wind and/or earthquake forces. High wind forces or rapid ground movement can cause a structure that is not properly anchored to lift, shift, or rotate slightly. When the house is properly anchored, the resistance to this movement produces stresses that are within the foundation wall. Consequently, the foundation wall must be designed to withstand these forces. In some cases, the combined vertical and horizontal forces acting on the foundation wall is great enough to require additional bracing and stiffening of the walls. In these cases, pilasters are often used to provide the additional support.

Settlement

All soils compress to some extent, and if the footing is not resting on bedrock, the foundation is subject to settlement. When the settlement is slight or uniform, or was anticipated in the design, it is of little concern. However, when the foundation settles unevenly (*differential* settlement), there is generally some cause for concern. Differential settlement introduces stresses that might seriously weaken the building. It often causes cracks in the foundation walls, unsightly cracks in the finished walls and ceilings, sloping floors and windows, and doors that bind. When the differential settlement is excessive, it can result in a structural failure of a portion of the foundation wall, a condition that can be quite costly to correct.

The principal cause of foundation settlement is a reduction of the volume of the voids in the soil supporting the foundation. The voids are the spaces between the soil particles and contain air and water. Sandy soil contains large granular particles with a relatively small volume of voids; clay soil contains fine-grained particles with a large volume of voids. The settlement of foundations built on a sandy soil tends to be quick and slight, whereas the settlement of foundations built on a clay soil tends to be greater and might occur over many years.

Usually you do not have to worry about foundation settlement in older homes because by the time you are considering buying the house, the settlement has already occurred, along with any accompanying problems. And what you see is what you get. However, if what you see is a house out of plumb, with sagging sections, it might have structural problems, and as a precautionary measure, you should have the house evaluated by a professional.

Some soils are considered poor for building-construction purposes. Homes built on a highly organic soil are vulnerable to excessive settlement when the soil dries out. Organic soil acts like a sponge, and as the soil dries, the organic matter shrinks. Depending on the amount of organic matter in the soil, the condition can cause differential foundation settlement.

Some soils on steep hillsides are subject to severe slippage. When the amount of buildable soil above an impervious subsoil layer is shallow, if the soil becomes saturated (because of an extended heavy rain), the entire upper layer of the hillside can slip to the bottom, along with any houses built on the side of the hill.

Some soils with a high clay content can swell or shrink up to 50 percent between wet and dry conditions. These soils are considered unstable, and unless special provisions are made during construction, the condition can cause excessive differential settlement cracks in the foundation.

If you have any questions about the soil the house is built on, you can usually get them answered at the local office of the U.S. Department of Agriculture, Soil Conservation Service, or at the office of the local county soil and water conservation district.

Inspection

Most houses built in the last seventy years have foundation walls of poured concrete, concrete blocks, or cinder blocks, whereas older homes generally have stone or brick foundation walls. When inspecting stone or brick foundations, especially in homes built before or around 1900, pay particular attention to the mortar joints. I have found that in homes of this vintage, many of the mortar joints in the foundation walls have deteriorated to a point where they are no longer functional. Some of the joints might have holes, and some may be filled with soft, crumbling mortar that can easily be raked out. In some cases, the joints between the foundation stones might have been filled with earth or mud rather than mortar.

Deteriorated mortar joints in foundation walls should be repointed, not so much because they represent a weakened structural condition (they do, but this is usually not a severe problem) but because they can allow water to penetrate into the basement (see FIG. 11-2) and enable mice and other pests to enter the area.

In some older homes with brick foundation walls, you might find soft, crumbled, and flaking bricks so badly deteriorated that they should be replaced. This condition is usually the result of using underburned bricks in the construction of the wall. Bricks that are underburned during the manufacturing process are softer and absorb water more readily. If you see this condition, it should be recorded on your worksheet, since masonry rehabilitation is needed.

Cracks Cracks in poured concrete or concrete-block foundation walls can result from shrinkage, differential settlement, lateral pressure on the wall by the soil, or poor-quality workmanship. It is not uncommon to find short cracks in foundation walls. These cracks might be vertical, horizontal, inclined, stepped, smooth, or irregular and are usually of no structural concern. All cracks, however, should be sealed as a precautionary measure

Fig. 11-2. Deteriorated mortar joints in fieldstone foundation wall. Note water seeping through the wall at base.

against water penetration into the basement. Otherwise, if there should be a hydrostatic pressure buildup in the soil against the foundation, water will seep through the cracks.

If there are long narrow cracks in the wall and both sides of the cracks line up so that there is no noticeable differential settlement, it is usually not a serious condition and can be controlled by sealing the cracks. However, when both sides of the cracks do not line up or there are long open cracks, a more serious condition of differential settlement exists. Since it is not possible to determine from a single inspection whether the differential settlement is active or dormant, the wall should be checked for incremental movement over a period of time, usually several months. In most cases, after some time, the differential settlement stabilizes with little effect on the house other than functional annoyances such as binding windows or a floor that might not be level. However, because of excessive settlement, an unstable condition can occur either in the foundation wall or the wood framing being supported by the wall. If there is any doubt in your mind about the condition of the foundation wall, you should have it checked by a professional.

Another type of crack of concern is a long, open horizontal crack in the foundation wall, especially if the wall shows signs of bowing. This condition is principally caused by an excessive horizontal pressure being exerted on the foundation wall by the earth backfill and indicates that the wall cannot adequately withstand these external lateral forces. Structurally the wall has failed; this condition should be recorded on your worksheet as a condition that requires further investigation and depending on the extent of bowing, reinforcement, or replacement.

Superior Walls In the 1980s, insulated, precast, ribbed concrete wall panels called *Superior Walls* were introduced to the market to be used for building foundations. (See FIG. 11-3.) The panels can be installed in a fraction of the time that it would take for poured concrete

Fig. 11-3. *Installing precast concrete "Superior Walls."*

foundations and traditional exterior walls. According to the manufacturer, when panels are delivered to the site, they are often erected in three or four hours. When installing a foundation, the Superior Walls are bolted together at the top and the bottom of each panel, and the joints between the panels are sealed with a polyurethane sealant. When inspecting the foundation, check for open joints and signs of wall tilt, and as with other foundations, check for cracks and signs of water penetration.

Structural support framing Because of the vulnerability to deterioration from rot and wood-destroying insects, all exposed wooden support members (girders, joists, posts, and sill plates) should be checked for structural integrity. Steel beams and columns, on the other hand, need be checked only for degree of rust or corrosion. Usually the rusting is only a surface defect that can be corrected by scraping, priming, and painting.

Wooden support members should be probed with a screwdriver or an ice pick, as described in chapter 8. If the wood is in good condition, the probe will not penetrate much below the surface. The portions of the joists and girders that are most vulnerable to deterioration are those resting on or adjacent to the foundation wall. While probing these areas, you should also check the sill plate (which is anchored to the top of the foundation) for structural integrity. Many older houses have wood support posts rather than steel columns. If there are wood posts, check their bases for decay. The base has a tendency to rot because of periodic dampness or seepage through the floor. A post with a rotted base should be replaced. Joists with deteriorated end sections, however, need not be replaced. They can be rehabilitated. In most cases, all that is needed is to place a similar size wood member alongside the affected joist and secure it to the portion of the joist that has not deteriorated.

Girders (the main support beams) are often supported at both ends by the foundation wall. If the girder is wood-constructed, the notch in the foundation wall in which the girder rests should be large enough to allow the sides and end of the girder to be ventilated. Otherwise, there can be a moisture buildup, which will promote decay. There should normally be at least a ½-inch clearance around the sides and end of the girder. For adequate support, there should be a minimum of a 3-inch end bearing in the foundation notch. (See FIG. 11-4.)

If you find that the end of the girder has deteriorated so that it is no longer providing adequate support, do not be overly concerned. The condition can be corrected at a reasonable cost without replacing the girder. In many cases, all that is needed is to support the girder by a column or pier located near the foundation wall. This repair, however, should be performed by a professional, since the column or pier must have an adequate size footing to spread the load.

When girders or joists have sagging sections, these are usually noticeable on the floors above. During your interior inspection, if you find floors that are not level, you should be alerted to the possibility of sagging support beams. Sagging sections can be braced with adjustable screw-type columns to prevent further sag. The adjustable columns can be used for releveling sagging sections. This procedure, however, should be approached with caution. The leveling process introduces new stresses that in turn can cause cracking.

During the wood-framing inspection, you might find joists or girders that have been notched at the top or bottom to accommodate the passage of pipes. Notching a beam at the

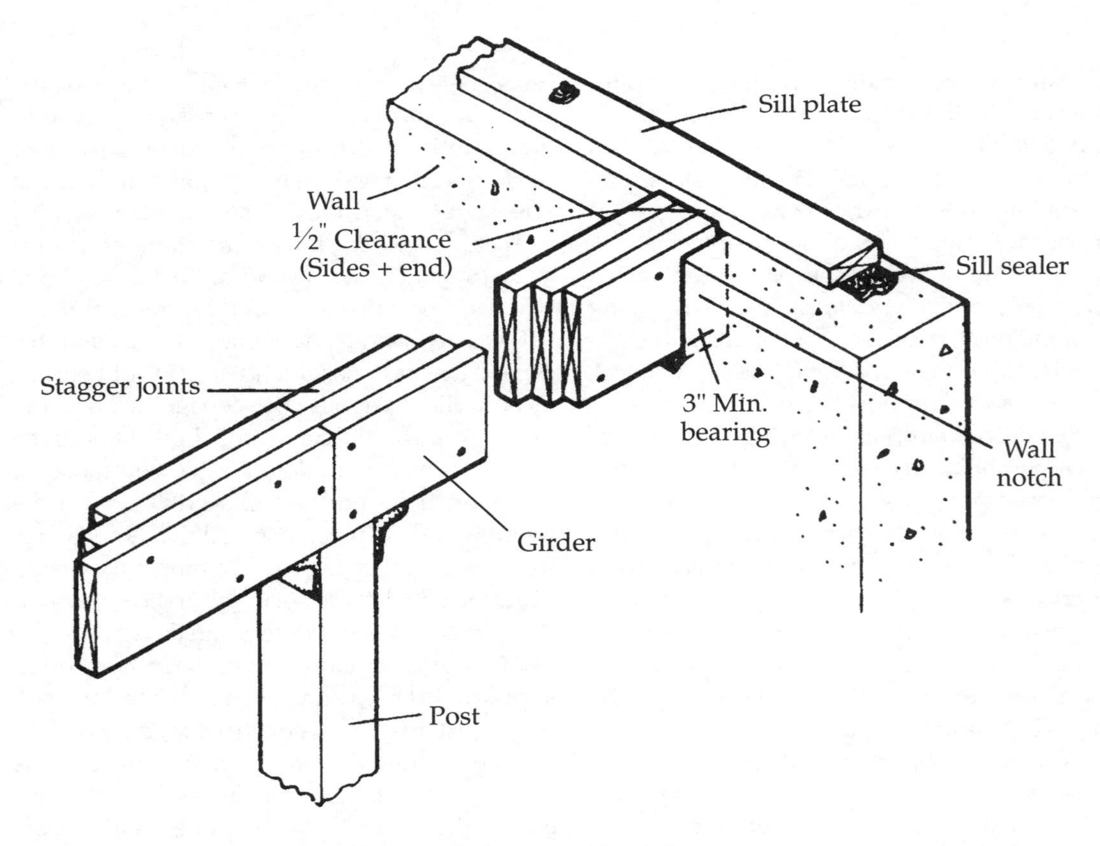

Fig. 11-4. Notch at top of foundation wall to support the girder.

top or bottom will reduce its strength. Just as a chain is as strong as its weakest link, a beam is as strong as its narrowest section. For example, if a 2-by-8-inch joist is notched to a depth of 2 inches, its strength will be reduced to that of a 2-by-6-inch joist. When a beam deflects, the top of the beam is in compression and the bottom is in tension. At a point approximately midway between the depth, the stresses change from one to the other so that at this point there is no compression or tension. Consequently, if a hole is cut midway between the top and bottom of a beam (as long as the hole is less than one-fourth the depth), there will be no effective reduction in strength. (See FIG. 11-5.) Using the above example, if a 2-by-8-inch joist has a 2-inch hole in its center, it will still be as strong as it was without the hole. If you find any notched beams, look specifically for signs of excessive deflection. This is an indication that reinforcement or bracing of the beam is necessary.

Fig. 11-5. *Hole in center of beam width does not effectively reduce beam strength.*

Engineered lumber Engineered lumber has replaced solid-sawn lumber as the framing and structural support member of choice by builders in many newer homes and in homes that are currently being built. The decision is based on the fact that although the lumber is initially more expensive, the total installed costs are less than that of dimensional lumber. The following products fall in the category of engineered lumber: glued laminated (GLULAM) beams, laminated veneer lumber (LVL), I-joists, and open web truss-joists. The advantage of engineered lumber is that it is stronger, stiffer, and more dimensionally stable than solid-sawn dimensional lumber. GLULAM beams and wood I-joists can carry greater loads over longer spans than is possible with solid-sawn lumber of the same size. Because of their "I" cross-sectional shape, wood I-joists weigh up to 60 percent less than solid lumber joists, making them easier to handle. They generally do not shrink, warp, cup, crown, or twist.

When inspecting the floor framing I-joists, look specifically for problem conditions such as notches in the flanges or holes in the webs that are larger than manufacturer's recommendations. Holes less than 1.5 inches in diameter are generally okay as long as they're not on or into the flange. In certain applications where joists line up under a load-bearing partition, depending on the load, there may be web stiffeners between the flanges of the I-joists. Web stiffeners are used to prevent problem conditions such as the web buckling out of plane or "knifing" through the flange. Also squash blocks are used in situations where there is a load-bearing wall above that is stacked over the wall below. The squash blocks transfer the load from the wall above to the wall below without relying on the joists for this support. (See FIG. 11-6.)

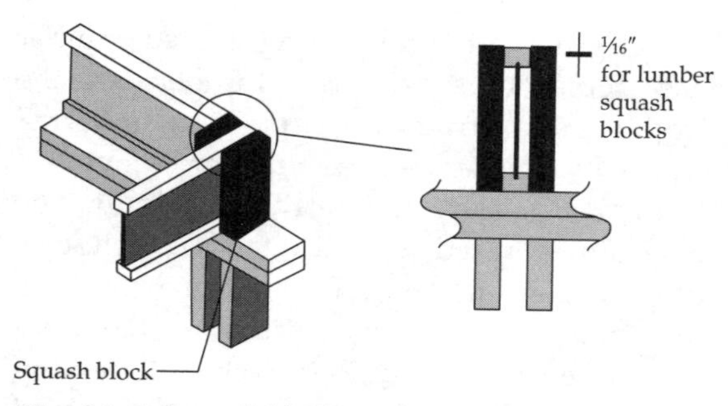

Fig. 11-6. Squash block installation.

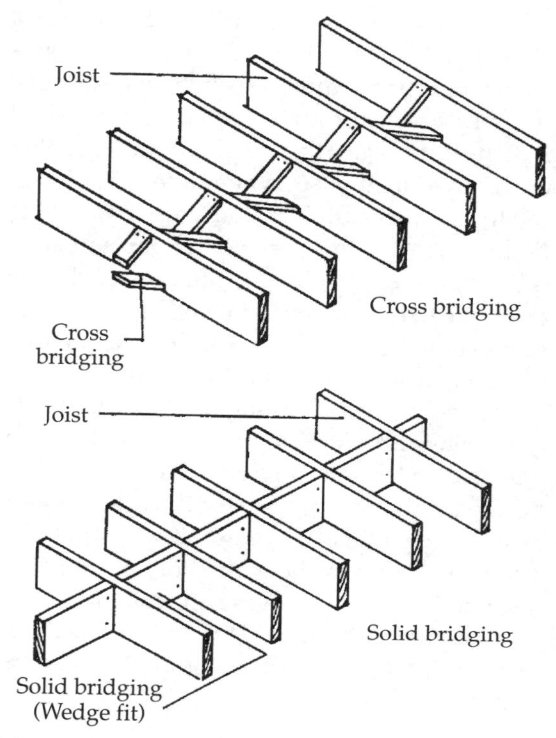

Fig. 11-7. *Bridging. Cross and solid bridging between floor joists. Cross bridging can be wood or metal.*

Check to see if the rim boards are solid lumber or engineered lumber. Only engineered lumber should be used with I-joists. Solid lumber typically has a higher moisture content than engineered lumber, and if used as rim boards, the shrinkage can cause the I-joists to carry the wall load, which could cause web failure.

Most building codes require cross or solid bridging between the floor joists. (See FIG. 11-7.) The purpose of bridging is to hold the joists in a vertical position and to transfer the floor load from one joist to another. Cross bridging consists of pieces of wood or preformed metal set in a diagonal position between the joists to form an X. Solid bridging consists of solid blocks set between the joists. While inspecting the overhead floor joists for rot, insect damage, notching, and sagging sections, you should also check for the presence of bridging. If it is missing, record the fact on your worksheet for later installation.

Dampness

In many parts of the country, the basement or lower level might be damp during portions of the late spring and summer. Dampness in a basement is a normal phenomenon that occurs because cool air cannot hold as much moisture as warm air. It does not necessarily indicate that the basement has a water problem. The temperature of the air in the basement or crawl space during the late spring and summer is always cooler than the outside air. Consequently, when outside air infiltrates into the basement through open windows, doors, cracks, or joints, the temperature of that air drops. This cooler air cannot hold as much moisture and results in a higher relative humidity of the air that entered the basement. Depending on the temperature and the amount of moisture present in the air, some moisture might condense on cool surfaces such as foundation walls and cold-water pipes.

Sometimes water droplets on the foundation wall caused by condensation are erroneously diagnosed as caused by seepage through the wall. If you see a damp-wet foundation wall, you can easily check whether the condition is caused by seepage or condensation. Simply fasten a small piece (4 by 8 inches) of aluminum foil to the foundation wall. Using wide strips of an adhesive tape, seal all the edges of the foil to the wall. After the foil has been on the wall for at least twenty-four hours, examine its surface. If it is moist, the condition is caused by condensation. However, if the foil surface is dry and the area behind the foil is damp, the condition is caused by moisture seeping through the foundation wall from the outside. Of course, it is possible for both the foil and the wall to be damp, indicating both seepage and condensation.

Dampness in a basement or crawl space should be controlled. It can produce conditions conducive to the growth of mold and decay fungi. Dampness can often be detected by musty odors or a clammy, close feeling. In the drier portions of the country, normal dampness (not caused by seepage) in the basement can be controlled by opening the windows and ventilating the area. However, in areas where the climate is hot and humid during the summer, the benefit gained by ventilating the area is lost by the introduction of moist air into the basement. In these areas, the dampness in the lower level can be controlled with one or more electric dehumidifiers. Most of these units have a humidity control that automatically shuts the dehumidifier off when the moisture content in the air reaches a preset level. Depending on the weather and the size of the dehumidifier, the unit might have to run for many hours during each day in order to wring out sufficient moisture from the air so that it is not uncomfortably damp.

Water seepage—causes and control

Depending on the topography, drainage conditions of the soil, and groundwater level (water table), the basement or crawl space might be vulnerable to water seepage. *Water seepage*, as used herein, is a general term that refers to water intrusion into the lower level of the structure. It might manifest itself as a small wet area, a puddle, or layer of water completely covering the floor. If the ground under and around the house is wet, water can seep into the basement through cracks and open joints in the foundation walls or floor slab. Since water seepage can be caused by a number of factors and water can leak into the basement at any number of locations, it is important to determine the cause and source of the seepage so that the proper corrective action can be taken.

For example, if water is entering the basement through the foundation walls, installing a sump pit and pump below the floor slab will not correct the problem. Similarly, if water is seeping into the basement through the floor slab, sealing the walls will not correct the water-infiltration problem. All too often the unsuspecting homeowner is talked into a full waterproofing job, which can cost several thousand dollars, when all that might be needed is to redirect the water discharging from the roof drainage system (gutters and downspouts) so that the water does not accumulate around the foundation.

High groundwater level

Water entering the basement through the floor slab is an indication that water pressure is being exerted on the underside of the floor. When the level of the water below the house is sufficiently high (due to a seasonal high water table or improper drainage) so that it pushes on the underside of the floor slab, it seeps into the basement through cracks, open joints, or porous sections of the slab. If the pressure is great enough, it can cause the floor to crack and heave.

If the seepage is minor, it can often be controlled by sealing cracks and open joints with a hydraulic cement and coating any porous areas of the slab with a cement-base or epoxy sealant. However, a better solution would be to lower the level of the water below the floor slab. This can be done by installing a sump pump below the slab. Subsurface water then flows into the sump pit in the manner of water flowing into a hole dug at the seashore. The water is then removed by the pump and discharged either into a storm drain or at a point sufficiently far from the house so that it will not be absorbed by the ground and flow back under the basement floor slab.

Depending on how the floor slab was constructed, a single sump pump might or might not be adequate to lower the level of the subsurface water. In areas with a seasonal high water table, a concrete floor slab should be installed over a gravel base. Water that accumulates below the slab can then flow through the voids between the gravel and drain away or flow into a sump pit. However, in many houses the floor slab has been installed directly over soil with poor drainage characteristics or over an inadequate gravel bed. In this case, water in the saturated area below the slab will not readily flow into a sump pit, and to control the water buildup, it is necessary to install a series of perforated drainpipes below the floor slab that terminate in the sump pit. Caution should be observed when lowering the level of the groundwater below the basement floor. With some slow-draining soils such as silts and clays, some soil can wash out from around the foundation footing. This can result in unequal settlement, which could crack the walls. Whether a sump pump or drainpipes are needed below the floor slab is an evaluation that should be determined by a professional.

If the house is located in an area with a high incidence of power failures, you should not depend solely on an electrically driven sump pump to control groundwater seepage. It is possible for the power to be knocked out when the water level below the floor slab is rising. As a precautionary measure, there should be an auxiliary backup sump pump in the sump pit. A backup system is particularly helpful in vacation homes where the house will be vacant for extended periods. One type of backup pump is a water-actuated (non-electrical) ejector pump. The Zoeller Pump Company, Louisville, Kentucky, manufactures this type of pump. The pump is connected to the house water supply and is activated by a float control. However, it will be of no help if the water to the house is supplied by an electrically driven well pump. If you install a water-actuated sump pump, it's important that you include a backflow preventer on the water supply because of the potential for contamination as a result of the cross connection. Another type of backup system is a battery-operated sump pump. This system will take over automatically to protect against flood damage when the power fails. The Zoeller Pump Company also manufactures this type of system.

Hydrostatic pressure—walls

Water seeping or leaking through the foundation walls into the basement is due to a hydrostatic pressure being exerted on the walls by saturated soil. This condition is the result of water accumulation around the foundation. The best way to control this type of problem is to minimize the amount of water that accumulates around the foundation. The following are some of the more common causes of water accumulation around the foundation, which can easily be detected and corrected by the homeowner.

- *Missing or defective gutters and downspouts to handle the rain runoff from the roof.* The downspouts must discharge the water away from the structure. All too often, an elbow or splash plate at the base of a downspout is missing, so that the water is discharged directly around the foundation.

- *Improper grading.* The ground immediately adjacent to the structure should be pitched so that it slopes away from the building. Around many homes this area is incorrectly pitched, resulting in surface water (rain or melting snow) collecting around the foundation.
- *Unprotected basement-window wells.* The area around basement windows, if not shielded from rain or serviced with a drain, can easily accumulate water that can leak through window joints or seep down around the foundation.
- *Uneven settlement of walkways or patio.* Occasionally I find that the walkways around the house or the patio have settled and are sloping toward the house. As with improper grading, this condition can cause surface water to collect around the foundation.
- *Leaky garden spigots.* Most homes have exterior-mounted spigots for connection to a garden hose. If the valve is faulty or is not tightened properly, water will drip or leak around the foundation. Water dripping at a rate that fills 1 cup per minute results in 90 gallons of water per day accumulating around the foundation. This water enters the basement through cracks or open joints in the foundation wall.

When the house is located on an inclined lot, surface and subsurface water flows toward the house from the higher portions of the lot. In this case, depending on the incline and the amount of water involved, water-flow control measures will include grading the lot on the high side, so that there is a swale to collect and redirect surface water around the house, and installing a French drain (curtain drain) below the ground to intercept subsurface water and direct it away from the house.

If the amount of water that accumulates around the foundation walls is not excessive, it can be prevented from penetrating into the interior by sealing cracks and open joints on the inside walls with a hydraulic cement and then coating the walls with a cement-base or epoxy sealant. Coating the wall is particularly helpful when the wall is porous, like a cinderblock wall. However, when an excessive amount of water accumulates around the foundation wall, as with a poorly drained soil such as clay, waterproofing the exterior surface of the basement walls might be more effective than treating the interior surface. In addition, a perforated drainpipe is normally installed near and parallel with the foundation footing. (See FIG. 6-5.) The purpose of this footing drain is to carry away water that is accumulating around the foundation and thereby reduce the hydrostatic pressure.

For the footing drain to operate properly, it must have a free-flowing outlet. I know of several cases where builders installed faulty footing drains around houses during construction. The problem was that the drains completely encircled the houses like a doughnut and had no free-flowing outlets. These footing drains were of absolutely no value. Even though initially a footing drain might function properly, over the years it can malfunction because the perforations in the drainpipe or the outlet become clogged. Also, many a footing drain has been damaged during a later modification of or addition to the structure. If the house has a footing drain, you should ask the owner to show you the location of the outlet. The drain outlet should be kept clear and should be checked occasionally during a heavy rain to ensure that it is operating properly.

Even though waterproofing the exterior surface of the foundation wall is more effective than treating the interior surface, quite often an interior treatment is chosen because of the costs involved in excavating around the foundation and temporarily relocating trees and shrubbery. For excessive water accumulation around the foundation, an interior treatment includes sealing the cracks and coating the walls to make them watertight and installing a drainpipe along the foundation footing below the floor slab that discharges into a sump pit.

Just a word about waterproofing the exterior surface of the foundation wall using a pressure-pumping process that requires no digging or relocation of plantings: *Caution*. In this process, a sealant, pumped through tubes that are inserted into the ground, is supposed to coat the wall and render it watertight. The effectiveness of this treatment depends on the condition and porosity of the ground around the foundation. Since contractors doing this work do not always take test borings and analyze the soil, the process is usually not effective, and additional measures are invariably necessary.

Fig. 11-8. *Water stains and mineral deposits (efflorescence) in the corner of a foundation wall.*

Inspection

The inspection for water seepage into the basement or crawl space should begin during your exterior inspection. As you walk around the house, record on your worksheet the location of those conditions that can cause water to accumulate around the foundation: faulty gutters and downspouts, improper grading, settlement of walkways, and so on. When you go into the basement, if there are problem conditions on the exterior, the walls and floor opposite those areas should be checked first for signs of water penetration. Figure 11-8 shows water stains and deposits in the corner of a foundation wall as a result of a faulty downspout. Even if there are no indications of past or current water seepage, the exterior problem conditions should be corrected.

A basement can have a water-seepage problem and be dry when you look at it. Many clients have told me that there were no seepage problems in the basement. After all, they looked at the basement during a heavy rain and found it "bone dry." So how could there be a problem? Well, as previously discussed, there are many causes for water seepage, and depending on the cause, a single rain might not result in seepage. For example, if water penetrates into the basement through the floor slab as a result of a seasonally high water table and the basement is inspected when the water table is 1 or 2 feet below the high level, a heavy rain will not raise the groundwater level sufficiently to cause water to seep through the floor slab. It takes time for rainwater to percolate into the ground and raise the water table.

Water puddles or flooded areas in the basement are obvious signs of a water problem. In most cases, however, you will not see standing water, and you must then make an evaluation of whether there is a condition of water intrusion based on other, more subtle signs. Water-seepage signs indicate only that water has seeped into the basement in the past. They do not indicate the frequency of the seepage or its extent. Consequently, if you see indications of water seepage, you should not engage a contractor to waterproof the house immediately upon taking possession. If you do, it could prove quite costly.

First talk with the homeowner about the condition. It is possible that whatever it was that caused the past seepage has already been corrected. (If the problem was corrected by installing buried drainpipes or coating the outside surface of the foundation wall, the correction would not be visible.) If the homeowner indicates that the problem has been corrected, you should ask to see a copy of the paid bill. Or get the name of the contractor so that you can call to find out exactly what corrective steps were taken. Quite often a contractor provides a guarantee against water seepage. If there is such a guarantee, you should find out whether it is transferable.

The possibility exists that even though there are signs of water seepage, the actual seepage might occur very infrequently—such as only after an excessively heavy rain as might occur every few years. In this case, depending on the extent of the seepage and the projected usage of the basement, costly waterproofing measures might not be justified. The best approach when considering the correction of water seepage is to correct immediately any obvious problem conditions such as faulty gutters and downspouts, improper grading, cracks through which water is actively leaking, and so on. However, before undertaking any major water-seepage control measures, such as excavating and coating the exterior surface of the foundation wall, inserting perforated drainpipes below the floor slab, or trenching and installing buried drainpipes in the yard, you should live in the house for at least one full year. This will enable you to evaluate the degree and extent of the seepage over a full weather cycle. If it turns out that the year is particularly dry so that there is no seepage, well and good. Wait another year. By not taking a "shotgun" approach and waterproofing everything, as recommended by many contractors, you might be able to resolve the problem at a cost that truly reflects the work needed to stop the seepage.

Seepage indications in an unfinished basement When looking for indications of water seepage, you should check the walls, the floor, the joint between the walls and the floor, and the bases of all the items stored or standing on the floor. Specifically, look for white powdery deposits on masonry foundation walls and floor. (See FIG. 11-9.) The

Fig. 11-9. *Efflorescence and water stains on foundation wall.*

deposits, called *efflorescence*, are mineral salts in the masonry that dissolve in the water as it passes through the walls or floor. When the water evaporates from the surface of the walls or floor, it deposits these salts. A thick layer of efflorescence is usually an indication of considerable seepage.

Walls Look for efflorescence, peeling and flaking paint, and scaling sections (surface deterioration) on the foundation wall. Any one of these items can indicate some degree of seepage. Porous walls, such as those made of cinder blocks, may have damp spots. Masonry-block walls are constructed with interior voids. When the hydrostatic pressure on the exterior portion of the wall is high, the voids often fill with water. As a result, the wall might be quite wet to the touch. (Caution: This might also be caused by condensation.) Vulnerable areas for seepage are cracks and the joints around pipes passing through the wall, such as the inlet water pipe and the drainpipe leading to the sewer. Look closely at these areas for water streak stains and efflorescence.

A poured concrete foundation wall is supposed to be more watertight than a concrete-block wall. This, however, assumes that the poured concrete wall is properly constructed. Quite often, it isn't. If the entire wall is not constructed with a single pouring, the joints between the sections constructed with each pour are vulnerable to water leakage. I inspected a new house just after construction was completed. The inspection was performed during a heavy rain, which was opportune, although not planned. While inspecting the basement, I found water leaking out of the joint at the seam between the individually poured sections of the foundation wall. The builder's mason apparently had not properly prepared the joint for a new pour, and consequently a cold joint with a poor bond was formed.

Look for seepage in a poured concrete wall around the tie-rod holes—holes in the concrete wall around the small-diameter metal rods that are used to hold (tie) the forms together when the wall is being poured. More often than not, these holes have been patched over. Also, these tie rods can corrode away over a period of time and when below grade are vulnerable areas for water intrusion. Sometimes you see efflorescence and water streaks just under the hole or patched sections. Occasionally I find these holes plugged with corks. This is not considered a permanent patch, and if seepage should develop, they should be plugged with hydraulic cement.

Floors and floor joints A vulnerable place for water seepage is the joint between the foundation wall and the floor. Look closely at this joint as you walk around the entire basement. Water stains and efflorescence are an indication of seepage. (See FIG. 11-10.) You might find silt deposits at the joint. This is also an indication of some degree of seepage. The fine silt is in suspension in the water as the water seeps in from the exterior. When the water evaporates, the silt is deposited. If you find evidence of water seepage through the joint between the floor slab and the foundation wall, you should record it on your worksheet for later correction. The joint should be sealed with either hot tar or a hydraulic cement.

Cracks tend to develop in the floor slab near the base of metal columns. Look for these cracks and any other cracks in the floor slab. Specifically, look for water stains, efflorescence, and silt deposits. Cracks in basement floors are a common phenomenon and are generally caused by slight settlement or shrinkage in the concrete. Usually they are of no concern other than the fact that water can seep through them. Therefore, they should be sealed. However, if the cracks are extremely wide or show evidence of heaving, they are of concern and should be checked further. In all

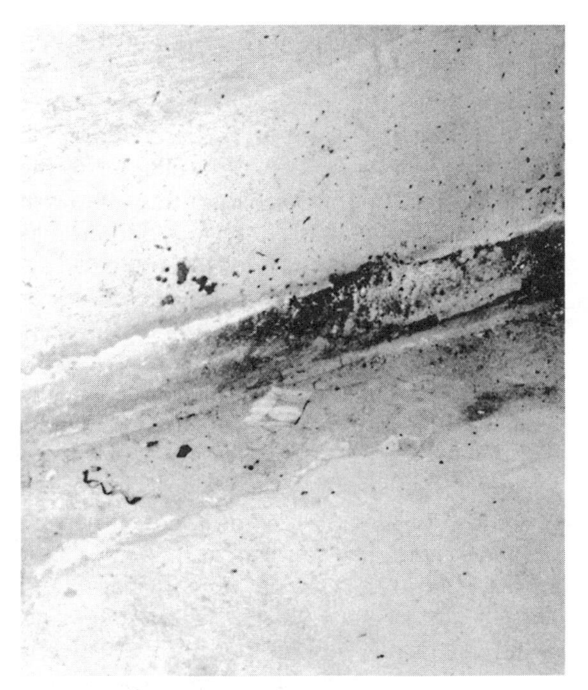

Fig. 11-10. *Signs of water seepage at joint between floor slab and foundation wall.*

probability, a cracked and heaved floor slab is the result of water pressure being exerted on the underside of the slab by a high water table.

Some homes have a cleanout and trap for the house waste line located in a pit below the basement floor slab. The bottom of the pit should be dry. If it is wet, it is an indication of a high groundwater level or a crack in the drain line. Occasionally I find that the top of the cleanout is open. (See FIG. 11-11.) It should normally be plugged. Some homeowners remove the plug so that the open cleanout can function as a drain if the basement becomes flooded. This is definitely not the way to eliminate the water in a flooded basement. If the basement periodically floods and there is no drain in the floor, a sump pump can be installed in the lowest section of the floor. The water can then be pumped out of the basement.

Fig. 11-11. *If the cleanout plug is removed, sewage can flood the basement.*

With the cleanout plug removed from the top of the house trap, the possibility exists that if the sewer line becomes overloaded (as is the case in some communities), sewage can back up and flood the basement. I know of several homes where this has happened. It was quite unpleasant.

Some people lose interest in a house when they find a sump pump in the basement. They think that the house has water problems. That is not necessarily so. There might have been periodic problems resulting from a seasonal high water table, but the sump pump might have controlled the water level. Or the pump might have been installed when the house was built in order to prevent a problem. To determine whether there still are problems, you must look beyond the sump pump. Look for signs of water seepage.

If the house has a sump pump, look down into the pit. If there is water in the pit and you are inspecting it during the dry season (the water table is usually highest during the spring), the sump pump will probably be operating continuously during the spring. Try to check the sump-pump operation. However, do not actuate a pump unless there is water in the pit. To check the operation of the pump and motor when there is no water in the pit, fill the pit using a garden hose or gently tip in a couple of large buckets of water. If you notice that the water is disappearing into the ground or under the slab before the pump has a chance to activate, wrap a plastic sheet around the wall of the pit. You may have to line the pit bottom with plastic as well. When you actuate the pump, watch the water level to see if it drops. You might find that the motor that drives the pump is operational but the coupling between the pump and the motor is broken. In this case, it will sound as if the pump is working, but it is not, and the water level will not drop. A sump pump is relatively inexpensive and can be repaired or replaced easily.

Although a sump pump is sometimes located in the low section of the floor (when it is being used to collect surface water), it should not be located where it will present a tripping hazard. (See FIG. 11-12.) The pump can be placed in a corner where it will not take up valuable floor space and can be connected to the low spot in the floor by a drainpipe placed below the floor slab that discharges into the sump pit.

Check to see where the sump pump effluent discharges. It should not discharge into the house drain, which in turn discharges into the sanitary sewer or, if there is one, the septic system. As discussed in chapter 6, the sump pump should discharge onto the lawn sufficiently far away from the foundation so as not to seep back into the house.

After inspecting the walls and the floor for signs of water seepage, you should check the bases of items stored or standing on the floor for water stains and rust. You might find that the walls and the floor of the basement appear to be freshly painted. If you see this, you should be suspicious because the new layer of paint will cover almost all the signs of a water-seepage problem. There are some areas, however, that are often omitted when painting the basement. These should be checked for water stains and rust. Specifically, look at the base of the steps leading to the basement and in particular the back of the bottom step. Also look at the base of columns. Wood columns might have stains or may be rotting, and metal columns might be rusting. The base of the inside portion of a

Fig. 11-12. *Sump pit in center of basement floor—a tripping hazard.*

furnace sheet-metal casing is often overlooked when painting. Look inside. Is there extensive rust? If there is, it might have been caused by a past flood. However, it could also have been caused by a faulty humidifier, so do not jump to a quick conclusion.

Seepage indications in a finished basement Usually the finished walls in a basement are a few inches or more away from the foundation walls. Therefore, seepage indications on the foundation wall will not be reflected in the middle and upper portions of the finished wall. Look at the bottom portion of the wall for signs of water intrusion. If the wall is paneled, look for rotted and warped sections and water stains. Sometimes there are grayish mold spots or mold fungi on the walls, a condition caused by excessive dampness. With a plasterboard wall, look for water stains and blackened areas and spots. The latter is mold and mildew. In some cases, the lower portion of the plasterboard wall might have deteriorated.

Next, look at the floor. If the floor is raised above the level of the concrete floor slab, the wood flooring and the wood-framing members used to raise the floor are vulnerable to rot in the event of seepage. The wood framing below the floor should be pressure-treated but often is not. Try to walk all over the floor, especially around the perimeter. If there are any rotting sections, they will feel soft and spongy beneath your feet. If the floor is covered with resilient tiles, the rotting sections might be visible, since the dampness in the wood tends to loosen the tiles. (See FIG. 10-4.)

An area that is particularly vulnerable to water seepage is the portion of the basement that faces a yard where the overall topography is inclined toward the house. Even if the ground adjacent to the house is graded properly, there will still be subsurface water moving toward the house. The walls and the floor of the vulnerable section should always be checked for signs of seepage. Sometimes this is quite difficult; some homeowners might inadvertently (or intentionally) block the area with furniture. If the area is blocked, ask the homeowner if it is all right to move the furniture.

If the floor slab is covered with resilient tiles and there is a problem with heavy seepage through the floor, it is usually noticeable by looking at the tiles. The joints between the tiles become swollen and filled with a white crusting of mineral deposits (efflorescence). Some tiles become loose, and efflorescence is noted below them. Occasionally the tiles are covered with wall-to-wall carpeting. In this case, the tiles cannot be examined for signs of seepage. However, some checking for seepage can be done. Ask the homeowner for permission to lift the edge of the carpet off the tacking strip along a section of the exterior wall. If there is seepage in this area, the tacks will be rusted and the wood strips water stained or rotted.

If the basement is heated with baseboard convectors mounted on an exterior wall, another place to look for signs of seepage is below the convector. If there is some seepage in that area (the joint between the floor slab and the wall), the base of the convector will be rusted.

On occasion a portion of the basement might become flooded as a result of faulty plumbing, an overflowing sink or tub, a malfunctioning water heater, and so on. Flooding from the above is basically a one-time affair and will not cause the type of damage that results from repeated wetting. There will be water stains, but there usually will not be rot, peeling or flaking paint, efflorescence, or heavy rusting.

Furnace room

The heating plant in a house is normally a furnace or a boiler. (See chapter 14.) In this chapter, the heating plant is called a furnace and the room in which it is located is called the furnace room.

In most homes, the furnace is located in the basement or lower level. It can be located in a large open area or confined in a relatively small room. In either case, the area around the furnace is considered a potential fire hazard. Consequently, there should not be any exposed wood framing in the ceiling or partition walls that are in close proximity to the heating plant. (See FIG. 11-13.) Exposed overhead floor joists or wall studs that are near the furnace should be covered with Type X fire-code plasterboard as a safety precaution. Sometimes I find that wood paneling has been installed around the furnace to make the area more attractive. However, it also makes the area more of a fire hazard, especially the portion of the wall around the chimney. (See FIG. 11-14.) If you find wood paneling near the furnace, you might consider covering it with fire-code plasterboard. There should be a minimum 2-inch clearance between the chimney and any wood framing or paneling. Note that the area around the furnace should not be used for storage. On many occasions I have seen combustible items actually stored against the furnace. Fortunately for the families involved, the items never ignited, but they could have!

Many heating systems have prefabricated chimneys that extend from the furnace room up through the interior portion of the structure, terminating above the roof. If the house has such a chimney, look at the joint between the chimney and the ceiling of the furnace room. If there are large openings, they represent a potential fire hazard and should be covered with a noncombustible material such as sheet metal. The open area around the chimney, if not properly blocked, can act as a flue and in the event of a fire in the furnace room, will draw the flames up to the attic.

While in the furnace room, also check for asbestos insulation around the furnace and heating pipes. (Asbestos as a health hazard is discussed in chapter 20.) The door to the furnace may be a louvered door. Depending on

Fig. 11-13. *Exposed wood joists above heating-system boiler—a potential fire hazard—should be covered with fire-code plasterboard.*

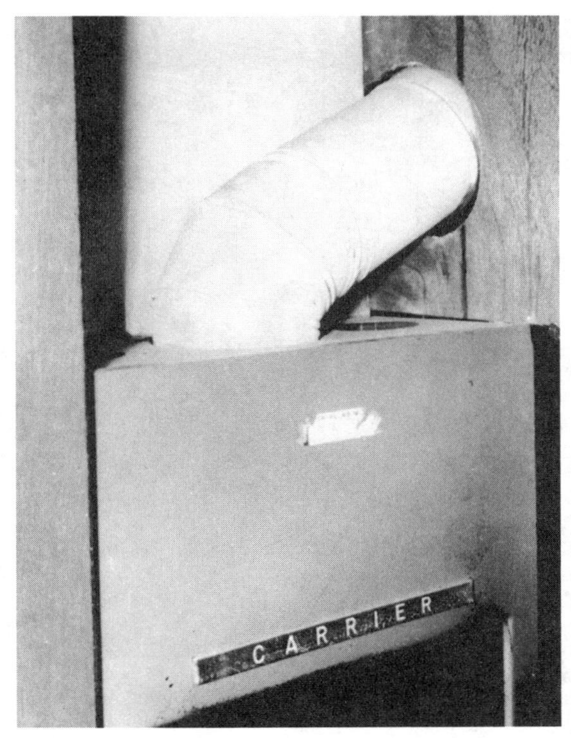

Fig. 11-14. *Wood paneling near furnace—a potential fire hazard.*

the local fire code requirements, louvered doors may not be permitted between the living space and the furnace room. The code may require that the door be fire rated. In this case check the adequacy of the ventilation; see below.

Ventilation

All fuel-burning heating systems must have adequate ventilation for proper operation. If the furnace is located in an unconfined space, the normal air infiltration into the area provides ventilation. However, when the furnace is located in a confined space such as a small room, inlet and outlet ventilation openings must be provided. The vent openings can lead directly to the outside or to a large unconfined area within the structure. The size of the openings depends on the total input (Btu/hour) rating of all the fuel-burning equipment in the enclosure. For most residential structures, an unobstructed inlet and outlet opening of 15 inches by 15 inches is sufficient.

If you would like to calculate the size of the vent openings needed, a safe formula to use is 1 square inch per 1,000 Btu/hour of input rating for both the inlet and outlet vents. The input rating will usually be found on a data plate mounted directly on the equipment. If the vent opening is covered by louvers, remember that metal louvers reduce the effective opening by about 25 percent and wood louvers by about 50 percent. An insect screen covering the louvers reduces the effective opening by another 25 percent.

Some homes have louvered entry doors to the furnace room, which provide the means for ventilation. Occasionally I find that for decorative reasons, the louvers have been covered over, blocking the effective ventilation opening. If the louvers have been covered, look for additional vent openings.

Crawl space

The foundation walls, piers, posts, and wood-support framing in a crawl space should be inspected (as described in the section on the unfinished basement) for deterioration, structural deficiencies, and evidence of water seepage. Pay particular attention to the wood-framing members that are very vulnerable to decay and termite infestation. (See FIG. 11-15.)

Many homes, however, have crawl spaces that are inaccessible and cannot be inspected. These homes were built in accordance with the Federal Housing Administration (FHA) Minimum Property Standards, which permit the ground level to be 18 inches below the bottom of the floor joists and 12 inches below the bottom of the girders. Even if there is an access opening to the area (see FIG. 11-16), the

Fig. 11-15. Cracked and rotting wood-frame members in crawl space.

Fig. 11-16. Access hatch to crawl area.

clearance in the crawl space is too low for a person to maneuver around easily and perform a detailed inspection. However, you should look into the area from the access opening (using your flashlight) to discover any obvious problems or problem conditions.

In most parts of the country, the crawl space will be quite damp, even though there are no problems with water seepage. The dampness is the result of the capillary rise of ground moisture. Unlike a basement where the ground is usually covered with a concrete floor slab, the floor in a crawl space is often bare earth. Even though the soil might appear dry and dusty, moisture can be present. In some soils, the capillary rise is more than 11 feet above the water table.

Dampness associated with capillary moisture can be effectively reduced by covering the ground with a vapor barrier, such as 4- to 6-mil polyethylene. Roll roofing is also a good vapor barrier, but it tends to deteriorate from fungi. If you do not notice a vapor-barrier ground cover, you should consider its installation.

To help minimize the dampness, the crawl space must be ventilated. There should be at least two vent openings on opposite sides of the foundation with a total free area of 1 square foot for each 1,500 square feet of crawl space area, providing there is a ground cover. When no vapor barrier is used, there should be at least four vent openings (one on each side) with ten times the total free area. Look for vent openings. If you do not see any or they have been permanently blocked, put a note on your worksheet to that effect.

Crawl spaces are usually not heated. Consequently, unless there is some insulation between the floor joists, there will be heat loss between the heated room above and the unheated crawl area. Since moisture from the house can travel down through the floor, the insulation should have a vapor barrier on one side to reduce further moisture entry into the crawl space. The vapor barrier should be located above the insulation, facing the heated room (rather than below, facing the crawl area). If this vapor barrier is located below the insulation, the vapor will condense on its surface during cool weather. Depending on the amount of vapor, the resulting condensation buildup can reduce the effectiveness of the insulation. Look for insulation. You might find missing, loose, or hanging sections, which should be replaced or resecured. (See FIG. 11-17.) If you find uninsulated heating ducts or pipes in the crawl space, record this fact on your worksheet as a reminder to insulate the exposed sections.

Some homes are constructed with both a basement and a crawl space. In this case, the crawl space need not be vented to the outside but can be vented to the basement. Look for evidence of water seepage in the crawl space. Even though there might be no signs of seepage in the basement, there might be some in the crawl area. I recently inspected a home that had a combination basement–crawl space. The basement had been painted, and there were no visible signs of a past water condition. The crawl space was separated from the basement by plywood doors that were painted on the basement side and looked good. However, when I inspected the crawl space, I found evidence of a previous water condition. Apparently, the backside of the plywood doors (facing the crawl area) had not been painted over. There were water stains on the lower section. (See FIG. 11-18.) While in the crawl space, check subflooring and support joists below kitchen and bathroom fixtures for evidence of decay and plumbing leaks.

Fig. 11-17. *Loose, hanging insulation in crawl space.*

Conditioned crawl space

During your crawl space inspection, you may notice that the foundation walls are insulated, the wall vents are all closed, the dirt floor is covered with large plastic sheets, and the area is heated or air-conditioned. Basically the area is dry and comfortable. In this case the traditional crawl space has been converted to a conditioned crawl space. In the past few years, the concept of a conditioned crawl space has gotten some traction in the building

Fig. 11-18. Water stains on back side of entry door to a crawl area.

industry. Limited research has shown that a conditioned crawl space is more energy efficient than the traditional vented crawl space and will eliminate some of its problems. With a vented crawl space, warm, moist summer air will enter the cool crawl space and condense. This could lead to a buildup of mold and wood rot and is conducive to termite activity. During the winter months, the cold air could contribute to heat loss and freezing pipes. With a few minor changes, the traditional vented crawl space can be retrofitted and upgraded to a conditioned crawl space. The main concern is that the area must be kept dry. Prior to upgrading, all possible sources of water penetration into the crawl space must be corrected.

Checkpoint summary

Foundation

- ❍ Note foundation-wall construction type: poured concrete, precast concrete panels, concrete block, brick, stone, and so on.
- ❍ Check for cracked areas of concrete; crumbled and flaking bricks; cracked, loose, missing, and eroding mortar joints.
- ❍ Check for long open cracks that do not line up and have shifted sections.
- ❍ Note long, open, horizontal cracks and signs of bowing in the foundation wall.
- ❍ Are sections of the structure sagging and no longer vertical? (Consult a professional.)

Wood-support framing

- ❍ Inspect all vulnerable wood-support members (sill plates, girders, joists) resting on the foundation wall for rot and insect damage.
- ❍ Note floor joists or girders that sag or have notched sections.
- ❍ Is there bridging or blocking between the floor joists?
- ❍ Inspect wood columns, support joists, and subflooring for cracked sections and evidence of rot.
- ❍ Are there any notches in the flanges of I-joists? Any holes in the webs larger than 1.5 inches in diameter?

Water seepage

- ❍ Is the ground adjacent to the house pitched so that it slopes away from the structure?
- ❍ Are there concrete patios and paths that are improperly pitched (toward the house)?
- ❍ Are there basement windows or stairwells that are vulnerable to flooding?
- ❍ Do downspouts
  - —have extensions?
  - —discharge against the foundation?
  - —terminate in the ground?
- ❍ Is there a sump pump?
- ❍ Is there water in the sump pit?
- ❍ Is the sump pump operational? Does the sump pump discharge into the house drain?
- ❍ Is the water being discharged away from the house or to a dry well?
- ❍ If the structure has been waterproofed, is there a guarantee or warranty available? Did you request a copy?

Basement walls

- ❍ Check for areas of scaling, peeling and flaking paint, damp spots, and signs of efflorescence.
- ❍ Check construction joints, tie-rod holes, and pipe openings for signs of seepage.
- ❍ Inspect wall paneling and base trim for stains, warped sections, and rot.
- ❍ Check underside of basement steps for water marks.
- ❍ Note areas of rust at base of metal columns and sheet-metal furnace casing.
- ❍ Check for dampness, noting musty odors and signs of mildew.

Basement floors

- ❍ Check for extensively cracked and heaved floor sections (usually the result of a high water table).
- ❍ Record all areas of active seepage and puddling.
- ❍ Check joint between foundation wall and floor slab for silt deposits.
- ❍ Check for porous areas and signs of efflorescence on floor and around perimeter.
- ❍ If floor is covered with tiles, are there swollen floor-tile joints?
- ❍ Is there efflorescence between joints?
- ❍ Inspect the house trap pit. Is it dry? Is the cleanout plug secure or loose?

Furnace room

- ❍ Check for exposed wood-frame members (wall studs, ceiling joists) that are in close proximity to the boiler or furnace.
- ❍ Check for large openings between the ceiling and the chimney.
- ❍ Is the room adequately ventilated?
- ❍ Check for asbestos insulation around the furnace and heating pipes.

Crawl space

- ❍ Inspect foundation walls, posts, and wood-support framing for deterioration and signs of water seepage.
- ❍ Check overhead subflooring and support joists for insect damage and/or rot.
- ❍ Check for adequate ventilation.
- ❍ Is this area damp?
- ❍ Is there a dirt floor? Is it covered with a vapor barrier?
- ❍ Is area insulated? Is insulation loose or incorrectly placed?
- ❍ Are there water-supply pipes that are vulnerable to freezing?
- ❍ Are there heat-supply ducts or pipes that should be insulated?
- ❍ Is the area conditioned crawl space?

12

Electrical system

The electrical system of a house can be compared to the nervous system of the human body. Just as every part of the body is supplied by nerves connected to the brain, every part of a house is (or at least should be) supplied by electrical wires connected to the inlet service panel box. These wires are called *branch circuits*. A proper electrical system is essential for a healthy house.

Inlet electrical service

The electrical service is provided through overhead wires or underground cables. If there are overhead wires, they can be seen around the outside of the house. These wires usually run from a utility pole to the house and are fastened to the structure at a point at least 10 feet above ground level. Count the number of wires coming into the house. If there are two wires (FIG. 12-1), the electrical service for the house is inadequate. Two-wire service provides only 110 volts, not 110/220 volts. There should be three wires coming into the house from the electrical service entry. (See FIG. 12-2.) Notice in FIG. 12-2 that the wires loop down before entering the weather-head. This is called a drip loop. It causes water that travels along the wires during a rain to drip down off the wire, thereby preventing it from entering the weather-head and flowing into the service entrance cable. Three-wire service provides 110/220 volts. In some cases, there might be four wires. Four-wire service also provides 110/220 volts but is

Fig. 12-1. *Two-wire inlet service provides 110 volts.*

Fig. 12-2. *Three-wire inlet service provides 110/220 volts.*

unusual in a residential structure. That service is three-phase and provides additional capacity, so that you may find it on some very large homes. It is usually found on a structure with heavy electrical demands such as an industrial or commercial building. The inlet service voltage might vary slightly so that in some areas it might be 120/240 rather than 110/220 volts. The difference is of no concern to the homeowner.

Before going into the house, look to see if any tree branches overhang or hit the inlet service wires. They should be pruned. Otherwise, the wire's outer insulation can be worn away, exposing the electrically hot conductors. This could be potentially dangerous if that section of conductors is within reach, as from an upper-level porch, or if it contacts a metal gutter.

While looking at the inlet service wires, check to see if they are securely fastened to the house and whether there are any frayed sections of the outer insulation. In many communities, the inlet service wire (from the attachment to the house to the electrical meter) is the responsibility of the homeowner, not the utility company. Service-entry wires that are badly frayed should be replaced because they can eventually result in a hazardous condition.

The inlet service wires, whether they are overhead or underground, terminate at the electrical meter, which can be mounted on the exterior or interior of the structure. The wires then run from the meter to a panel box, sometimes called the *service switch*. The panel box is basically a distribution center. The branch circuits throughout the house terminate in this box. See FIG. 12-3 for a generalized residential wiring diagram.

Electrical capacity

The unit of electrical power is called a *watt*. Watts are equivalent to volts times amps ($W = V \times A$). The electrical service that I have

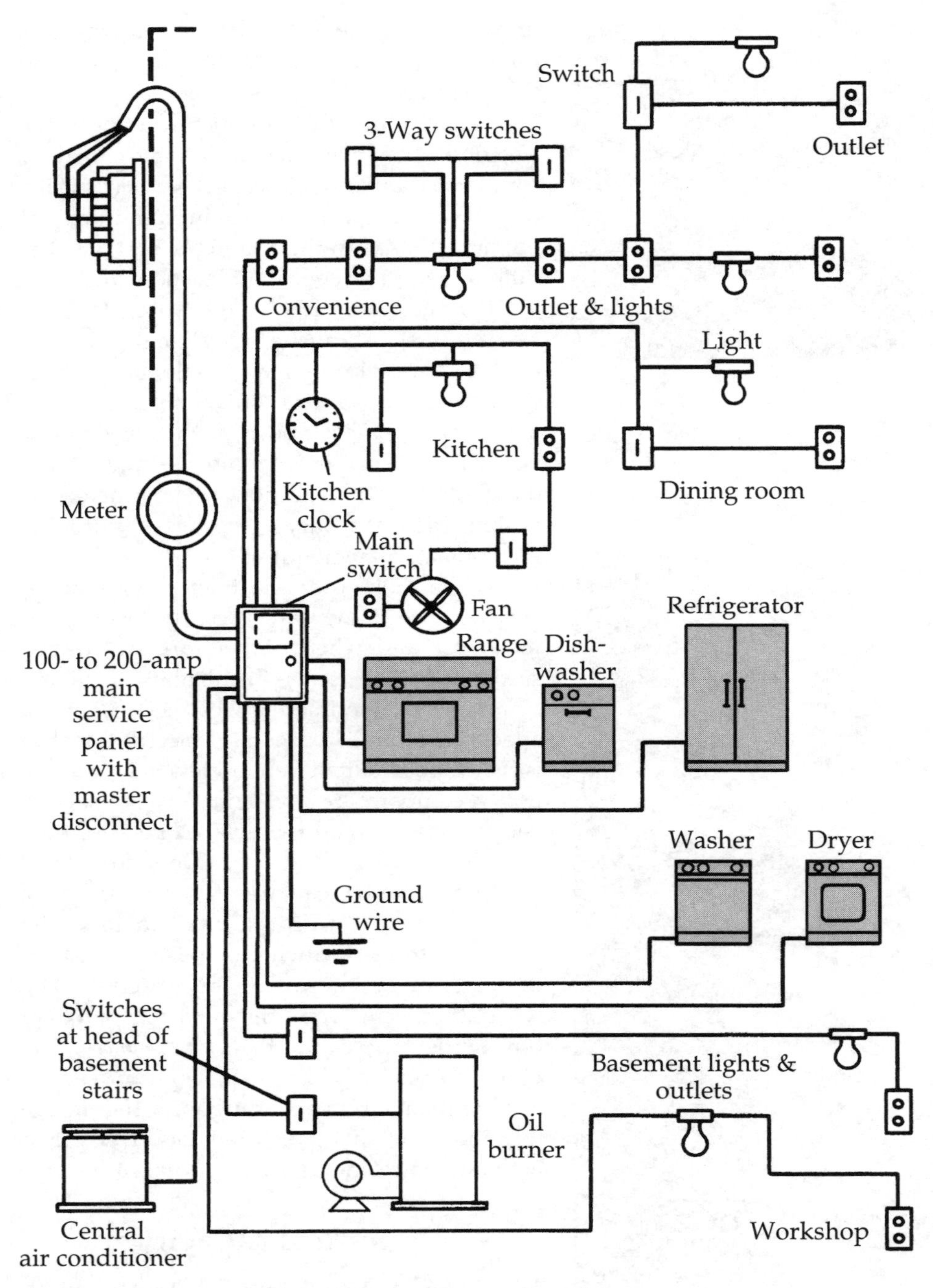

Fig. 12-3. Generalized wiring diagram for a residential structure.

found in houses and the corresponding equivalent power capacity is shown in TABLE 12-1.

In older houses, unless the electrical service has been upgraded, you are likely to find 30 and 60 amps at 110 or 110/220 volts. Any service less than 60 amps at 110/220 volts is considered inadequate. Homeowners with such electrical service might indicate that they have lived in the house for over forty years and have found the service totally acceptable. That might be true—for them—because they have adjusted to the lower electrical capacity by not using any major electrical appliances. Nevertheless, if you buy a house with low electrical service, you will probably have to upgrade the service after you move in.

The capacity of the electrical service should be great enough to satisfy the power requirements of the various electrical appliances to be used. Table 12-2 shows some typical

Table 12-1. Power capacity.

Amps	Volts	Watts	Evaluation
30	110	3,300	Inadequate
30	110/220	6,600	Inadequate
60	110	6,600	Inadequate
60	110/220	13,200	Marginal (small house only; no major appliances)
100	110/220	22,000	Minimum
150	110/220	33,000	Good
200	110/220	44,000	Very good

Table 12-2. Power requirements of appliances.

Appliance	Power requirement (watts)
Attic fan	400
Central air conditioning	6,000
Clothes dryer	4,500
Computer	200–400
Copy machine	1,300
Dishwasher	1,500
Fax machine	65
Forced-air furnace (electric heat)	28,000
Freezer	575
Garbage disposal	900
Hand iron	1,000
Instant hot water dispenser	1,000
Lamp (each bulb)	25–150
Microwave oven	500
Range, electric	8,000
Portable room heater	1,600
Printer	23
Room air conditioner	1,100
Sauna	8,000
Steam bath generator	7,500
Television (color)	150–450
Water heater	2,500–4,500

electrical appliances and their associated power requirements.

Considering the many electrical appliances that are available to the homeowner, the minimum electrical service is 100 amps at 110/220 volts. However, if the house is small and the intention is not to use many electrical appliances, 60 amps at 110/220 volts will probably suffice. When a house is equipped with an electric water heater, electric range, electric clothes dryer, and a central air-conditioning system, it should have at least 150-amp service. If electric heat is used in addition to the above appliances, the house should have 200-amp service.

Fuses and circuit breakers

The panel box contains either circuit breakers or fuses to protect individual branch circuits from an overload. Overloaded circuits are one of the chief causes of home electrical fires, and proper protection of branch circuits is essential. FIG. 12-4 shows a circuit-breaker panel box, and FIG. 12-5 shows a fuse panel box. Circuit breakers are more convenient than fuses. Once they have been tripped, they can be reset like a switch, whereas once a fuse has blown, it must be replaced. On the other hand, circuit breakers are somewhat less reliable than fuses. Circuit breakers have been known to "freeze" in the *on* position and should be manually tripped periodically to ensure operational integrity.

Is the panel box a Federal Pacific Electric (FPE) service panel? There have been numerous reports questioning the safety of this panel box and its Stab Lok circuit breakers. Some of the breakers have failed to shut off, or "trip," in response to an overload and arcing has occurred in the panel box. As a safety precaution, replacing the panel box is recommended.

It is important that the capacity of a fuse be matched with the current-carrying capacity of a branch circuit. When a fuse has been blown,

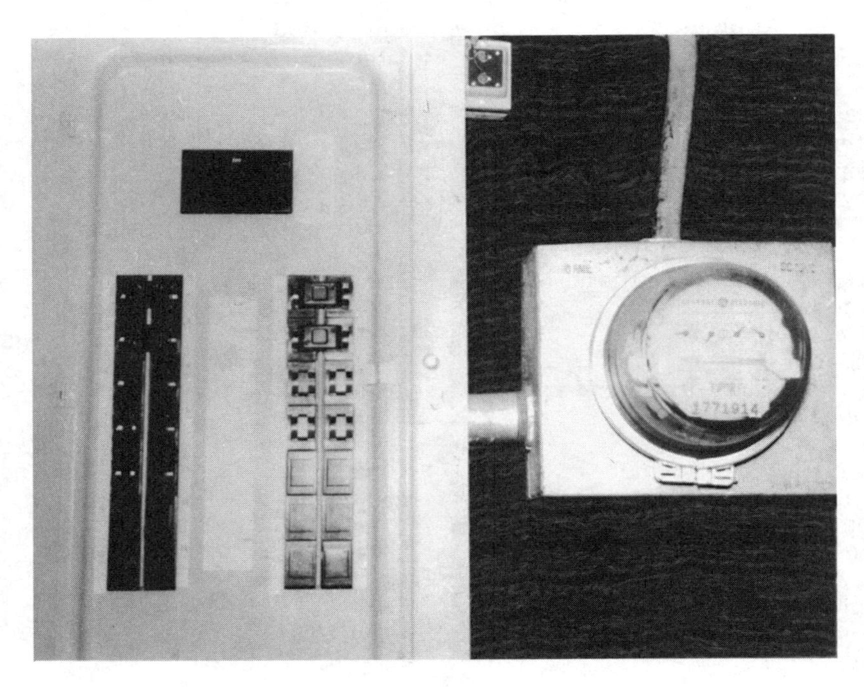

Fig. 12-4. *Circuit-breaker panel box with master disconnect.*

Fig. 12-5. Fuse panel box with master disconnect.

it must be replaced by a new fuse with the same current-carrying capacity. Too often, a homeowner replaces a 15-amp fuse with a 20- or 30-amp fuse, not realizing that all three are physically but not electrically interchangeable. In this case, if there is an overload on that branch circuit, the fuse will not blow, but the wires will become excessively hot and possibly cause a fire. This type of problem can be circumvented by replacing the fuse with a Fustat. (See FIG. 12-6.) A Fustat is basically a fuse with an adapter that fits into the fuse holder in the panel box. Once the adapter is inserted, it usually cannot be removed. The adapters are sized so that they accept only fuses of a specific current capacity and are not interchangeable.

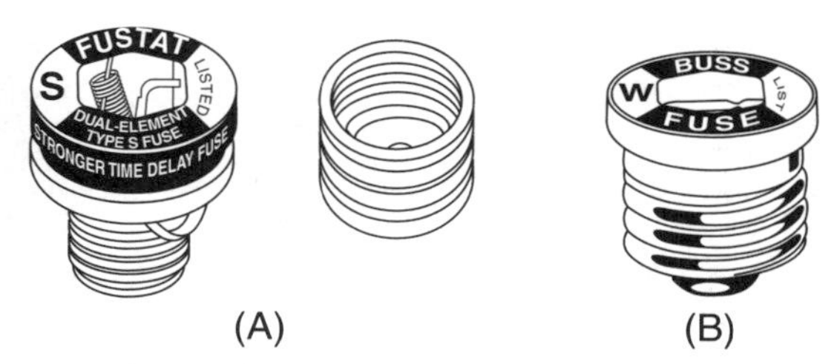

Fig. 12-6. A—Fustat with adapter; B—plug fuse. Bussman Mfg.

Fig. 12-7. Minibreaker/circuit breaker fuse.

There are also fuses that act like circuit breakers. (See FIG. 12-7.) Cooper-Bussmann manufactures 15- and 20-amp mini-circuit-breaker plug fuses that fit into any standard fuse socket. The minibreakers have a little push button that pops out when the circuit is overloaded. To reset the circuit breaker/fuse, simply push the button back in. The minibreakers are also designed for time delay to handle temporary starting loads so they do not trip unnecessarily when motors or appliances start up.

Inlet service panel box

Some panel-box covers have a door. If there is a door on the panel box, lift or swing it open. Warning: Do not remove the panel cover. The cover should be removed only by a professional inspector or electrician because of the danger involved with exposed electrically hot wires. Some homeowners remove the panel cover when doing home wiring and then forget to replace it. Others remove the cover and then cover the panel box with a picture. They think this provides easy access. It does—for children too—and is very dangerous. Panel-box covers should be mounted and securely in place at all times.

When looking at the panel-box cover, note whether any knockout plates or fuses are missing. They represent a potential hazard; it is possible for a child to stick his or her finger in these openings and be electrocuted. If there are openings, they must be permanently blocked off. Check the size and number of the fuses or circuit breakers. There should be at least two 20-amp appliance circuits and one 15-amp lighting circuit for each 500 square feet of floor area. Anything less than this is not considered adequate. Also, if the branch circuits are protected by fuses, note if there are many blown fuses in the area indicating that one or more of the circuits might often be overloaded, a condition that might require installing additional branch circuits.

Panel-box interior

Because of the hazards involved, certain items should not be checked by the home buyer-owner, specifically those items found inside the main panel box. A professional home inspector or licensed electrical contractor checking the interior of the panel box can tell you whether the fuses or circuit breakers are properly sized for their respective branch circuits. Improperly protected branch circuits are found fairly often.

Once the panel cover has been removed, the actual inlet service capacity can be determined. It depends on the size of the inlet service wire only, not on the rated capacity of the panel box or the size of the fuse or circuit breaker used for the master disconnect, if there is one. Some municipalities require that the main electrical disconnect be located on the exterior of the house. Figure 12-4 shows a circuit-breaker panel box with a 200-amp master disconnect. A home buyer seeing this would normally believe that the house is supplied with 200-amp electrical service, and rightly so. In this case, however, the inlet service wire is too small to supply 200 amps. When the cover was removed, it was determined that the inlet service was 100 amps and that the master disconnect circuit breaker did not provide the overload protection.

Subpanel When finishing a basement or constructing an addition on a house, a subpanel is generally used to provide the needed branch circuits, rather than running those circuits from the main panel. During your inspection, you might see a subpanel. As with the main panel, because of the potential hazard, the home buyer-owner should not remove the subpanel cover. A professional home inspector or licensed electrician should check the interior to determine whether or not the grounding wire and the neutral are bonded together as in the main panel. They should definitely **not** be bonded in the subpanel. If you bond the neutral and ground on the same bus or ground bar at the subpanel, you create a parallel neutral current path that could energize all metal parts of the electrical installation. Whereas, by isolating the neutral from ground at the subpanel, any current would then go back to the main panel and go to the service ground.

Aluminum wiring

Inspecting the interior of the panel box also reveals whether aluminum wiring was used for the branch circuits rather than copper wiring. Aluminum wiring is considered a potential fire hazard. Between 1965 and 1973, about 1.5 million homes were wired with aluminum, which at the time was approved by the National Electrical Code. It was later found that dangerous overheating occurred at some of the connections in the 15- and 20-amp branch circuits, which resulted in fires.

If aluminum wiring was used, the electrical connections to the receptacles, switches, and light fixtures throughout the structure should be checked by a competent electrical contractor to determine if they have been properly made or show evidence of problems. If the house does have aluminum branch circuits, you should be aware of the following trouble signs:

- Unusually warm cover plates on switches and outlet receptacles
- A distinctive or strange odor in the vicinity of a receptacle or switch
- Sparks or arcing at switches and outlets
- Periodic flickering of lights (sometimes traceable to faulty appliances or fixtures)

Correcting the problem does not require rewiring the house. Switches and outlet receptacles that are unmarked or marked AL/CU should be replaced with devices that are marked CO/ALR. Or existing switches and outlets can be used, provided short copper pigtails are attached to the ends of the aluminum wires and the devices. It's also necessary to connect light fixtures with copper pigtails. The home buyer-owner should not attempt to correct this condition.

Grounding

The electrical system must be grounded as a safety precaution. This means that a portion of the wiring in the main panel box must be deliberately connected to the ground. This is done by connecting the wiring to a grounding wire that in turn is clamped to a metallic inlet water pipe or to a rod driven into the ground.

Check to see if the electrical ground has been properly connected. There should be a wire coming out of the main panel box that runs to the inlet water pipe. Sometimes the wire is not visible, so go directly to the inlet water pipe. This pipe is located by the water meter (if there is one) and usually protrudes through the foundation wall or lower-level floor slab. The electrical ground wire should be clamped to the water pipe on the street side of the water meter. (See FIG. 12-8.) It might also be connected on the house side, but in such cases there must be a jumper cable running around the meter to the street side.

Improper ground connections

It is surprising how many houses have loose clamps on the water pipes, resulting in an improperly grounded electrical system. Many homeowners do not realize that the wire clamped to the water pipe is a grounding wire for the electrical system, and they loosen the clamp when making a repair or finishing the basement, often forgetting to reclamp the wire because nothing "apparent" happened when it was disconnected.

Look at the clamp. Is it loose? If it is, it must be resecured. This can usually be done with a screwdriver. Sometimes the clamp or screws have corroded and must be replaced. If there is no clamp or ground wire visible, you should question the integrity of the ground system. Sometimes the ground wire is clamped to a water pipe other than the inlet pipe. This procedure provides a false sense of security because a section of the pipe

Fig. 12-8. *Ground wire clamped to inlet water pipe.*

can be removed, causing an open circuit in the electrical ground. Some states require a jumper wire between the hot- and cold-water pipes that are connected to the tank-type water heater. This provides electrical bonding between the two pipes and prevents an interrupt in the ground circuit during that period in which the water heater is being replaced. These states may also require a jumper wire around the water meter (if there is one) and also around a plastic water filter in the main water line (if there is one). These requirements can be checked with the local authorities.

The inlet pipe to which the electrical ground is connected must be a functioning water pipe. This ensures that the pipe extends a considerable distance into the ground. According to the electrical code, the pipe should extend at least 10 feet into the ground. If a new inlet water pipe has been installed, the ground wire should be moved to the new pipe. Otherwise, there is no way of knowing how far the old pipe extends into the ground. Often, when replacing an old water pipe, the portion outside the structure is cut off at the foundation wall. This cannot be determined when looking at the pipe from inside the house.

When the inlet water pipe is plastic, as is often the case with a well, the electrical system is grounded by connecting the grounding wire to a rod driven into the ground. If this is the case, check the clamp connection and the rod. If either the clamp or rod is loose, the integrity of the electrical ground should be questioned.

Interior electrical inspection

Now that the basic system has been checked, walk around the interior portion of the house and determine the adequacy of the electrical distribution and possible violations.

Electrical outlets

Check each room for electrical outlets. You might feel that the number of outlets in each room is not very important. This is not so. If there are an insufficient number of outlets, the homeowner will tend to use extension cords. Most extension cords have a lower electrical current capacity than the outlet; and an overload on the extension cord can result in a fire rather than a blown fuse. Figure 12-9 shows one arrangement that is a potential hazard. The convenience outlet in the kitchen is capable of carrying 20 amps, whereas the extension cord can safely carry only 10 amps. Kitchens should ideally have 20-amp appliance outlets that are conveniently located (about every 24 inches) above a working counter.

The number of outlets needed for each room depends on the room's usage and the position of the furniture. Generally speaking, one receptacle outlet per wall for an average-size room (10 feet by 12 feet) is adequate. The outlets in larger rooms should be spaced so that no point along the wall is more than 6 feet (measured horizontally) from an outlet.

In addition to checking whether there are a sufficient number of outlets in each room, you should inspect the outlets to determine if they are functional and whether they are grounded properly. You can check the outlets with a simple plug-in tester available at electrical supply stores. Do the outlet receptacles have two or three slots? Newer outlets have three slots and are intended for use with appliances that have three-prong plugs, although two-prong plugs can also be used. The third slot is a grounding connection for grounding appliances. This is particularly important for appliances that are not double-insulated or have a metal casing. (See the discussion on electrical outlets in chapter 10.) Even though an appliance has a three-prong plug and requires grounding, it can be used with a properly grounded two-slot receptacle

Fig. 12-9. The home wiring "octopus"—a potential fire hazard.

with an adapter. There are, however, many electrical appliances that are double-insulated against shock hazards and thus do not require a grounding connection. These appliances can be safely used with outlets that have only two slots.

Bathrooms should have at least one, preferably two, outlets that are readily accessible. Often there is only a single outlet in the wall-mounted light fixture located above the medicine chest, which usually cannot be conveniently reached by anyone less than 6 feet tall. As a safety feature, the outlets and switches in the bathroom must not be reachable from the tub or shower.

According to the National Electrical Code, receptacle outlets in bathrooms, kitchens, garages, crawl spaces, and unfinished basements of new construction must have ground-fault circuit protection. The protection can be achieved by using a special receptacle or a circuit breaker that has been equipped with a ground-fault interrupter (GFI). A GFI is an electronic device that trips (opens) the circuit when it senses a potentially hazardous condition. It is very sensitive and operates very quickly. The GFI interrupts the power in less than $\frac{1}{40}$ second if it senses an imbalance in the electrical current of as little as 0.005 amps. The quick response time in interrupting the power is fast enough to prevent injury to anyone in normal health. Ground-fault protection, however, is not a retroactive requirement and thus will probably not be found in most homes. If there is a GFI circuit breaker or receptacle, it should be tested to see if the fault-sensing function is operational. These units are equipped with manually operated test buttons that trip the circuit when they are operating properly. The GFI should then be reset. Underwriter's Laboratories, Inc., recommends that all GFIs be tested monthly.

In addition to circuit breakers that protect against ground faults, there is a new circuit breaker that protects against electrical arcing conditions, referred to as arc faults. Arc faults are considered one of the major causes of fires in the home. The circuit breaker is an arc fault circuit interrupter (AFCI). It is designed to detect unexpected arcing in electrical wiring and disconnect the power before the arc generates a high temperature that can ignite nearby combustibles such as wood, paper, and carpets. Arc faults can occur in wiring or electrical cords where the insulation has become brittle or cracked, or where it has been punctured by staples or nails. It can also occur at a loose connection where wires are attached to switches and outlets, and where furniture has been pushed up against plugs. As with a GFCI, the unit is equipped with manually operated test button that will trip the circuit when the unit is operating properly. It is recommended that all AFCIs be tested monthly.

When walking through the hallways, rooms, and up and down steps, it should be possible to light the path ahead and to turn off the light without retracing one's steps. This can be done with three-way switches. Also, there should be a convenience outlet in hall areas for nightlights and cleaning equipment.

Knob-and-tube wiring

Knob-and-tube wiring is no longer used in new construction. However, since you might be considering an older house, the building might have knob-and-tube wiring throughout or in portions of the structure. A knob-and-tube wiring system uses porcelain insulating knobs, tubes, and flexible nonmetallic tubing for the protection and support of single-insulated conductors. (See FIG. 12-10.)

Obviously you cannot see the type of wiring behind the walls. However, very often there are exposed wires in the unfinished attic or basement. If there are exposed sections of knob-and-tube wiring, the outer insulation covering should be checked for broken and open sections. The insulation is often dry and brittle and chips easily. If there are exposed conductors, the exposed areas must be covered with electrical tape as a safety measure. Knob-and-tube wiring, although obsolete, is considered safe, providing that no modifications are made to that portion of the electrical system. If any changes or extensions are made, rewiring of that branch circuit might be necessary. All modifications to knob-and-tube wiring must be made by a licensed electrician. In addition, the outlet receptacles of a knob-and-tube-wired system are not grounded. Consequently, as a precautionary measure, they should be used only with appliances that do not require grounding.

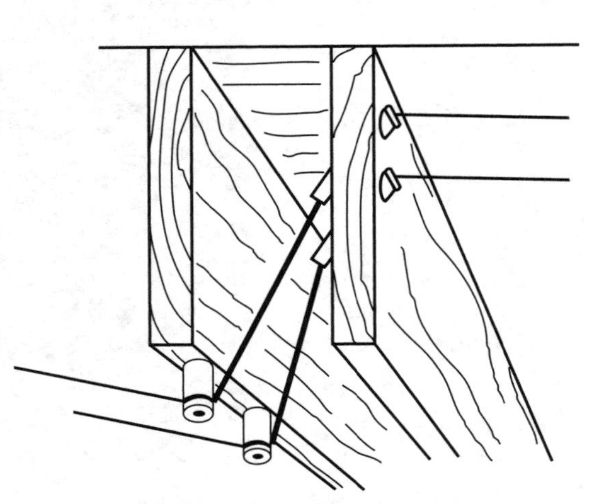

Fig. 12-10. *Knob-and-tube wiring, which is obsolete.*

Low-voltage switching systems

Light fixtures are normally controlled by switches that interrupt the electrical flow in a 110-volt circuit. However, in some houses, the light switches operate at about 24 volts. This type of wiring is generally installed so that the lights can be controlled from three or more locations. With this system, it is possible to control all the lights in the house from a master panel located anywhere in the house. The circuit for a low-voltage system includes a transformer and an electrically operated switch (relay) that is usually mounted near the fixture. If your house has this type of system, make sure that there are replacement relays. Many electrical-supply houses do not stock these relays, and they would have to be specially ordered from the supplier. If you do not have replacement relays and a relay breaks down, you might be without light from that fixture for several weeks before the condition can be repaired.

Violations

Electrical violations should concern the home buyer-owner because they represent potential safety or fire hazards. As you walk around the interior and exterior of the structure, keep in mind those items that are considered common electrical violations. If there is an outside pole lamp, are the wires leading to the lamp buried? They should be. If a portion of the wire is exposed, is it stamped *UF—Sunlight Resistant*? This indicates that it can be used as exterior wire. Many times, homeowners unknowingly use interior-type wire as exterior wire. This is quite dangerous because sun, rain, or soil conditions can cause the insulation covering the wire to deteriorate, exposing the conductors. Incidentally, if there is an outside pole lamp, it should be turned on to see whether it is operational. If it is not, the problem may be as simple to correct as changing the bulb or replacing the switch. However, there might be faulty underground wiring. Do all of the outside electrical outlets have exterior-type covers that provide protection against water penetration? Although not a retroactive requirement, it is recommended that all outside lights and outlets be protected with ground-fault interrupters.

Inside the house, particularly in the basement, look for open junction boxes, loose wires, and exposed wiring and splices (unless the wires are low voltage). Recently, an inspection of a four-story multifamily brownstone walkup in New York City revealed several violations in one circuit located in the basement. The wire coming from the panel box was rated at 15 amps, but it was protected in the panel box by a 30-amp fuse. The wire was hanging in loops from the basement ceiling and had open splices with exposed conductors (splices in a branch circuit should be contained in a closed junction box). One of the hanging loops with exposed conductors was resting on the inlet water pipe leading to the heating-system boiler, and the floor below the pipe was wet. You seldom run into a more dangerous condition. Yet there were ten families living in that house.

If the wiring looks makeshift or appears to have been modified by a nonprofessional, you should request that the seller provide you at contract closing with the Board of Fire Underwriters' certificate of approval for the electrical wiring currently in the house. In your area, the Board of Fire Underwriters might be called by a different name, or a private commercial service might be used.

Finally, when walking through the house, see if all of the receptacle outlets and switches have cover plates. They should. Also, when using an extension cord, the homeowner sometimes runs the cord through the inside of a partition wall. This is a violation of the electrical code. If an outlet is needed, a permanent one should be installed.

Checkpoint summary

Exterior

❍ Is electrical service provided by underground cables or overhead wires?

❍ Are overhead service wires securely fastened to the house?

❍ Count the number of service wires. Note that a two-wire service provides only 110 volts, not 220 volts.

- ❍ Inspect inlet service wires for cracked, missing, and frayed sections of insulation.
- ❍ Note overhanging dead tree limbs or branches in contact with service wires.
- ❍ Inspect for exterior-mounted main panel boxes. (Replacement on the interior is recommended.)
- ❍ Check outside electrical outlets for weather protection.
- ❍ Inspect exterior wiring for proper type. It should be marked *UF—Sunlight Resistant*.
- ❍ Inspect exterior lights and outlets for operation.
- ❍ Note inoperative fixtures or fixtures that are missing, loose, or hanging by wires.

Interior

- ❍ Inspect main panel box. Is it a fuse or circuit-breaker type?
- ❍ Are the GFCIs operational, and are there any AFCIs?
- ❍ Is it a Federal Pacific Electric service panel box?
- ❍ Does system contain a main disconnect?
- ❍ Note the following (do not remove the panel cover):
  - — Loose or missing cover.
  - — Missing knockout plates or fuses.
  - — Are there at least two 20-amp appliance circuits?
  - — Is there at least one 15-amp lighting circuit for each 500 square feet of floor area?
  - — Are there many spare or burned-out fuses present?
- ❍ If panel-box cover is removed by an electrician or an inspector, he should inform you of
  - — the amount of service;
  - — circuits that are improperly protected (overfused);
  - — circuits that have aluminum wiring;
  - — evidence of water leakage or corrosion deposits.

Grounding

- ❍ Inspect electrical system for proper ground protection.
- ❍ If there is a municipal water supply, is the ground wire from the main panel box fastened on the street side of the water meter?
- ❍ Inspect the connection for tightness of fit and corrosion.
- ❍ If the inlet water pipe is plastic (often in a well-pumping system), check on the exterior for a rod or pipe to which the ground wire should be clamped.
- ❍ Note whether the ground wire is missing or has a loose or corroding section.

Interior wiring, outlets/switches, violations

- ❍ Does house contain old or obsolete wiring (i.e., knob-and-tube-type wiring)?
- ❍ Inspect for cracked and open sections of insulation.
- ❍ Inspect wiring in basement and attic areas for
  - — loose and hanging sections;
  - — extension-cord-type outlets;
  - — open junction boxes;
  - — exposed splices;
  - — makeshift or nonprofessional alterations.
- ❍ Inspect each room for electrical outlets
  - — at least one outlet in the bathroom.
  - — at least one outlet per wall for an average size room (10 by 12 feet).
- ❍ Are outlets functioning (electrically hot)?
- ❍ Check for loose outlets, switches, and missing cover plates.
- ❍ Are stairways and hallways adequately lit?
- ❍ Are there three-way switches?
- ❍ Are there outlets in the hallways for nightlights and cleaning equipment?
- ❍ Note violations such as
  - — open splices;
  - — fixtures hanging by wires;
  - — extension-cord wiring that passes through partitions or around door openings.

13 Plumbing

As with the electrical system, the major portion of the plumbing system is concealed behind the walls and below the floors. Nevertheless, the part of the plumbing system that is accessible for inspection permits you to make a meaningful evaluation of its condition. A basic plumbing system consists of a water-supply source, distribution piping, fixtures, drainage piping, and a waste-disposal system. Figure 13-1 shows the layout of a plumbing system for a typical one-family, two-story house.

Water supply and distribution

Water is supplied to the property line from the street water mains of the local utility company. At the property line, there is a shutoff valve called a *curb valve* that can be used to control the water supply to the

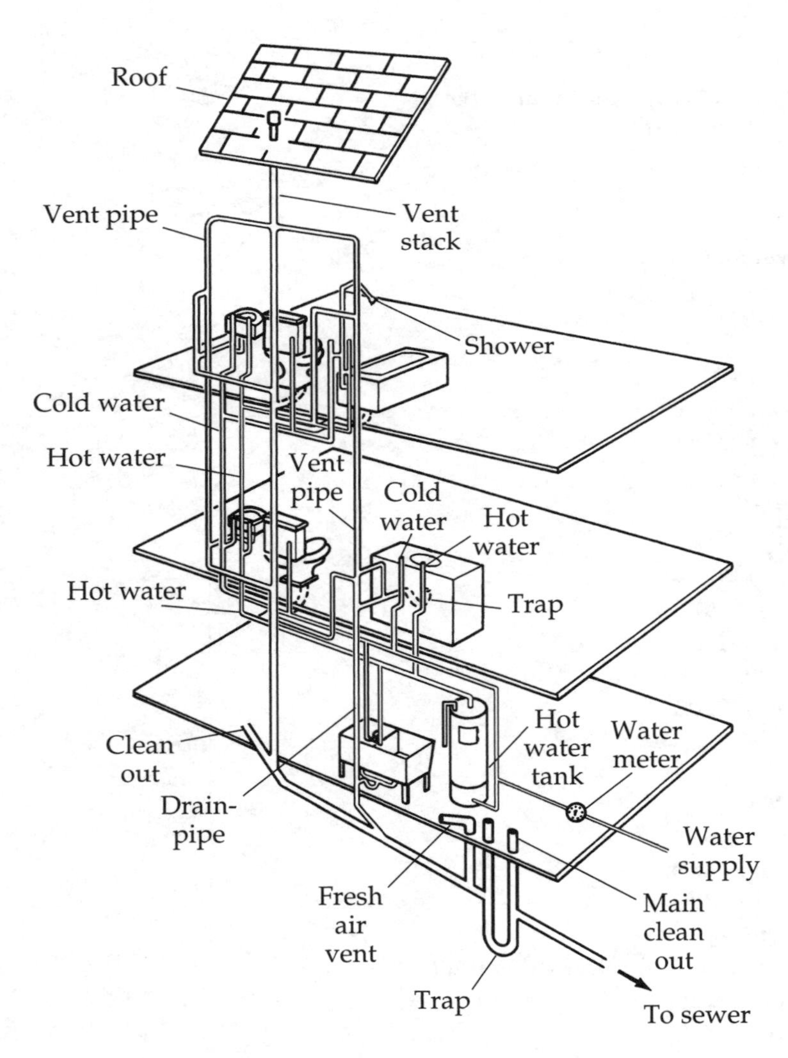

Fig. 13-1. Plumbing system layout, showing water supply, drain and vent pipes for a typical one-family, two-story house.

house. If you do not pay your water bill on time, the water company can close this valve. The pipe that actually delivers the water to the house, the *house service main*, runs from the curb valve to just inside the house and is the responsibility of the homeowner. Should this pipe need repair or replacement, it is at the homeowner's expense. If water is not supplied by a utility company, it will be supplied by a well-pumping system.

Inside the house, the cold water is distributed to the various fixtures located throughout the house. At the house inlet, there is a shutoff valve that the homeowner can use to close the water supply for the entire house. If there is a water meter in the system, it will usually be located inside the house near the inlet before any branch connections. Sometimes it is located near the curb valve. The cold-water pipe has a branch connection (usually near the inlet) that directs a portion of the water to a heater for generating the domestic hot water. The water heater can be a separate stand-alone unit or part of the heating system and is discussed in chapter 16. Distribution piping for the hot water will run from the heater to the various fixtures. The hot-water pipes are very often near and parallel to the cold-water pipes. However, they should be at least 6 inches apart so that the heat from the hot-water pipes does not affect the temperature of the cold-water system.

Fixtures

Plumbing fixtures are located at the end of the water supply and the beginning of the drainage system. They provide a means by which the water brought into the house can be used. Depending on their purposes, the fixtures have either hot water, cold water, or both. They also have a drain for the removal of the waste water. This waste water is channeled away from the fixtures through drainpipes to a sanitary sewer or a private sewage-disposal system such as a septic tank or cesspool.

Drainage system

The drainage system is more complex than the water distribution system, since it consists of three parts—traps, drainpipes, and vents. The drainage system begins just below the fixture with a water-filled trap. The trap is generally U-shaped and should have water at the bottom portion. The water in the trap forms a seal to prevent sewer gases usually in the drain line from entering the room.

The wastes flow from the fixture trap down the drain line and out to the sewer or private sewage-disposal system. Unlike the water distribution system, where the flow is under pressure, the drainage-system flow is entirely by gravity. Consequently, the drainpipes are larger in diameter than the water pipes, varying from 1½ inches to 4 inches, compared to ⅜ inch to 1 inch for the water pipes. In some communities, the house drain line leading to the sewer must have a house trap. The trap is usually located inside the house near the foundation wall. Its purpose is to provide a seal and prevent the gases that occur in the sanitary sewer from circulating back through the plumbing system. When there is a house trap, there should also be a fresh-air inlet pipe connected to the main drain. This air inlet pipe is located on the house side of the drain approximately 1 foot from the trap. (See FIG. 13-2.) In cold climates, it is located about 5 feet from the house trap to prevent the water seal from freezing during the winter. When the outer end of the fresh-air inlet terminates on the outside of the foundation wall, it should be covered with a perforated metal plate that admits the air and prevents obstruction. When it is freestanding, it should be covered with a cowl or gooseneck. (See FIG. 13-3.) The

Fig. 13-2. *House trap on main drain line leading to a sewer. Fresh-air inlet pipe on house side of the trap terminates on the outside of the foundation wall.*

function of the fresh-air inlet is to maintain atmospheric pressure at the house trap and to ensure complete air movement within the drainage system. With a private sewage system (septic tank), a house trap on the drain line is not needed. The gases that are generated in septic tanks are usually discharged to the atmosphere through the house drainage-vent system.

Venting is needed in the drainage system, since it provides a means to discharge to the atmosphere gases that develop in the system. It equalizes the air pressure in the drainage system by allowing air to flow into and out of the drainpipes. This free air movement maintains atmospheric pressure at the various fixture traps, which prevents the waste water from siphoning the water seal out of the drain trap. (See discussion on fixture traps in chapter 10, page 140.) Venting in the drainage system is achieved by vent pipes connected to the drain line near each fixture trap and to a pipe that terminates above the roof line. This pipe, called the *vent stack*, is visible from the outside. Vent pipes must be unobstructed. They carry no water or wastes.

Waste-disposal system

Waste disposal from a residential structure will be either through sewers connected to a community waste-treatment plant or through a private disposal system such as a septic tank or cesspool. When all other items are equal, a house with a sewer is more desirable than one with a private disposal system. Sewers are relatively maintenance-free. On occasion, there might be a blockage, which can usually be cleared at low cost by using a drain auger ("plumber's snake"). Maintaining a private disposal system, on the other hand, can be quite costly.

Whether the house is serviced by a sewer or is connected to a septic tank usually cannot be determined during an inspection. The house drain line passing through the foundation wall is the same regardless of whether there is an exterior connection to a sewer or to a septic tank. Since not all municipalities require a house trap on the main drain, the absence of a trap does not mean that there are no sewers. And just because there is a sanitary sewer in the street, it should not be assumed that the house is connected to the sewer line. Tying into the sanitary sewer (if the connection is permitted) is at the homeowner's expense. I know of several instances where homeowners elected to stay with their septic systems rather than go to the expense of tying into the sewer. In some communities, however, once a sewer

Fig. 13-3. Terminations of fresh air vent: top left—perforated plate covers opening in foundation wall; bottom left—perforated cover plate for vent opening is missing, a common condition; right—a freestanding "gooseneck" cover.

line is installed, all the homes on that street are legally obliged to connect. Your best bet is to contact the local municipal building department and ask them if this house is connected to the sanitary sewer system. It might save you a lot of aggravation later on.

Cesspool

A cesspool is basically a hole in the ground that has been lined with stone, brick, or some other material. It is constructed to allow raw contaminated liquid sewage to leach into the soil while retaining the organic matter and solids. Because of environmental and health considerations, most communities no longer allow cesspools in new construction. The older homes that have cesspool disposal systems are not required to upgrade them to septic systems as long as they are functioning properly. However, when problems develop, the homeowner is often legally obligated to replace the cesspool with a septic system rather than repair it. Depending on soil conditions, topography, and available space, the installation of a new septic system may cost over ten thousand dollars.

Septic system

The conventional septic system described below is the most common method of on-site waste water treatment and disposal in the

United States. It consists of a watertight container that functions as a detention tank for sewage sludge and a disposal field for the absorption of the liquid wastes. A septic tank is usually made of concrete but might be made of steel or fiberglass. Raw sewage from the house is discharged into the septic tank through the house drain line. (See FIG. 13-4.) After the sewage settles, the solids are decomposed by bacterial action and are converted into a liquid and a sludge that accumulates at the bottom of the tank. Several types of gases are by-products of the decomposition process, the most common of which is methane, an odorless and highly inflammable gas. The gases generated in the septic tank usually flow back through the house drain and are discharged harmlessly to the atmosphere at the roof-mounted vent stack.

When the level of the liquid (*effluent*) in the septic tank rises to the outlet port, the effluent flows through the outlet pipe to a drainage field. The drainage field, also called the *leaching field*, consists of a series of perforated pipes set into a bed of gravel. As the effluent flows through these pipes, it trickles through the perforations and is absorbed into the ground. The rate at which the ground absorbs the effluent (*percolation rate*) determines the size of the leaching field. When the topography changes abruptly or the area available for a leaching field is too small for adequate absorption, a seepage pit is used. It is basically a covered pit with an open jointed or perforated lining through which the effluent will seep or leach into the surrounding soil.

To determine the size of the leaching area needed for a house, percolation tests are taken in the area of the proposed sewage-disposal system. The rate of water absorption will depend upon, among other things, the type of soil and the level of the water table in that area. If the percolation tests are taken during a drought or when the water table is low (the

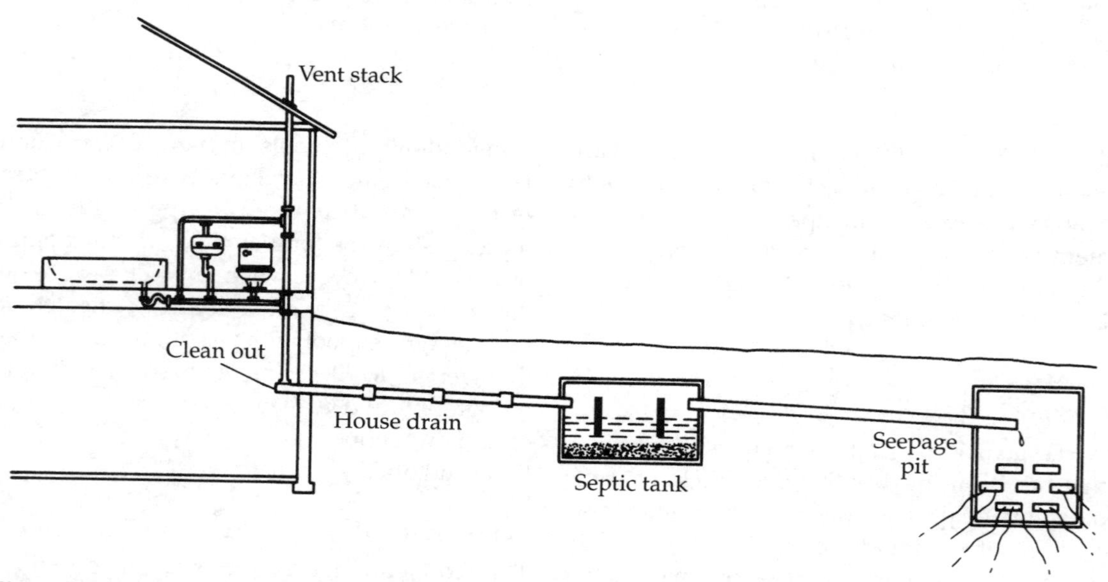

Fig. 13-4. *Typical septic system with seepage pit. Depending on the topography and amount of available land, the pit may be replaced by a leaching field.*

water table rises and drops during the year), the concluding data might result in designing an undersized leaching field that will cause premature failure of the septic system. I know of a village just northwest of New York City where the septic systems in hundreds of homes failed within eight years of installation, some of them within two years. A properly designed and maintained septic system should last between twenty and thirty years; indeed, fifty or more years is not uncommon.

Failure of the septic system will usually show up in the area of the leaching field and not necessarily over the septic tank, although there might be indications in both areas. Look for wet spots or a lush growth of grass. Both can be accompanied by an objectionable odor. When the ground in the seepage field becomes saturated and can no longer absorb the effluent, the liquid will build up and accumulate on the surface. The effluent contains nitrogen and other compounds that are natural fertilizers. When the effluent surfaces, it causes vegetation in the area, particularly grass, to thrive and have a lush green color. However, a healthy-looking patch of green grass over the leaching field is not necessarily an indication of a septic failure. During dry weather, when grass is apt to grow very slowly and turn brownish, the ground over the leaching field (depending on the depth of the field) often contains sufficient moisture to promote the growth of the grass and maintain the green color.

For the leaching field to function properly, there must be sufficient voids in the soil so that the effluent will be absorbed by the ground. Over the years, the voids can be filled with suspended solids, reducing the rate at which the effluent is absorbed into the soil to the point where the effluent surfaces. When this occurs, it is necessary to install a new leaching field. Aside from the cost, a serious problem can arise concerning how to handle the effluent when there is no more room on the property for a new leaching field or seepage pit. If the house has a septic system, find out if there is sufficient room for expansion of the leaching field should it be necessary at a later date.

Premature failure of the septic system can also occur as a result of neglect and abuse by the homeowner. A septic tank should be cleaned or at least inspected for sludge buildup every two to four years. If the tank is not cleaned periodically, the sludge will build up to such a level that the solids are carried out of the tank and into the leaching field. Eventually these solids will clog the voids in the soil or the perforations in the leaching-field pipes, blocking the normal flow of effluent. When this happens, the leaching field requires replacement. This type of problem can be avoided. Unfortunately, many homeowners neglect to inspect and clean the septic tank periodically.

On many occasions I have been told by an owner that the septic tank has never been cleaned. The owner knew from "experience" that if the septic system functioned properly, it was best not to disturb it. After all, his neighbor across the street had had his septic tank cleaned after fifteen years, and two weeks later the system failed and had to be replaced. If he had not touched the system, it would have been all right, wouldn't it? No! That system apparently had been on the verge of failure before the cleaning. The cleaning had nothing to do with the failure.

Occasionally septic systems can operate satisfactorily for many years without the tank being cleaned. In those cases, the tank might have been grossly oversized for the number of people living in the house, or, perhaps the house was used as a vacation home and occupied only part of the time. One method of extending the life of a septic system is to reduce the volume of water that passes through the tank and leaching field. This is often done by installing a separate drain line for the waste water from fixtures such as the washing machine and connecting it to a dry well or seepage pit that is located away from the leaching field.

Sometimes the homeowner indicates that he has not cleaned the septic tank because he has been using a chemical compound or septic-tank cleaner that he pours down the drain. This cleaner is supposed to improve bacterial activity and eliminate the need for periodic cleaning. The advertising claims for these cleaners are not well founded; there will always be a sludge buildup that must be removed. In addition, some of the cleaners might contain compounds that will actually reduce the bacterial process and can cause the system to fail.

To reduce the possibility of premature failure of the septic system, some homes are equipped with a grease trap in the waste-disposal system. The trap separates grease from the kitchen waste line, thus preventing it from entering the septic system. A buildup of grease in the system can result in clogging or reducing the porosity of the leaching field and can also affect the bacterial action in the septic tank. If there is a grease trap in the waste-disposal system of the house, you should be advised that for maximum effectiveness, it should be frequently cleaned or at least inspected for grease buildup.

Discharging large volumes of water into the septic tank, such as rain runoff from roof gutters or storm drains can over a period of time adversely affect the tank's operation. Large volumes of water can flood the tank, forcing suspended solids into the leaching field where they can eventually block the perforations in the pipes and clog the voids in the soil.

Depending on the soil condition, a water softener can be detrimental to a septic system. If the seepage field consists of a clay-type soil, the waste water from the water-softener regeneration process must not be discharged into the septic tank. The salt brine in the water-softener waste water is not broken down by bacterial action as it passes through the septic tank to the leaching field and can clog the voids in the fine-textured clay soil.

Even though a septic system might be faulty, there might not be any visual indications of a problem when you inspect the house. Whether there is a water (effluent) accumulation over the leaching area or septic tank will depend upon the dryness of the season, the usage of the plumbing system prior to the inspection, and the degree of deterioration of the septic system. One way to check the operation of the system is to "push" it by turning on the water in the tub and letting it run for about one hour. (Before you do this test, ask the owner for permission.) The tub drain must be open so that the water will flow into the septic tank. Assuming an adequate water supply, this pushing should introduce about 275 gallons into the septic system, which very often is enough to cause the effluent in a faulty system to surface and be visible when you reinspect the areas over the tank and leaching field. This test, however, can be misleading if the homeowner had the tank pumped out a few weeks prior to your inspection.

In addition to the conventional septic system described above, there are alternative septic systems available, which are installed when warranted by specific soil conditions, topography, and available space. Alternative septic systems are more complex and costly than conventional systems because of the use of pumps, alarms, and diverter valves. Two such systems are the alternating drainfields septic system and the mound septic system.

Alternating drainfields septic system This system is similar to the conventional septic system's drainfield design. However, as a result of alternating the drainfields, the overall system can be designed so that it takes up less space than a conventional system. Because many municipalities have a requirement that there must be room for a 100-percent drainfield expansion, a smaller drainfield area is a desirable feature, since it will allow construction on small pieces of property, which were considered unusable.

When the drainfield is unsaturated, the treatment of septic effluent is enhanced. As the effluent passes into and through the soil of an unsaturated field, the aerobic bacteria digest the organics in the waste stream and form a filtering layer (a bio-mat) at the soil/gravel edge. On the other hand, saturated conditions and excessive amounts of organics in the effluent reduce the treatment capacity of the system. Excessive amounts of organics escaping from the septic tank can result in a bio-mat growing so thick it will actually prevent adequate water flow into the soil. Consequently, depending on the wastewater flow and the load of suspended solids, some drainfields become clogged and less effective, and in some cases, the fields fail. The alternating drainfields septic system resolves this problem by having two or more drainfields. This system is used primarily to overcome soils with restrictive percolation rates, such as clay. By switching from one field to the other every six months to a year, each drainfield can alternately be used and rested. The resting period causes the bio-mat, which eventually clogs the field, to deteriorate due to lack of nourishment, and allows the drainfield to completely drain, thus renewing the field. Switching from one drainfield to the other is done through a diverter valve, which is housed in a valve box on the effluent line from the septic tank. The valve provides for independent flow to one drainfield or the other, but not both. It must have a watertight lid or cover, which permits unobstructed access for maintenance, inspection, and operation.

Mound septic system Mound systems are used when there are problems with the site for the drainfield, such as a high water table, inadequate percolation rates, and a shallow soil cover over creviced or porous bedrock. This system consists of a septic tank, a dosing chamber or tank with a pump and a high water alarm, and a drainage field constructed in a mound that is raised above the natural grade to maintain a proper distance from the water table or bedrock. (See FIG. 13-5.) The dosing tank is placed between the septic tank and the mound and accumulates septic tank effluent. Float-type control switches inside the tank turn the pump on and off. Once the accumulated effluent in the tank reaches a predetermined volume, the pump delivers a "dose" of effluent to the laterals in the mound by an electrical control system. The mound can be 3 to 5 feet at its highest.

An adequate evaluation of the plumbing system requires an interior and an exterior inspection. Although most of the plumbing inspection is performed in the interior, an exterior inspection can reveal venting problems, violations, and septic-system problems.

Exterior inspection

Vent stack

As you walk around the house looking at the roof, look for a plumbing vent stack. If you do not see one, it indicates either that the plumbing system is not properly vented or that the vent stack terminates in the attic. Both possibilities are violations of the plumbing code. If the construction of the roof is such that there are sections that are not visible from the ground, a vent stack should be looked for during the attic inspection. If there is a vent stack, you will see a pipe coming up from the floor and going out through the roof. Sometimes it is

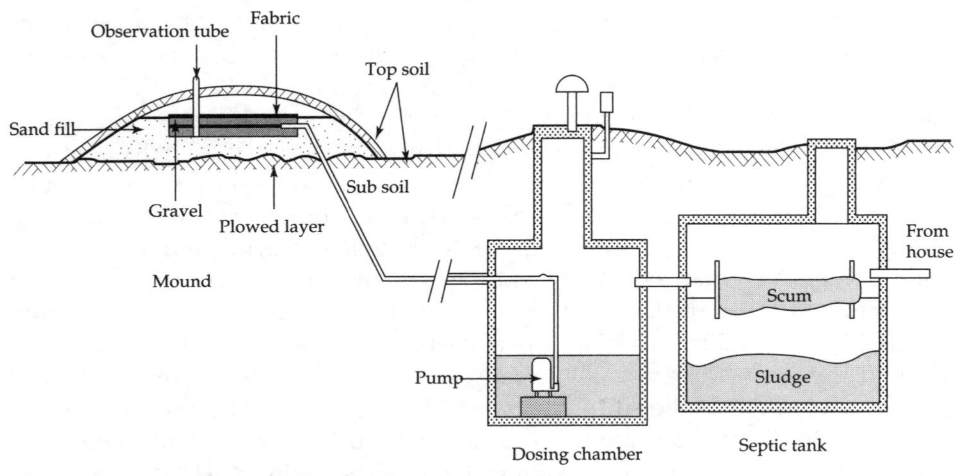

Fig. 13-5. *Schematic of a typical mound septic system.*

difficult to check for the vent stack in the attic because of the restricted space and the lack of adequate lighting.

The vent stack should terminate above the roof. If you see one that terminates near a window or one that runs up along the side of the building (in the northern portion of the country), it is in violation of the National Plumbing Code (as explained in chapter 3).

Fresh-air vent

When there is a fresh-air vent for the house trap, if the outer end terminates on the outside of the foundation wall, the opening should be covered with a screen or perforated metal plate. The covering is needed to prevent children from stuffing toys, balls, or other objects into the opening and blocking air movement. Technically, an unprotected fresh-air vent opening is also a violation of the plumbing code. If you do not see a cover over the fresh-air inlet vent, record the fact on your worksheet as a reminder for later installation. You should also use your flashlight to determine if there is any blockage.

Lawn sprinkler systems

Does the house have an underground lawn sprinkler system? If it does, look specifically for a vacuum breaker (antisiphon device) on the water-supply line for each zone. When the piping arrangement is such that the water is supplied to each zone through a manifold, the vacuum breaker should be on the pipe supplying water to the manifold.

A vacuum breaker is recommended to prevent dirty water or foreign matter that normally accumulates around pop-up spray heads from flowing back into the potable water supply as a result of a negative pressure in the line. It is a relatively simple device and is usually no larger than a few inches in diameter and a few inches high. (See FIG. 13-6.) The vacuum breaker has an atmospheric vent that is sealed when water is flowing to the spray heads. When the water pressure drops to below atmospheric pressure (vacuum) because of a problem in the supply line, a mechanical float drops, opening the atmospheric vent. This allows air to enter the piping system upstream of the spray heads

Fig. 13-6. Typical atmospheric vacuum breaker used in lawn irrigation systems. Rain Bird Sprinkle Mfg. Corp.

and thereby prevents dirty water at the spray head from being siphoned back into the supply piping.

Vacuum breakers must be located at least 6 inches above the level of the highest sprinkler head. Look for them near the foundation wall. Sometimes the piping is located in a well adjacent to the foundation, similar to a basement window well. Usually, the vacuum breakers and the electric solenoid valves that control the water flow to the various zones are in the same general area.

Septic system

If the house has a septic system, find out the location of the tank and the leaching area. As you walk around the house performing the general exterior inspection, look specifically for puddles over the leaching area and septic tank. If the puddles have a film on top and there is a foul odor in the area, you can be sure that there is a septic problem. If the puddle is a clear liquid and there is no odor, there might still be a septic problem. The best way to find out whether the liquid is septic effluent or surface water as a result of rain is to collect a sample and have it analyzed. If no puddles are noted during your initial exterior inspection, reinspect the area after "pushing" the septic system as described above.

Interior inspection

Fixtures

Since you will start your general interior inspection at the attic level and work your way down to the basement, the interior plumbing inspection will begin in the first room that has fixtures. (See the "Bathroom" section of chapter 10 for further discussion of plumbing.) The fixtures should be checked for general condition—cracked, chipped, and stained sections—and for operation. Do the faucets function properly, or are there leaks around the handles or spout? Is there an air gap between the spout and the top level of the water when the sink or tub is filled? There should be a gap to prevent back siphonage. Is the sink drain leaking or show signs of past leakage? Occasionally, you might find a pot below the drainage trap to catch dripping water. Sometimes you might find rubber-hose-type connections on the drain or a drain that has been taped up. These are makeshift corrections and require proper attention. Are there individual shutoff valves for the water supply to the various fixtures? Shutoff valves are not necessary but are desirable when making repairs or replacing the faucet. When you open and close the faucet rapidly, do you hear a "water hammer" noise? You shouldn't. But if you do, it can usually be corrected with an antiknock coil or air chamber. Are the toilet bowl flushing and fill valves operating properly? If you hear a whistling noise while the bowl is filling up, the fill valve needs adjustment. After the water closet (tank) has been filled, do you still hear water running? If you do, minor maintenance is needed.

Cross–connection

A cross-connection is where the potable water supply can become contaminated because it is connected to another source of water, which may be nonpotable. Look for situations that could result in a cross-connection. Occasionally, in old homes I find that the faucet spout on the laundry sink extends below the sink's flood rim. This is a contamination problem waiting to happen. If the sink is filled with waste water from a washing machine discharge or some other source, and at the same time a partial vacuum is created in the municipal water supply, the waste water will be drawn back into the potable water, contaminating it. A vacuum can be created because of a number of reasons, such as when a work crew opens a water main to make repairs or when water is drawn from a fire hydrant to fight a fire. If you see this condition, recommend that the faucet be replaced with one where the spout is above the flood rim. The new faucet would create an air gap between the spout and any water in the sink, thereby preventing a cross-connection.

Water pressure, flow

After inspecting the fixtures, faucets, and associated valves, pay particular attention to the water flow and drainage. When there is a considerable flow of water from a faucet, most people say the pressure is good, and when the flow is merely a trickle, they say the pressure is bad. This is a popular misuse of the word *pressure*. It is true that if the pressure is low, the water flow will be low. However, a low-flow condition is not usually caused by low pressure. It is caused by a constriction in the inside of the supply pipes. Depending on the quality of the water, mineral or corrosion deposits can form along the inside of a pipe, reducing the effective pipe opening to that of a straw. In this case, even when the source pressure is good, the flow will be less than minimal.

Check the cold- and the hot-water flow separately by opening two faucets and flushing the bowl at the same time. If the flow appears to be less than adequate, record the fact on your worksheet as a reminder to check further. The low flow might be caused by a kink in the supply pipe, small-diameter distribution piping, or low water pressure at the service entry. However, the probability is greatest that the low flow is caused by a decrease in the inside diameter of one or more sections of pipe supplying the fixtures.

Old galvanized iron piping is particularly bad for water flow. In addition to mineral deposits, there is often a buildup of rust that further constricts the flow. If you find low water flow, especially in an older house, look for galvanized piping. You might find that the house has been partially repiped. Some iron sections might have been replaced with copper or brass. Mixing iron in a copper plumbing system is not desirable, for two reasons. First, just as a chain is as strong as its weakest link, a plumbing system is as good as the weakest section of pipe. Even a small section of iron pipe whose inside diameter has been reduced by a buildup of rust and mineral deposits will lower the flow of water discharging from a faucet. Second, when ferrous (iron) and nonferrous (copper) metals are in contact with each other, a galvanic action takes place between the metals that accelerates the corrosion of the plumbing system at the point of contact.

On occasion, in those homes that have water supplied by a well-pumping system rather than a local utility, you might find that the water discharging from the faucet has a pulsating flow. This condition is caused by a rapidly fluctuating water pressure and is the result of a waterlogged pressure tank. The condition, relatively easy to correct, is discussed on page 205.

Plumbing wall hatch

In many older homes, an opening in the wall of a hallway provides access to the plumbing pipes for the tub. The opening is usually no larger than 2 feet by 3 feet and often has a wooden cover that has been painted like the rest of the wall so that it is not very noticeable. Sometimes this access hatch is in a closet or bedroom. If there is such a hatch, remove the cover. You should be able to see the water-supply pipes, the overflow, and the drainpipes. (See FIG. 13-7.) In newer homes a wall hatch that covers plumbing could also indicate modular construction.

Pipes

Water-distribution pipes can be made of copper, brass, galvanized iron, and more recently plastic (PEX). PEX piping will be discussed separately. Once you become familiar with the types of metal pipe, you will be able to differentiate between them very easily. Here are some pointers. Brass and galvanized pipe have threaded joints; copper joints are soldered. So if the pipe joints are threaded, you will know that they are not copper pipes. Do not try to determine whether the pipe is brass or galvanized by its color. There might be a dirt film on the pipe, or it might be painted. The easiest and most foolproof method is to use your magnet. If the pipe is made of galvanized iron, it will attract the magnet; if it is made of brass, it will not.

Fig. 13-7. Wall hatch in hallway provides access to pipes for bathtub.

Now look to see if the piping is a mixture of copper, brass, and galvanized sections. Even if there is no noticeable drop in water flow, mixed plumbing is not desirable because the resultant galvanic corrosion at the joints will eventually cause leakage. Figure 13-8 shows such a joint. The white encrustations on the pipe are mineral deposits left when the water oozing out of the fitting evaporated. Although deposits on this joint self-sealed past leakage, it is only a temporary correction, for the deposits can come loose at any time. Properly correcting this condition requires replacing the iron section with pipe of copper or brass.

PEX (plastic) pipe PEX piping (tubing) is made from cross-linked high-density polyethylene. Cross-linking is a chemical reaction that permanently links together polymer chains of the polyethylene. PEX was introduced in the United States in the 1980s, but was slow to catch on in the industry. It is now accepted by most municipalities and can be used for all potable water-supply lines, both hot and cold. It has a number of advantages over metal pipe. It is flexible, won't corrode or develop pin holes, has fewer connections and fittings, and is easier and faster to install. Because of its flexibility it is very freeze-break

Fig. 13-8. Mixed plumbing as seen from a hallway wall hatch. Some of the fittings are deteriorating (note the rust and mineral deposits).

resistant, and a 90-degree corner bend can be made without the need for elbow fittings. Also, PEX is considerably less expensive than copper pipe. PEX comes in three colors: standard white (unpigmented), red, and blue. The red and blue colors are used to help plumbers and homeowners distinguish between hot and cold water-supply lines. Unlike copper pipe, PEX piping does not require soldering at connection joints. There are two popular methods for making connections with PEX: one uses a crimping tool and a crimping ring to secure the tubing over a brass fitting, and the other uses an expansion tool, special expansion fittings, and plastic rings to secure the joints. As you walk around inspecting the pipes, check the exposed joints for signs of past or current leakage. If there is any doubt about the use of PEX tubing for plumbing applications in your house, contact the local authority with jurisdiction over plumbing to verify the acceptance.

Polybutylene pipe There is another plastic pipe that you should be aware of: polybutylene pipe. It's a gray plastic water-supply pipe that was used in the construction of millions of homes throughout the United States between 1978 and 1995. In the 1980s, complaints started to come in that leaks were developing around the fittings and in some cases the pipes were splitting. Some of the leaks were the result of improper installation such as overstressing the pipe at joints and fittings, and overbending the pipe into a tight curve. However, the majority of the leaks were caused by the deterioration of the pipe and/or the plastic fittings. It was determined that polybutylene pipe reacts with oxidants, such as chlorine, normally found in tap water. The reaction, which is internal to the pipe, causes it to scale, flake, and to become brittle, with micro fractures developing. As the pipe ages and reacts with water-soluble oxidants, the condition becomes more severe. Although from a visual inspection, assuming that there are no indications of leakage, the pipe may look to be in good condition. The problem is that since the deterioration is occurring within the pipe, it is very difficult to determine if the pipe is truly in good condition.

The best a home inspector can do if he/she finds polybutylene piping is to alert the buyer about its past history and to anticipate future problems.

Basement inspection

After inspecting all the fixtures, associated pipes, and fittings in the interior rooms, the rest of the plumbing inspection is carried out in the basement. Some homes are built on ground level and do not have a basement. In those cases, this portion of the inspection will be performed as part of the interior room inspection.

Water-supply pipes

Look for the entry of the water-supply pipe. It will be located near the foundation wall and will usually have a meter near the inlet. If you cannot find the water-service entry, ask the homeowner. Sometimes it is concealed behind boxes or storage shelves. The inlet service pipe will generally be made of copper, brass, or galvanized iron. However, in some older homes, you might find that the pipe is made of lead. A lead pipe can be detected by the type of joint between the sections. Lead pipes have wiped lead joints that appear in a horizontal section as a spherical bulge. (See FIG. 13-9.) There is usually a joint near the foundation wall. If you do not see a joint, you can gently scratch the surface of the pipe. If the pipe is made of lead, the surface will be relatively soft, and the scratch will expose an area with a silver-gray color.

Although a lead water pipe might be acceptable in a plumbing sense, it can be a potential health hazard. Depending on the quality of the water, some of the lead might dissolve out. Since the amount of lead that a person can safely absorb is limited and cumulative, by drinking this water over an extended period of time, the maximum tolerance level can be reached. If you find a lead inlet pipe, you should have the water analyzed for lead content. In many communities, the local health department will do the analysis free or for a very nominal charge. If the lead content is high, the pipe should be replaced. Again, remember that replacing the inlet pipe is not the responsibility of the water company; the cost must be borne by the homeowner.

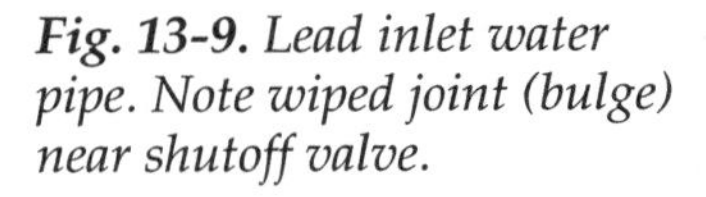

Fig. 13-9. *Lead inlet water pipe. Note wiped joint (bulge) near shutoff valve.*

During my inspections, I periodically find a lead inlet pipe. When I do, I always recommend to the prospective home buyer that as a precautionary measure the water should be analyzed. In one home, located in White Plains, New York, an analysis of the water revealed that it had seven times the allowable concentration of lead. Needless to say, that pipe was replaced.

If the water flow from the plumbing fixtures is low and the piping in the house is good (all copper or brass pipes and fittings with no leaks), the problem will be a constriction in the inlet water supply or low pressure at the street main. On several occasions, I have inspected houses where the old water pipes were completely replaced with new copper pipes and still the water flow was low. Further inspection revealed that even though the house had been repiped, the old galvanized inlet pipe had not been replaced. Under normal soil and water conditions, this pipe should last about forty years. As these pipes age, rust deposits build up on the inside, restricting the flow. Also, a galvanized iron pipe will eventually corrode from the outside and leak. Replacement can cost anywhere from several hundred to several thousand dollars, depending on the length of the line and ground conditions.

When domestic water is supplied by a utility company, the inlet pipe will lead directly to the house distribution piping. In some areas, depending on the quality of the water, there might be a water softener or a filter between the supply and distribution piping. Normally, however, you will not find a storage tank (similar to that needed in a well-pumping system) between the supply and distribution piping. If you do find such a tank, you should suspect a low-pressure condition at the street main. When the water pressure in the street main is low and there are simultaneous demands for water by the houses on the street, the flow to each house may be inadequate. To compensate for this condition, a storage tank is often installed to provide a reservoir to supply water during these periods. Water drawn from the tank is then replaced when the plumbing fixtures in that house are not being used. If you see such a tank between the inlet supply and distribution piping, record the fact on your worksheet.

When domestic water is not supplied by a utility company, it will be supplied by a well-pumping system. Such systems are discussed in the last section of this chapter.

At the house inlet side of the water-supply pipe, there will be a master shutoff valve that can close the water supply to the entire house. Sometimes there are two valves, one on each side of the water meter. See whether the valve is operational. Close and open it. Over the years, because of lack of use, the valve often freezes in the open position.

Distribution piping

Water is supplied to the various fixtures throughout the house by distribution piping. The distribution system begins by the inlet supply pipe just after the water meter and consists of two components: supply mains and fixture risers. The supply mains are usually suspended from the basement ceiling and can be readily inspected in an unfinished basement. The fixture risers run between the supply mains and the fixtures and are usually concealed behind the walls. For the most part, the risers cannot be inspected.

In addition to copper, brass, and galvanized iron pipes, most communities also allow the use of plastic (PEX) pipes for both hot- and cold-water distribution. Trace and inspect all of the exposed water pipes in the basement. You will find that there will be a branch takeoff pipe that leads to the domestic water heater, and if the house is heated by steam or hot water, there will be a cold or hot takeoff leading to the boiler. As a point of interest, over the years, cold-water copper and brass pipes take on a darker color than the hot-water pipes. Sometimes you might find sections of cold piping "sweating" profusely. This is not a problem condition. It is merely condensation and can easily be eliminated by insulating the pipes or reducing the amount of moisture in the air with a dehumidifier.

Faulty plumbing does not necessarily mean that there is a steady stream of water leaking from a pipe or fitting, although if that is the case, immediate correction is necessary. Of particular concern are indications of aging and deterioration. Look specifically for signs of past and current leakage around fittings and valves. Look for mineral deposits, corrosion, and patched sections. The presence of galvanized iron pipes and fittings in a copper and brass plumbing system is a potential problem, as discussed previously. If you see iron pipes, you should make an estimate of the amount and anticipate their replacement. Copper pipes often take on a greenish cast, particularly around the fittings. Although this condition looks as if it could have been caused by water leaking from the joint, it is not. It is usually caused by the soldering flux. A leak, on the other hand, usually shows up as an encrustation of mineral deposits around the joint.

Brass pipes found in residential structures will usually be "red brass" or "yellow brass." You can often tell the difference by the color. Red brass, however, is not really red but a yellowish brown. Yellow-brass water pipes are more vulnerable to corrosion and dezincification (zinc leaching out of the brass into the water) than red-brass pipes. The projected life for yellow-brass pipes is about forty years; seventy-five years for red brass is not uncommon. The weakest part of a brass pipe is the threaded joint. With some older yellow-brass pipes, the threads are paper thin. If force is applied to one of the pipes (with a wrench during a repair or even by leaning on the pipe), the joint could easily rupture. Usually weak joints can be detected by a slight encrustation of mineral deposits. If you find encrusted joints on brass pipes, you should anticipate repair or replacement of those sections.

When the distribution pipes are brass, look along the length of the pipes for signs of leaks. Brass pipes are vulnerable to pinhole leaks along their length. Depending on its chemical quality, the water in the pipes can cause some of the zinc in the brass to dissolve. When this occurs, pinhole openings can be seen along the length of the pipe. Because of the small size of the openings, water drips from the holes very slowly. In many instances, the water will evaporate before it drips, leaving whitish mineral deposits around the opening. (See FIG. 13-10.) Eventually the deposits can self-seal the leak, although the pinhole openings will get larger.

I once inspected a house that had this problem. The owners had moved out in the beginning of the winter before they found a buyer. As a precautionary measure, water was drained from all of the pipes to keep them from freezing over the winter. My inspection took place the following spring when the water was turned on. The sudden surge of pressure in the pipes was enough to loosen all of the deposits, and water started to leak out all along the pipes so that it looked like a sprinkler system. If you see brass pipes with mineral encrustations along their length, even though there might not be any current leakage, those pipes should be replaced.

Drainage pipes

The wastes discharging from toilets and sinks flow from the fixtures down to the sewer or septic tank by means of drainage pipes. As with the distribution piping, only the portion

Fig. 13-10. Brass water pipe with pinhole leaks, a condition caused by leaching zinc. Note mineral deposits on underside of pipe.

of the drain line that is in the basement will be visible for inspection. In those houses built on ground level or those with finished basements, the drainpipes will probably not be visible and cannot be inspected.

Drainpipes are generally made of cast iron, galvanized iron, copper, lead, or plastic. Very often the drainage system will consist of a combination of the different types of pipes. This is acceptable. The problems encountered in the distribution system when using iron and copper pipes together or lead pipes do not exist in the drainage system.

For most of the day, the drainpipes are empty. The wastes flow down the drain by means of gravity. Consequently, the main house drain in the basement must have a steady downward pitch leading directly to the sewer or septic tank. Look at the drain line. If there is a low point along the length of the pipe, there is a problem. (See FIG. 13-11.) The low point in the pipe will allow grease and sewage

Fig. 13-11. Low point in drainpipe can cause solid wastes to build up and block the flow.

solids to settle and eventually block the pipe. This condition is a violation of the plumbing code and must be corrected. When inspecting the drainpipes, look for signs of current and past problems such as cracked and patched sections, improper pitch, and leakage.

One question that I am asked fairly often by prospective buyers is "Can we put a bathroom in the basement?" The only difficulty in installing a bathroom is how to handle the drainage, which must flow by gravity. If the house drain is connected to the sewer at a level above the basement floor, the wastes from the fixtures will have to be pumped up to the house drain so that they can flow out to the sewer. If there is a bathroom in the basement, check to see if the wastewater drainage discharges into a sewer ejector tank. The tank is normally located in an unfinished area of the basement. Since the tank is installed below the floor slab, the only portions visible are the top cover and the associated discharge and vent pipes. (See FIG. 13-12.) The cover should be gasketed and

Fig. 13-12. *Sewage ejector pumping system below the floor slab.*

tightly secured to the tank, and there should be adequate seals around the pipe penetrations. There should also be a gate valve and a check valve on the discharge line. A float switch controls the ejector pump. You can check the operation of the pump by flushing the toilet and then letting water run in the sink until the pump is activated. Another approach, although one that is not approved by most building codes, is to use a specially designed toilet that can lift the wastes about 10 feet without the use of a pump. It uses water pressure. This toilet is moderately priced. However, it is quite sensitive and if used for anything other than human wastes and toilet paper, it can become clogged. If only a sink is desired in the basement, a simple, inexpensive lift pump can be used to lift the waste water to the house drain.

When the house drain is located below the basement floor slab, connecting the fixture drains to the house drain will require breaking up sections of the floor. To minimize the installation cost, the proposed bathroom should be located near the existing drain line. In some cases, when the drain line is at or just below the basement floor surface, the toilet is located on a platform to facilitate the connection.

Well-pumping systems

When domestic water is not supplied by a utility company, it will be supplied by a private well-pumping system that includes a well pump, storage tank, and pressure switch.

Wells

As explained in chapter 6, part of the water hitting the surface of the earth as rain, snow, hail, or sleet seeps into the ground and percolates down until it hits an impermeable rock strata through which it cannot penetrate. The water then flows along the strata until it eventually reaches the ocean or a river, which can be more than a thousand miles away. The underground flow is not like a running stream but more like a turtle climbing a rock pile. The water flows through the pores and cracks of rock formations, sometimes surfacing along the way as a river or lake. The water composing this underground flow is *groundwater*, the top surface of which is commonly called the *water table*.

Subsurface rock formations that readily yield the groundwater to wells are *aquifers*. There are two types of wells—shallow and deep wells. A well that draws from an aquifer located less than 25 feet below the earth's surface is a shallow well. When the aquifer is more than 25 feet below the earth's surface, the well is a deep well. Wells over 500 feet deep are not uncommon.

Because of the proximity of the surface, shallow wells are vulnerable to contamination from cesspools, malfunctioning septic systems, barnyard manure, and industrial waste disposal. Deep wells, although less vulnerable to contamination, can also become polluted. Bacterial and chemical pollutants move downward in the soil until they reach the water table and then flow with the groundwater. To a large extent, the soil acts as a natural water purifier for bacterial contamination by filtering small suspended solids and allowing large pollutant particles to settle out. In addition, bacterial pollutants tend to die after a period of time; their life spans are usually short in the unfavorable conditions found in the soil.

Chemical pollution of the water source, however, can persist for years. I recently read a newspaper article about a toxic chemical solvent, trichlorethylene (TCE), that was contaminating the water supply of seventeen private wells. The solvent, which is used for thinning paint or removing grease, can cause neurological problems if inhaled or ingested in high concentrations. The source for the contamination could not be determined. The local health commissioner, however, indicated that he thought it was the result of TCE being dumped in the area many years ago, prior to the homes being built.

Water that has a foul taste or odor and appears dirty may be completely potable, whereas water that is very clear and has a good taste may be polluted. You cannot tell by looking at it or tasting it whether the water is contaminated. As a precautionary measure, well water should be analyzed once a year for both bacterial and chemical pollutants. If the house has a well, have the water analyzed prior to contract closing.

Well pumps

The purpose of the pump is to draw water from the well and push it through the distribution piping with sufficient force that the water overcomes the frictional resistance of the pipes and provides an adequate flow at fixtures.

There are three basic types of well pumps used for residential structures: *submersible*, *jet*, and *piston*. All three can be used for shallow or deep wells. The submersible pump, however, is most frequently used for deep wells. In shallow wells, the jet or piston pumping mechanism is not located in the water. It is located on top of the well. The water is drawn up to the pump by a suction action, not unlike drinking through a straw. A suction will result in a pipe immersed in a body of water when the pressure inside the pipe is reduced below atmospheric pressure (vacuum). Under ideal conditions, the maximum suction lift is 34 feet. However, because of pump inefficiencies and frictional resistance of the pipe walls, the practical limit of suction lift is 25 feet, which is used in defining a shallow well. A deep well, therefore, is one in which water is pumped from a depth that exceeds 25 feet.

Well pumps and their accessory equipment are usually very reliable. Nevertheless, all well-pumping systems require occasional repair or replacement. The projected life expectancy of a pump is seven to ten years, although many pumps run without trouble for twenty to thirty years.

Piston pump These pumps are no longer in general use, although they might be found in older homes. Basically, they are motorized versions of the old hand pump. A motor drives the piston that alternately sucks water into the cylinder and then discharges it on every other stroke. In a shallow well, the pump (motor-piston assembly) is above the ground. In a deep well, the motor is above the ground, and the piston assembly is located in the well. Usually the motor is connected to the piston assembly by a belt and pulley. Inspect the belt for partially torn and frayed sections and adequate tension. Look for signs of leakage around casing joints and the piston rod. There should be none. An overall evaluation of any pumping system must include an inspection of the accessory equipment, which is discussed later in this chapter.

Jet pump The jet pump consists of a jet assembly and a centrifugal pump. The centrifugal pump can be thought of as a small paddle wheel driven by a motor. As the wheel turns, it imparts energy to the water, increasing its velocity and pressure. A portion of the water discharging from the centrifugal pump is diverted to the jet assembly, which has no moving parts. However, it uses this recirculated water to perform two functions. It creates a suction that draws well water into the assembly and pushes this water back up to the centrifugal pump. After passing through the pump, some of the water is again rediverted to the jet assembly and the remainder directed to the plumbing system for distribution.

You can tell whether the pump is a shallow-well or deep-well jet pump by the number of pipes extending into the well. The basic difference between the two pumps is the location of the jet assembly. In a shallow-well jet pump, the jet assembly is built into the centrifugal pump casing and has only one pipe extending into the well. In a deep-well jet pump, the jet assembly must be located within the well (so that the suction lift does not exceed 25 feet). In this case, there are two pipes extending into the well. (See FIG. 13-13.) In areas where the temperature drops below freezing, proper weather protection of jet pumps is important. According to Gould Pumps, Inc., frozen pumps represent one of the most common reasons for pump replacement.

Submersible pump A submersible pump consists of an electrically driven centrifugal pump designed so that both the electric motor and the pump can operate under water. This pump is intended for placement directly in the well and is used primarily for deep wells. However, it can also be used for shallow wells. Water is drawn into the unit through screened openings located between the motor and the pump. A single discharge pipe is connected to the top of the pump and runs to the storage tank, which is usually found in the lower level of the house. When inspecting the pumping system, you will not see the pump, only the accessory equipment. (See FIG. 13-14.)

Because the electric motor is located in the well, in those areas where electrical storms are frequent, it is advisable to have a lightning arrester at the motor power supply. This will conduct high-voltage surges from the line to the ground before they enter and damage the motor. Submersible pumps have the advantage of quiet, dependable operation. They are relatively maintenance-free and are more efficient than jet or piston pumps. However, if a problem develops with the pump or motor, the entire unit must be withdrawn from the well.

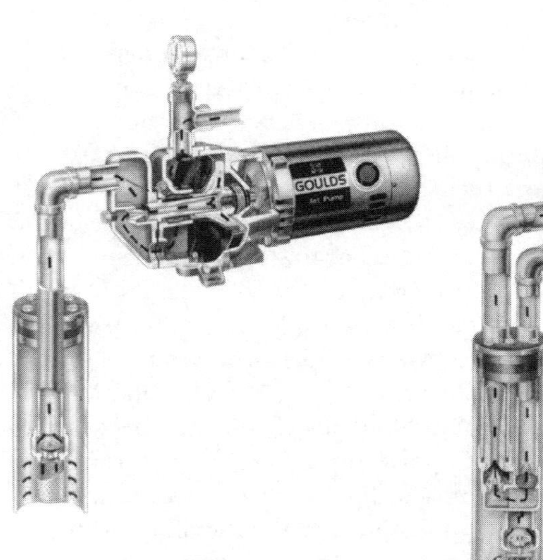

Fig. 13-13. *Jet pumps: left—shallow-well jet pump, one pipe extending into well; right—deep-well jet pump, two pipes extending into well. Gould Pumps, Inc.*

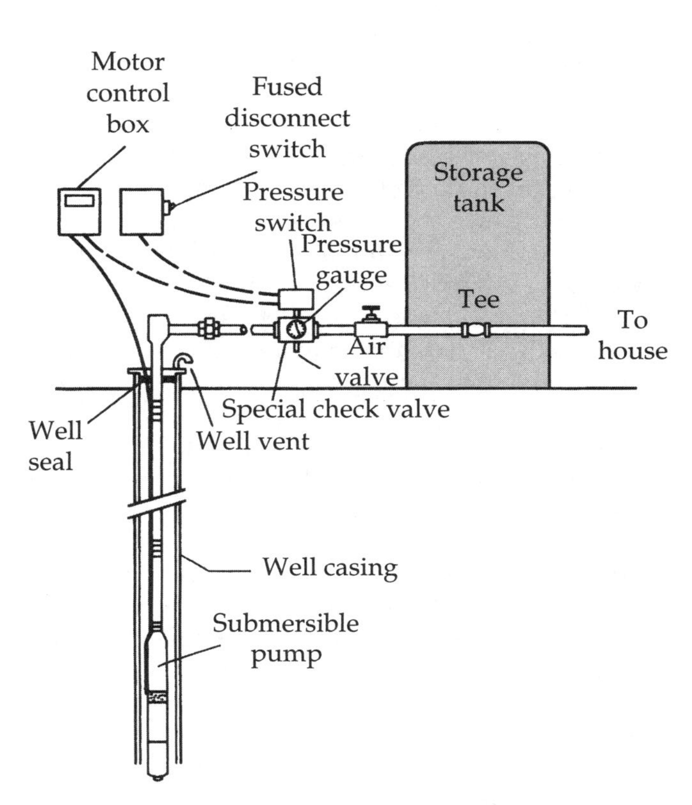

Fig. 13-14. Typical submersible pump installation. Only the controls and storage tank are visible, since the pump is located in the well.

Accessory equipment

For a private well-pumping system to provide water service comparable to that offered by a utility company, accessory equipment is needed.

Storage tanks The pump discharge line must be connected to a storage tank. The tank, also called a *pressure tank*, is generally located on the lower level of the house but might also be located in an outside pump house. Water from the tank is forced into the house supply pipe whenever there is a demand at one of the plumbing fixtures. A properly functioning tank provides a water reservoir that balances the capacity of the pump against the usage demand. It prevents excessive short-cycling (too rapid starting and stopping), which can cause switch and motor trouble.

The water in the storage tank is under pressure. Since water cannot be compressed, the tank must be partially filled with air to function properly. Over time, the water in the tank absorbs the air, so that the tank eventually becomes completely filled with water. The tank is then "waterlogged," and the pump performs as if no tank were used. Any small request for water, such as filling a glass, will cause the pump to cycle rapidly. This in turn will cause premature wear on the pump, motor, and switch.

The pressure range normally used for well-pumping systems is between 20 to 40 or 30 to 50 pounds per square inch (psi). If a waterlogged condition exists and there is a demand for water, you will hear the pump starting and stopping rapidly if the pump is the jet or piston type. However, when a submersible pump is used, you will not hear the pump. In this case, you can tell that a waterlogged condition exists because the pressure switch will be clicking on and off. Also, the pointer on the pressure gauge will be fluctuating between the high- and low-pressure limits. If you find a waterlogged tank during your inspection, record it on your worksheet. The condition can be easily corrected by draining and then injecting air into the tank.

Some pumping systems have an air-charger apparatus that introduces air into the tank with each cycle to avoid waterlogging. Some prepressurized tanks claim to achieve a permanent separation of air and water by means of a plastic or rubber diaphragm or bag. These tanks should also be checked to determine whether they are waterlogged. I have found that on occasion they are, a condition that indicates a faulty diaphragm or bag.

In most areas of the country, the storage tank should be insulated to prevent condensation during the summer. Some tanks show signs of deterioration because of excessive rusting, a condition brought about over the years by moisture condensation on the tank. Depending on the degree of deterioration, the tank might have to be replaced or scraped, repainted, and insulated.

Pressure switch, gauge, and relief valve All systems must have a pressure switch and pressure gauge. The switch automatically starts and stops the pump at predetermined pressures. The pressure differential between start and stop is usually about 20 psi. The normal pressure range, as mentioned earlier, is 20–40 psi or 30–50 psi, sometimes 40–60 psi. Pressure in excess of 65 psi is abnormal and should be checked out by a pump service company. It might be caused by a faulty pressure switch.

The switch can easily be checked by turning on a faucet. Look at the pressure gauge and watch the pressure drop until the pump is activated by the low limit. Have someone turn off the faucet while you watch the pressure building up on the pressure gauge. When the upper limit is reached, the pump should stop. If it does not or it cuts out at too high a pressure, there is a problem. If the pressure gauge is broken, this test cannot be performed. All too often, I have found inoperative pressure gauges. If you find one, it should be replaced.

Standard tanks are normally rated for a maximum pressure of 75 psi. As a precautionary measure, there should be an automatic pressure relief valve on the storage tank or associated piping. In many jurisdictions it is a code requirement. The relief valve will prevent problems associated with excessive pressure buildup if the pressure switch malfunctions and allows the pump to continue running.

Look at the pressure gauge to see if the system will hold pressure when the pump is not running and no water is being used. If the pointer on the gauge drops, it indicates a leak. If no signs of leaks were noted during the plumbing inspection, the leak is probably between the storage tank and the well. Note this item on your worksheet, for it must be corrected.

General considerations

When the yield of a well is less than 5 gallons per minute (gpm), some municipalities require an auxiliary storage tank from which water can be drawn during periods of peak demand. The rate at which water will be used in a home can vary from 1 gpm (rinsing hands) to a peak rate of 12 gpm or more, depending on personal habits and plumbing fixtures available. The approximate rates at which the various home fixtures use water is shown in TABLE 13-1. This assumes adequate-size distribution piping.

Table 13-1. Water use rates (gpm).

Bathroom sink (lavatory) faucet	3
Water closet (toilet)	4
Bathtub	5
Shower	5
Dishwasher	2
Washing machine (laundry)	5
Garden hose	3
Lawn sprinkler	2

Table 13-2 shows the approximate water-supply requirements of home fixtures.

Obviously, all of the fixtures will not be in operation at the same time. Nevertheless, for a home with two full bathrooms, the pumping system should be designed so that it can supply a peak of about 10 gpm, even if this is greater than the yield of the well. If you find an auxiliary storage tank or the controls for a storage tank (the tank might be buried), you should try to find out the design criteria for the well-pumping system. You might find that the water flow at the design peak demand is less than you require. The design criteria might be known by the seller or might be available through local health department records or the company that installed the system.

Table 13-2. Water supply requirements.

Filling bathroom sink (lavatory)	2 gallons
Filling average bathtub	30 gallons
Each shower	Up to 60 gallons
Older water closet (toilet)	4–5 gallons per flush
Newer toilets	1.6 gallons per flush
Dishwater	3 gallons per load
Washing machine (laundry)	Up to 50 gallons per load

*This figure will, of course, vary with each individual. Also, shower heads are available that will reduce the flow rate.

Checkpoint summary

Exterior inspection

❍ Did you note any vent stacks that
  — terminate near windows?
  — run up an exterior side of the house (in northern climates)?
  — have TV antennas and so on strapped to them?

❍ Is the drainage system connected to a municipal sewer, a septic tank, or a cesspool?

❍ Do you know where the septic system or cesspool is located?

❍ If house is connected to a septic tank, has the tank ever been cleaned? When?

❍ Did you note any wet spots or any foul odors in the area of the septic system?

❍ Are there any areas where liquids are oozing from the ground?

❍ Does property contain a lawn sprinkler system?

❍ Is sprinkler water-supply line protected by a vacuum breaker?

Interior inspection

Fixtures (operation and condition)

❍ Check all plumbing fixtures for operation.

❍ Note cracked, chipped, or stained areas.

❍ Are sinks or bowls loose?

❍ Do faucets leak around handles or spouts?

❍ Do sinks, bowls, tubs, and showers drain properly, or are they sluggish?

❍ Do sink and tub drains open and close properly?

❍ Are there any missing or inoperative "pop-up" units?

❍ Does toilet bowl fill and shut off properly?

❍ Do any fixture drain lines leak, have makeshift patches or missing traps?

❍ Do fixtures have individual shutoff valves on supply lines?

Water pressure, flow

- ❍ Check individual fixtures for low hot- or cold-water flow.
- ❍ Is water flow adequate?
- ❍ Note if there are knocks (water hammer) when faucets are opened and closed rapidly.
- ❍ Note any fixtures with galvanized iron piping or kinked lines (copper).

Piping

Inlet service

- ❍ If water is supplied by a utility company, locate the inlet pipe and the water meter if any.
- ❍ Is the inlet pipe made of iron, brass, copper, or lead?
- ❍ If inlet pipe is lead, take a water sample for analysis.
- ❍ Is there a master shutoff valve? Check its operation.

Distribution piping (supply mains, fixture risers)

- ❍ Are these pipes copper, brass, galvanized iron, plastic, or a combination?
- ❍ Are there signs of leakage, patched or corroding pipe sections or valves?
- ❍ If system is basically brass, note any mineral deposits along the undersides of pipes or around threaded joints.
- ❍ Pipes located in an unheated area such as a crawl space, garage, and so on may be vulnerable to freezing and should be insulated.
- ❍ Are any pipes sweating?
- ❍ Are any pipes improperly supported?
- ❍ Are hot- and cold-water lines adequately spaced apart?

Drainage pipes

- ❍ These pipes are generally made out of cast iron, galvanized iron, copper, lead, or plastic.
- ❍ Look for low points or sagging sections where solid wastes can accumulate.
- ❍ Are any visible drainage lines improperly pitched?
- ❍ Note any signs of leaking, cracked or patched sections.
- ❍ Is there a sewage ejector tank?
- ❍ If there is, is it operational?

Well-pumping systems

- ❍ Have a water sample analyzed for contamination.
- ❍ Is there a deep- or shallow-type well?
- ❍ Is well pump a piston, jet, or submersible type?
- ❍ Do you know the design criteria (gallons/minute) of the system?
- ❍ Are installation records and recorded flow available?

Accessory equipment

- ❍ Is storage tank insulated?
- ❍ Are there signs of rust or corroding areas?
- ❍ Does tank contain a pressure-relief valve?
- ❍ Is pressure gauge operational?
- ❍ When system is active (pumping), note the pressure differential.
- ❍ Does the pressure exceed 65 psi?
- ❍ Does gauge fluctuate rapidly, or does pump cycle on and off?
- ❍ Does system hold its pressure when all faucets are shut and there are no interior plumbing leaks?

14

Heating systems I

Heating systems fall into two principal categories—central heating and area heaters. In a central heating system, warm air, hot water, or steam is generated in one location of the house and is distributed through ducts or pipes to heat other portions of the house. An area heater is basically a space heater and is used to provide warmth to the room in which it is located, as with a fireplace or potbellied stove. The area heaters of today, however, are much more sophisticated and are equipped with temperature and safety controls.

The principal energy sources used in heating systems are gas, oil, and electricity. To some extent, coal, wood, and solar energy are also used for heating residential structures.

Central heating systems

Most of the homes in the United States have central heating systems. The basic components of these systems are

- A burner for converting gas or oil to heat or a resistance coil for converting electrical energy to heat.
- A heat exchanger for transferring this heat to the air or water. When the heat exchanger is used to produce warm air, it is a *furnace*. When it produces hot water or steam, it is a *boiler*. Because many people have unknowingly been calling a boiler a furnace, the term *furnace* is pretty much the generic name for the heat exchanger.
- A distribution system, consisting of ducts or pipes for conveying the warm air, hot water, or steam to the various parts of the house.
- Heat outlets such as registers (vents) or radiators for transferring heat into the room.
- Automatic safety and temperature controls.

A central heating system provides heat to the rooms throughout the house and in some cases to the nonhabitable areas such as the garage or unfinished basement. In many cases, the system can be extended to provide heat to additions or modifications to the house such as a finished attic or a new dormer. Whether the heating system can be extended depends on its heating capacity and the configuration of the distribution system. If you are thinking about extending the heating system of the house, you should consult a professional to determine its feasibility.

One advantage of central heating is that the distribution system can be designed so that the house is divided into separate heating areas, *zones*. A multizoned house is more economical and more efficient to heat. Zoning is used to maintain the same or different temperatures in various parts of the house. Only those rooms that require it need be heated. Zone control is automatic. Each zone has a thermostat control that opens and closes valves for hot-water and steam systems or dampers for a warm-air system. If the house was originally designed with only one heating zone and you are considering converting it to a multizone system, have the conversion plans checked out by a professional. The configuration of the distribution system might be such that conversion to a multizone system is not economically justifiable. Reaching the break-even point on your investment (cost versus fuel savings) can take many years.

One of the problems of a central heating system is distributing the heat evenly to all parts of the house. The larger the house, the more difficult it is to obtain an even distribution. Often the registers or radiators farthest from the furnace or boiler do not supply as much heat as those that are closest. Minimizing this problem is called "balancing the heating system" and requires the use of dampers for warm-air systems, throttling valves for hot-water systems, and certain types of air valves for steam systems. Balancing the heating system to your family's requirements is best performed after living in the house for a while. It might be that the system is already properly balanced for your needs and requires no further adjustment. Balancing is discussed in this chapter in the sections pertaining to the various systems.

Heating outlets: registers and radiators

The most effective location in a room for forced-warm-air registers and hot-water and steam radiators is along the exterior wall near windows or doors. This enables the heated air to mix with the cold air that very often infiltrates into the interior through the joints around windows and doors. Of course, if the joints are properly caulked and weatherstripped, the air infiltration on a cold, windy day will be minimized. Most homes, however, are not adequately caulked and weatherstripped. (See the section on caulking and weatherstripping in chapter 19.) The mixing of warm air with the cold air around the exterior walls will eliminate cold spots, reduce drafts, and produce a more uniform heat distribution in the room. If the registers or radiators are not located along the exterior walls in any of the rooms, the overall heat distribution in those rooms might be less than what you consider desirable.

All central heating systems have advantages and limitations. If you feel that the type of heating system is crucial to your decision on whether to buy the house, these advantages and limitations are important. Your decision, however, should be based on fact and not hearsay. I have had many clients tell me that they would consider only a house heated with a hot-water system because a warm-air system is "too dry." Well, all heating systems are "too dry" unless the air is intentionally humidified. I have been in homes heated with a hot-water system where the humidity was less than 10 percent, and I have been in homes heated with warm humidified air where the humidity was over 30 percent. Since a hot-water system does not have ducts, the house cannot be humidified from a central location as with a warm-air system. The advantages and limitations of various heating systems are discussed later in this chapter.

Thermostat and master shutoff

There are many types of controls for heating systems. The ones most familiar to homeowners are the thermostat and the master shutoff. The thermostat is used to turn on or shut down the heating system on an as-needed basis automatically. It is a temperature-sensitive switch that normally operates at low voltage (24 volts), although some operate at line voltage (110 volts). As the temperature drops below the thermostat setting, contacts within the thermostat close, activating the heating system. When the temperature in the room containing the thermostat rises above the setting, the contacts open, shutting down the heating system.

In some thermostats, the contacts are exposed to air and dust and should be cleaned periodically. Otherwise, a dust layer can form on the contacts that prevents them from operating properly. Over the years, the contacts in some thermostats become worn because of cleaning and no longer close properly. These thermostats require replacement. In newer thermostats, the contacts are encased in a glass enclosure or have been replaced by a sealed mercury switch.

In a house with a multizoned heating system, each zone will be controlled by a thermostat. The placement of the thermostat is quite important. Since the thermostat senses only the temperature in the surrounding area, the placement of the thermostat must be such that the temperature in that area is representative of the temperature for the entire house or for that zone. The thermostat must never be placed in a draft or in an area where air circulation is blocked.

Research has indicated that by lowering the thermostat setting 5 to 10 degrees prior to going to sleep and resetting it in the morning, you can save 5 to 15 percent of your fuel bill, depending on your geographical location. If you want to take advantage of these savings but feel that you might forget to lower the thermostat setting each night, you can replace the regular thermostat with a clock thermostat. A clock thermostat will automatically lower the temperature setting each night and raise it each morning. From a convenience point of view, a clock thermostat is very worthwhile. If the house does not have a clock thermostat, you should consider its installation. Some units allow for a double setback, which is useful in houses that are empty during the day.

Every heating system should have at least one master shutoff switch. Usually, the switch is located near the furnace or boiler. Sometimes the switch is located at the top of the stairs that lead to the basement. In the event of a problem with the heating system, this switch can be used as an emergency shutoff for the burner. The switch is also used by a repairman when servicing the system. By turning the master switch off, no one can inadvertently turn the heating system on by raising the thermostat.

Warm-air systems

Air heated in a furnace travels via supply ducts to the rooms. The warm air enters the rooms and is discharged through wall or floor registers or ceiling diffusers. The cooler air in the room, being displaced by the heated air, travels through return ducts back to the furnace where it is reheated and recirculated. If the system does not have cold-air return ducts, the cold air travels back to the furnace via gravity. Usually the stairway allows the cold air to travel to the first floor, and grilles in the floor of the first level then allow it to recirculate to the furnace. Of course, doors must be open to permit the air movement. Where doors are closed, the heat rises in a very restricted manner.

The air in a warm-air system can be supplied to the furnace by recirculating heated air through ducts or by drawing in cool air from the basement. The latter method is no longer used in modern warm-air systems because of the potential danger of exposure to poisonous gas and because of the heating inefficiency. If the chimney is clogged, the exhaust gas, which contains poisonous carbon monoxide, backs up into the basement. It is then drawn into the furnace and distributed throughout the house. Also, since the temperature of the basement air is lower than the temperature of the recirculated air, more fuel will be required to heat it to the desired temperature. Although this method of supplying air to the system is for the most part no longer used, it will still be found in many older systems. You can easily spot it. Look for a large opening in the furnace casing.

There is only one condition under which a furnace or its heat exchanger must be replaced: when the walls in the heat exchanger that separate the circulating air from the hot exhaust gases deteriorate because of age, premature corrosion, or cracks and thus allow the exhaust gases to mix with the circulating air. Included in the exhaust gases is carbon monoxide which is poisonous. The mixture of air and exhaust gases circulating around the house is quite dangerous. The "life expectancy" of a furnace refers to the average number of years of usage that can be expected before the walls of the heat exchanger deteriorate. For many modern furnaces, the projected life is between fifteen and twenty years, although some older ones have been safely operational for well over thirty years. A number of manufacturers are now providing replacement heat exchangers so that the overall life expectancy of the furnace can be extended indefinitely.

Advantages

Warm-air systems have an advantage over other types of heating systems in that the air in the house can be cleaned (dust particles removed by filtering) and humidified. Most systems use either inexpensive disposable filters or permanent-type filters that require periodic washing. Some systems utilize an electronic filter, which is very effective in removing dust and pollen from the air.

Not all warm-air furnaces are equipped with a humidifier for adding moisture to the circulating air. If you do not see one, you should consider installing one. The humidifier may be mounted in the main return or supply duct and is usually located near the furnace. Humidifiers such as the evaporative-plate or wick type add some moisture to the circulating air. However, they are not totally effective. A more positive introduction of moisture into the airstream can be achieved with a power spray humidifier that is controlled by a humidity-sensing device.

Additional advantages of a warm-air system:

- Adaptability to a central air-conditioning system. (See chapter 17.) The distribution ducts and the furnace blower can be used to circulate the cool air. This results in considerable cost savings when installing a central cooling system.
- There are no distribution pipes to freeze and burst. Consequently, if the heating system is not operational for several days during the winter, as would be the case in the event of an extended power failure, there would be no need to worry about the distribution system freezing (a condition that can occur with a hot-water heating system). Of course, regardless of the type of heating system, during an extended power failure, the domestic water pipes are vulnerable to freezing.
- The replacement cost for a new warm-air furnace is less than the replacement cost for a new hot-water or steam boiler.

Disadvantages

The major disadvantage of a warm-air system is that in the event of a faulty heat exchanger, the exhaust gases will mix with the circulating air and be distributed around the house.

In a multizoned warm-air heating system, the zones are not totally independent of one another. The zones are controlled by motorized dampers located in the ducts. When the dampers are closed, they block the airflow in their respective ducts, preventing that portion of the house from being heated. However, some air will always flow around the closed damper, decreasing the overall efficiency of operation. In quality-constructed homes, zone control is often obtained by using two separate furnaces with separate distribution systems rather than by using motorized dampers.

Gravity warm air

A gravity warm-air system is a very simple system that is often found in older homes. Generally, the furnace is quite large and often looks like a mechanical "octopus" with many ducts sticking out of the upper portion. (See FIG. 14-1.) There are no moving parts, motors, or electrical connections other than those required for the thermostat and burner control. The basic principle of operation is the fact that warm air rises and as it does

Fig. 14-1. *Left—Old "octopus" warm-air furnace; system was converted from gravity to forced warm air by installing a blower unit, which is located in the black casing on left side; right—Heat-supply ducts at top portion of "octopus" furnace.*

displaces the cool air, which in turn results in air circulation.

The heat distribution in a house heated by a gravity system is often less than desirable. To obtain a uniform heat distribution, there must be good air circulation. And in order to have good circulation, there must be a considerable temperature difference between the warm air entering the room and the cooler air in the room. As the room begins to warm, the temperature difference decreases, thereby reducing the air circulation.

The cooler air returning to the furnace is drawn into the return duct by a natural draft that is the result of thermal air currents. The moving air does not have much force. Consequently, there will usually not be a filter in this system for cleaning the air. The resistance offered by a filter (especially a dirty one) is often enough to block the airflow.

Although a gravity system is considered obsolete, it can be updated to a forced warm-air system by installing a blower unit (fan and motor assembly). For the most part, the old gravity systems are quite rugged and can last for many years. I have seen many units that are over fifty years old. They are usually made of heavy-gauge cast iron or steel and can probably be used for many more years, assuming that efficiency of operation is not a consideration. As mentioned earlier, the only time a furnace must be replaced is when there is a crack in the heat exchanger that allows the exhaust gases to mix with the circulating air.

Forced warm air

Because it is economical to install and is versatile, the forced warm-air heating system is found in more homes than any other central heating system. The basic difference between this system and the gravity warm-air system is a blower in the heat exchanger that circulates the warm air. Since the warm air is distributed under a draft, comfort heating can be achieved at lower furnace temperatures with lower fuel consumption. In addition, the supply and return ducts need not be as large as those of a gravity system.

Controls There are three basic controls for this system: a thermostat, a fan control, and a high-temperature limit control. The thermostat has been discussed earlier in this chapter. The purpose of the fan control is to prevent the fan from circulating cool air around the house. The fan control is a temperature-sensitive switch that turns the blower on and off at preset air temperatures. It is independent of the thermostat. When the thermostat calls for heat, only the burners should fire. The fan should not begin to operate. If it does, the fan control is either faulty or in need of adjustment. After the heat exchanger warms up to a temperature of about 110° F to 120° F, the fan should begin to operate. When the temperature setting on the thermostat is satisfied, the thermostat will shut off the burner but not the fan. The fan will continue to operate until the temperature in the heat exchanger drops to about 85° F.

If the heat exchanger gets too hot, the high-temperature limit control will shut off the burner. The limit control is usually set at about 175° F. For proper operation, the fan should begin to operate before the burners are shut off by the limit control. Otherwise, the temperature of the air discharging from the registers will be too high for comfort heating.

Condensing furnaces Because of the need to conserve energy and the escalating cost of fuel after the energy crunch of the 1970s, the heating industry developed a new generation of gas-fired forced-warm-air furnaces. These furnaces have been able to achieve an overall operating efficiency of 90 to 97 percent, whereas many of the conventional furnaces found in most homes today have an overall operating efficiency of around 60 to 65 percent. Because the U.S. Department of Energy set a standard in 1992 that requires furnaces to have an overall efficiency in the 80 percent range, conventional furnaces are no longer installed in new homes or as replacement units.

The new furnaces are called *condensing furnaces*. The increase in efficiency is the result of the installation of a secondary heat exchanger that extracts heat from the exhaust gases that normally flow up the chimney with conventional furnaces. During this process, the temperature of the exhaust gases drops to a point where the water vapor in the exhaust gases condenses, thereby releasing additional heat. With the conventional furnace the temperature of the flue gases is about 450–550° F. With the high-efficiency units, it's about 120–130° F.

To achieve the increased efficiency, it was necessary to incorporate additional components into the overall furnace package. Included with the furnace are a power vent fan (also called an *induced draft blower*), plastic piping to vent the flue gases through a side wall or the roof, condensate drainage piping, and an intake air duct for those units that have a sealed combustion system using outside air.

A power vent fan is needed to overcome the additional resistance to the flow of the exhaust gases caused by the secondary heat exchanger. Also, with the low temperature of the exhaust gases, it is not necessary to use the conventional chimney.

Pulse-combustion furnace Another type of high-efficiency furnace is the pulse-combustion furnace. As with the condensing furnaces, the high efficiency of a pulse unit results from the extraction of heat from the exhaust gases until the associated water vapor condenses. However, the combustion process is completely different. Whereas the heat in a condensing furnace results from a continuous burning of fuel, in a pulse combustion furnace, it results from sixty to seventy tiny explosions of a gas-air mixture per second in the combustion chamber.

Distribution systems There are two basic configurations used for the distribution of a forced-warm-air heating system—the *extended-plenum* and the *radial* configuration. (See FIG. 14-2.) Regardless of the configuration, there should always be a physical separation from the furnace at the beginning of the duct. The separate sections are usually connected by a heavy canvas fabric. (See FIG. 14-3.) The separation is intended to isolate the distribution system from the furnace so that blower noise will not be transmitted throughout the house. The isolation fabric in many older homes contains asbestos. If you suspect the fabric contains asbestos, recommend that it be tested and replaced if necessary. (See chapter 20, Environmental concerns.)

In the extended-plenum configuration, a large rectangular supply duct extends in a straight line from the plenum mounted on the furnace. From this main duct, branch ducts supply warm air to the various rooms. The large supply duct results in less resistance to airflow and produces a more effective heat distribution for those rooms farthest from the furnace.

In the radial configuration, there is no main supply duct. Each branch duct takes off directly from the plenum and runs to the individual room registers. This system is usually found in smaller houses. A variation on the radial configuration is the perimeter-loop-duct arrangement, which is intended for houses built on a slab. This configuration uses a duct that encircles the perimeter of the floor slab and is connected to the furnace by feeder ducts. (See FIG. 14-4.)

The branch ducts for both the extended-plenum and the radial configuration will be either rectangular or round. The round ducts are usually relatively small in diameter (4 to 6 inches), have a higher resistance to airflow, and are normally not used for air-conditioning. (Do not confuse these ducts with the insulated flexible ducts used in many central air-conditioning systems.)

The return duct for both configurations is usually made of sheet metal and has a rectangular cross section. However, in some homes with basements, you might find that a portion of the return duct has been formed using a section of the overhead wood framing. This is done by covering the channel that is formed by adjacent joists with sheet metal. (See FIG. 14-5.)

Supply registers and return grille As discussed earlier, the most effective location for the warm-air supply registers is along the outside walls. If the supply register is not located along the outside wall, and often it is not, the

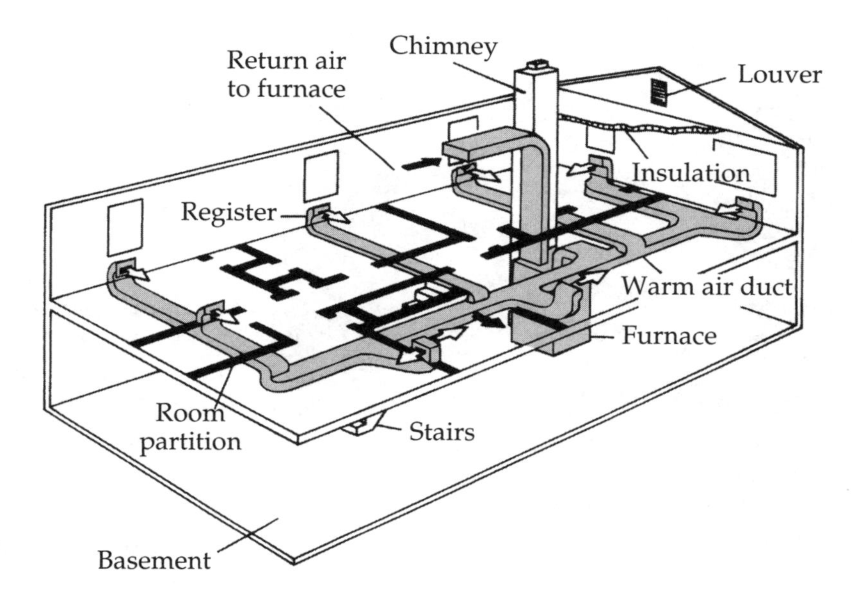

Fig. 14-2. *Warm-air distribution systems. Top—extended plenum. Bottom—radial.*

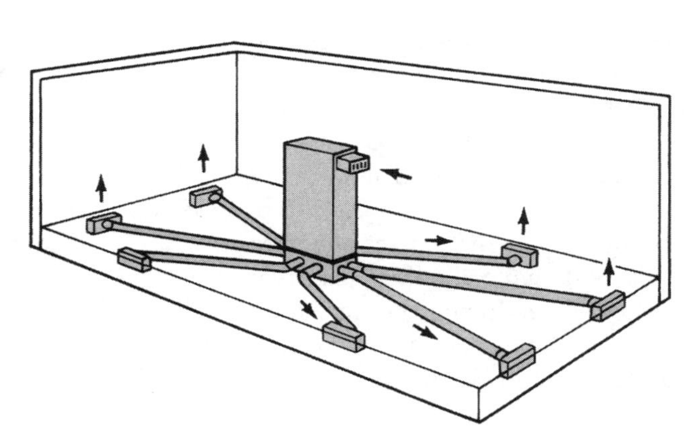

location of the return grille will be important for developing a uniform heat distribution. For optimum distribution, warm air entering the room from the supply register circulates around the room and then leaves through the return grille. If the return grille is located on the same wall as or a wall adjacent to the supply register, there will be a short cycle of the air circulation, which reduces its effectiveness. In this case, warm air discharging from the supply register can be drawn into the return grille before it has a chance to circulate.

You should look for separate return grilles for each room or a centrally located return grille for each floor. When each room has a separate return, the grille should be located on the wall opposite the supply register. If there is a centrally located return, the supply register should be located on the wall farthest from the door. This will provide a better

Fig. 14-3. Canvas fabric cover over the separation between the furnace and the main duct. The separation isolates the distribution system from noises and vibrations that may develop in the furnace.

circulation, since the air will flow through the door opening. Also, about 1 inch should be cut off the bottom of the door to allow for air movement while the door is closed.

Balancing the warm-air flow for your personal needs can best be accomplished after you move into the house. The various supply ducts will have dampers that can vary the airflow. By reducing the flow to the registers closest to the furnace, more warm air will flow to the registers farthest from the furnace. In addition, each register will have dampers that can be used for further restricting and "fine-tuning" the airflow.

Heat pump

Central heating using a heat pump is basically the same as a forced-warm-air system. However, the means by which the furnace is heated differs. The heating element is not a gas or oil burner but a component of a reversed-cycle air-conditioning system. The heat-pump system is discussed in detail in chapter 17.

Geothermal heating/cooling

A geothermal heating/cooling system is basically a modified heat pump system. It is often referred to as a ground source heat pump. Unlike an air source heat pump where the compressor/condenser is located on the outside of the house, the geothermal heat pump is generally located in the basement or utility room. It is packaged in a single cabinet that includes the compressor, refrigerant heat exchanger, and various controls. The system uses the earth as a heat source and a heat sink. It takes advantage of the earth's relatively constant temperature, at depths below 4 feet, of 40 to 55 degrees year round. The system circulates a water/antifreeze solution through a

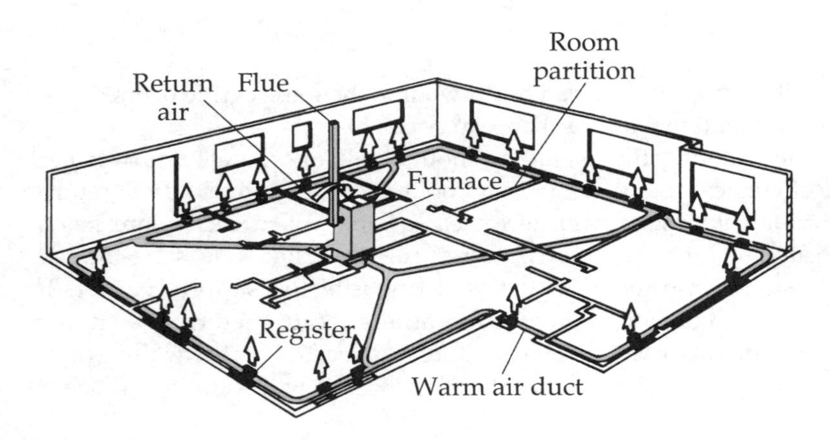

Fig. 14-4. Perimeter-loop duct arrangement, a variation on the radial configuration.

Fig. 14-5. Return duct for warm-air heating system. The duct was made by enclosing the channel formed by adjacent ceiling joists with a sheetmetal covering. Usually found in older systems.

closed loop of polyethylene pipe that is buried in the ground or set beneath the water. Depending on which cycle the system is operating under (heating or cooling), the solution adds or extracts heat from the heat exchanger in the heat pump cabinet. Geothermal heating/cooling is discussed in chapter 21.

Hot-water systems

This system operates on the principle of circulation and recirculation. Water heated in a boiler is transmitted through pipes to radiators located throughout the house. At the radiators, the hot water gives up some of its heat. The cooler water then continues to flow back to the boiler where it is reheated and recirculated.

In this system, the boiler, distribution piping, and all the radiators are completely filled with water. When heated, water expands, increasing its overall volume. Consequently, all hot-water heating systems must be equipped with an expansion tank to store the increased water volume temporarily. When the system is shut and the circulating water cools, the volume decreases, drawing the water back from the expansion tank. Without such a tank, excessive pressures could be built up in the system that could rupture the distribution pipes and fittings.

The water circulating within the heating system operates under a pressure that normally ranges from 12 to 22 psi. Although the water is constantly recirculating and there is no need for additional water, an automatic water-feed device is provided with all systems as a precautionary measure. The automatic water-feed device is a pressure-reducing valve. The water supply to the boiler is taken from the house water supply. Since the house supply pressure is normally in a range from 30 to 60 psi, it must be reduced before being introduced into the boiler. The reducing valve is usually preset by the manufacturer to 12 psi.

There are two basic types of hot-water heating systems—gravity and forced. They are classified according to the means by which the water within the system circulates.

Gravity hot water

As with the gravity warm-air heating system, the gravity hot-water system is inefficient, not very responsive to changing demands for heat, and no longer installed in new construction. However, it might be found in many older homes. The principle of operation is similar to that of a gravity warm-air system: As the water is heated, it becomes lighter than the cooler water and tends to rise. Since the system is filled with water, as the hot water rises, it displaces the cooler water, forcing it to return to the boiler for reheating, and thus induces circulation. To keep the resistance to flow at a minimum, the size of the distribution piping is relatively large—about 3 inches in diameter—compared to the distribution piping in a forced system—about 1 inch in diameter.

Gravity systems might also be classified by whether they are *open* or *closed*. In an open system, the expansion tank has an overflow pipe that is open to the atmosphere. The expansion tank must be located above the highest radiator to ensure that the radiator will be filled with water. It is usually found in the attic with the overflow pipe extending through the roof or side of the building. (See FIG. 14-6.) If the attic has been partitioned off into rooms, the expansion tank might be found in a closet or corner. When the tank is located in an unfinished, unheated area, it must be insulated to minimize heat loss and to protect against freezing should the system malfunction.

Hot-water and steam systems are normally equipped with automatic pressure-relief valves that discharge when the pressure exceeds 30 psi and 15 psi, respectively. A relief valve, however, is not needed in an open gravity hot-water system because if the pressure should build up to a point where it exceeds the design pressure, the water will simply discharge through the overflow pipe in the expansion tank.

In the closed system, no portion of the expansion tank is open to the atmosphere. In this case, the expansion tank can be located anywhere in the system, usually near the boiler. Because the system is closed, it can operate at higher pressures, and therefore higher temperatures, without turning to steam. The higher temperatures permit the use of smaller radiators.

There are two types of expansion tanks used in closed systems—air cushion and diaphragm. In the air-cushion type, air is initially trapped in the tank to provide a cushion, which is compressed as the water in the system expands and enters the tank. Over the years, the air in the tank can be absorbed by the water resulting in a "waterlogged" expansion tank. In this case, when the heating system is turned on, since there is no room for expansion, the pressure will increase rapidly until the relief valve discharges. When this happens, the expansion tank must be flushed and the air cushion reinstalled.

Fig. 14-6. *Expansion tank for "open" gravity hotwater system—located in attic.*

To eliminate waterlogging, the diaphragm expansion tank was developed. This tank uses a rubberized diaphragm to separate the air cushion from the water. For the most part, the diaphragm does eliminate waterlogging. However, where the diaphragm is faulty, waterlogging does occur.

If the heating system is gravity hot water, do not fret. It can be converted into a forced-hot-water system by installing a centrifugal pump with the associated controls to circulate the water through the distribution pipes and radiators. If the system is the open type, the expansion tank will have to be replaced with a closed tank.

Forced hot water

Quite often my clients comment on the small size of the boiler for the forced-hot-water system. They apparently are used to seeing a boiler originally designed for use in a gravity system or one that was converted from coal-fired to oil- or gas-fired, both of which are considerably larger—two to six times as large, depending on the manufacturer. Regardless of the size, you can easily recognize a forced-hot-water system by the presence of a circulating pump in the distribution return pipe just before the connection to the boiler. (See FIG. 14-7.)

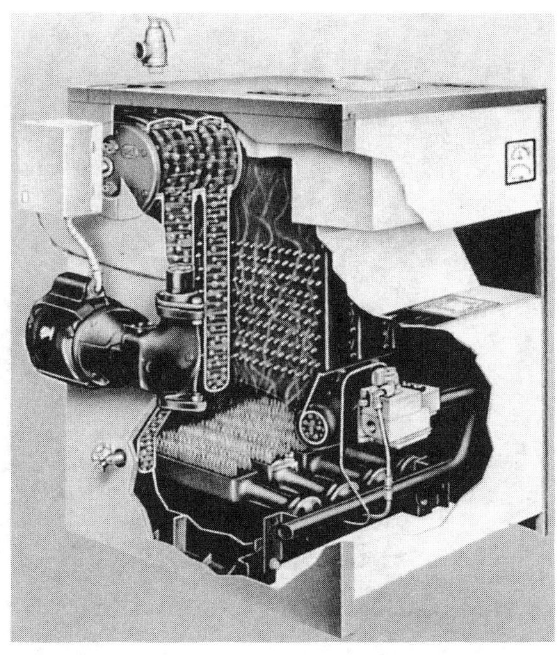

Fig. 14-7. *Boiler for forced-hot-water heating system. Note the circulating pump.*

Boilers The boilers used in these systems will be made of either cast iron or steel. Cast-iron boilers are more resistant to corrosion than steel boilers and thus have a longer projected life. The projected life span of a modern cast-iron boiler is about twenty-five to thirty years. However, many of the boiler manufacturers will provide only a twenty-year warranty. The older gravity-type boilers were probably made with a heavier-gauge metal. I have inspected many that were over fifty years old and still going strong. (See FIG. 14-8.) Steel boilers, being vulnerable to corrosion, have a shorter projected life, usually about twenty years. I have seen steel boilers that required replacement after fifteen years.

Other than for reasons of efficiency and economy, the only time a boiler must be replaced is when a leak has developed that cannot be effectively patched. Sometimes, just after firing a boiler that has not been operational for a day or more, you might see a slight amount of water dripping into the firebox (assuming that the firebox is accessible for visual inspection). This is often the result of condensation caused by cold water circulating into the boiler or some slight joint movement. As the boiler heats up, the various sections tend to move slightly. In some cases, this results in a slightly open joint that allows water to drip out. However, as the sections continue to heat up, they expand, compressing the joint and sealing the leak. Although this condition is usually not a problem, if you see water dripping into the firebox, it would be best to have the condition checked by a professional.

Fig. 14-8. *Old cast-iron boiler with new oil burner. Boiler was converted from a coal burner.*

On occasion, I have found a regular tank-type domestic water heater (of the type described in chapter 16) being used in the heating system in place of a boiler. From a safety point of view, this is acceptable because there is a high-temperature and pressure control along with a thermocouple control for the gas valve. However, water heaters have a projected life of seven to ten years and are often guaranteed by the manufacturer for only five years. In addition, with the exception of those homes located in the sunbelt, the Btu (heat) output of these units per hour is less than that needed to heat most homes adequately. This type of setup may be effective for heating an addition to a house where the existing heating system cannot be extended. The only drawback is the short projected life of the water heater.

Condensing boilers In addition to developing high-efficiency condensing furnaces, some manufacturers also developed high-efficiency boilers. As with the furnaces, the increase in efficiency results from extracting the heat from the exhaust gases that normally flow up the chimney with conventional boilers. By redesigning the heat exchanger within the high-efficiency boiler, more of the heat extracted from the exhaust (flue) gases is transferred to the water. In the process, the temperature of the flue gases drops to a point where the water vapor in the gases condenses, thereby giving up more heat. This process raises the overall operating efficiency of the boiler to about 87 percent, compared to about 60 percent for the conventional boiler.

Although boilers heat water and furnaces heat air, there are many similarities between high-efficiency boilers and furnaces. Neither requires a chimney for venting hot exhaust gases. They can vent the relatively cool flue gases through a side wall using a plastic pipe. Both require an intake air duct for those units that use outside air for combustion. Both require an induced draft blower (power vent). Both also require provisions for condensate drainage.

Pulse boiler Another type of high-efficiency boiler is the gas-fired Hydro-Pulse boiler. Its overall operating efficiency is about 90 percent. As with the boiler discussed above, the high operating efficiency is the result of extracting so much heat from the exhaust gases that the water vapor contained within condenses. However, the combustion process differs. With the gas-fired high-efficiency boiler, as well as the conventional boiler, heat is generated as a result of a continuous burning of the gas-air mixture, whereas with the Pulse boiler, heat is generated as a result of sixty to seventy mini explosions of the gas-air mixture per second. This results in a noisier operation than with a conventional or high-efficiency boiler. Consequently, vibration isolators are usually installed between the boiler and connecting pipes to minimize noise transmission.

Advantages and disadvantages Because the forced-hot-water system operates under pressure with the water being circulated by a pump, it is very flexible. It can be used to heat an area below the level of the boiler. Additional, completely independent heating zones can be readily installed. Because the hot water remains in the pipes after the boiler is no longer being heated, there is less heat fluctuation and a more even temperature distribution. Also, the operation of this system is relatively quiet.

One of the disadvantages of this system is that with water in the pipes, the distribution system is vulnerable to freezing temperatures. In the event of an extended power failure or if an oil-fired boiler runs out of oil for several days during a cold spell, the pipes can freeze and burst. I know of several cases where antifreeze was introduced into the circulating water to prevent that type of a problem. In addition, this system is not adaptable to central conditioning of the air such as cooling, humidifying, and filtering.

Distribution piping There are three basic types of distribution piping arrangements for forced-hot-water systems: series loop, one pipe, and two pipe. The specific distribution system installed usually depends on the size and cost of the house.

Series loop This is the simplest and least expensive piping arrangement to install. It is usually used in smaller homes where room-by-room radiator adjustments are not needed for balancing the heat distribution. In a series loop, the radiators, usually baseboard convectors, are an integral part of the supply piping. (See FIG. 14-9.) If a radiator is shut off, the flow throughout the system is stopped. In this arrangement, the temperature of the water entering the last radiator will be considerably less than it was when it entered the first radiator. To minimize the difference in temperature and produce a more uniform heat distribution, the heating system for larger homes is often designed so that the house is divided into two or more heating zones. Each zone has a separate piping configuration and either has a separate circulating pump or shares a common circulating pump with the other zones, but has a separate thermostatically controlled valve in the main supply pipe.

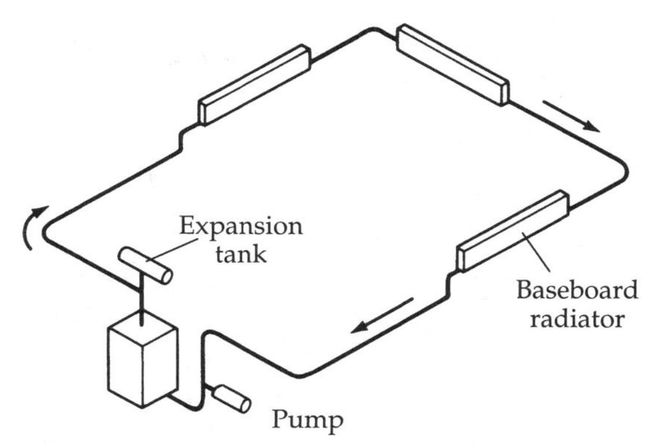

Fig. 14-9. *Series-loop piping configuration for forced-hot-water heating system.*

One pipe In a one-pipe distribution configuration, as with the series loop, a single pipe makes a complete circuit from the boiler and back again, serving as both the supply and the return. In this case, however, rather than the radiators being integral with the supply pipe, they are attached to it by two risers, one connected to each end of the radiator. (See FIG. 14-10.) Each radiator will also have a shutoff valve located at the inlet riser. Each radiator can be shut without affecting the water flow in the supply main. Consequently, this system can provide room-by-room heat control. However, as with the series loop, there is a considerable temperature difference between the water entering the first and last radiators. To compensate for the cooler water entering

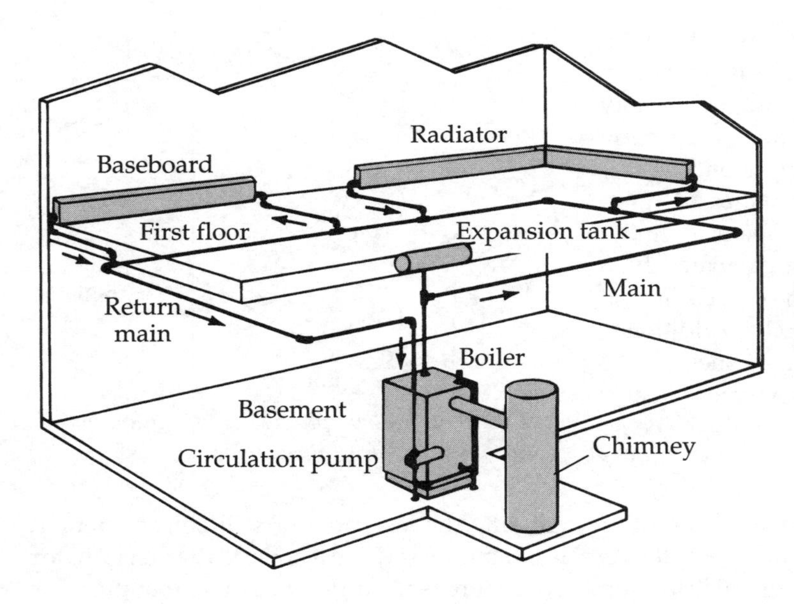

Fig. 14-10. *One-pipe distribution configuration for forced-hotwater heating system.*

the radiators downstream, larger radiators are often installed. They emit heat comparable to the smaller ones closer to the boiler.

Two pipe The two-pipe distribution configuration is the most costly to install, but it overcomes the deficiencies of the other configurations. There are two main pipes, one for the supply and the other for the return. The inlet and outlet ports of the radiators are attached to the mains by risers. (See FIG. 14-11.) The cool water leaving the radiator does not mix with the hot water flowing through the supply main, so the temperature difference between the water entering the first and last radiator is small.

Even though both the one- and two-pipe configurations will provide heat control for individual rooms if the radiator shutoff valve is manually closed or partially closed, the configurations are also often used in homes with zone heating. Zone heating is automatic. All that has to be done is to set the thermostat to the desired temperature for that portion of the house.

Radiators As discussed previously, the optimum location for a radiator is along an outside wall, preferably near or under a window. There are three basic types of radiators found in hot-water heating systems: freestanding cast iron, freestanding convector, and baseboard convector. Freestanding cast-iron radiators are found mostly in older homes. Unless an old home is being restored, a freestanding radiator is usually considered undesirable. Many clients have asked whether the radiators can be replaced with the newer finned-tube baseboard radiators. Yes, they can. Baseboard convector radiators are usually preferred because they distribute the heat better than cast-iron or freestanding convector radiators and are less conspicuous. They achieve an even temperature throughout the room because they distribute the heat near the floor. Natural convection causes the heated air to rise and warm the outside wall.

Panel heating Another method of heating using a forced-hot-water system is to embed the distribution piping into the walls, floors, or ceilings and let those areas (panels) function as radiators. (See FIG. 14-12.) The heat from the distribution piping is conducted to

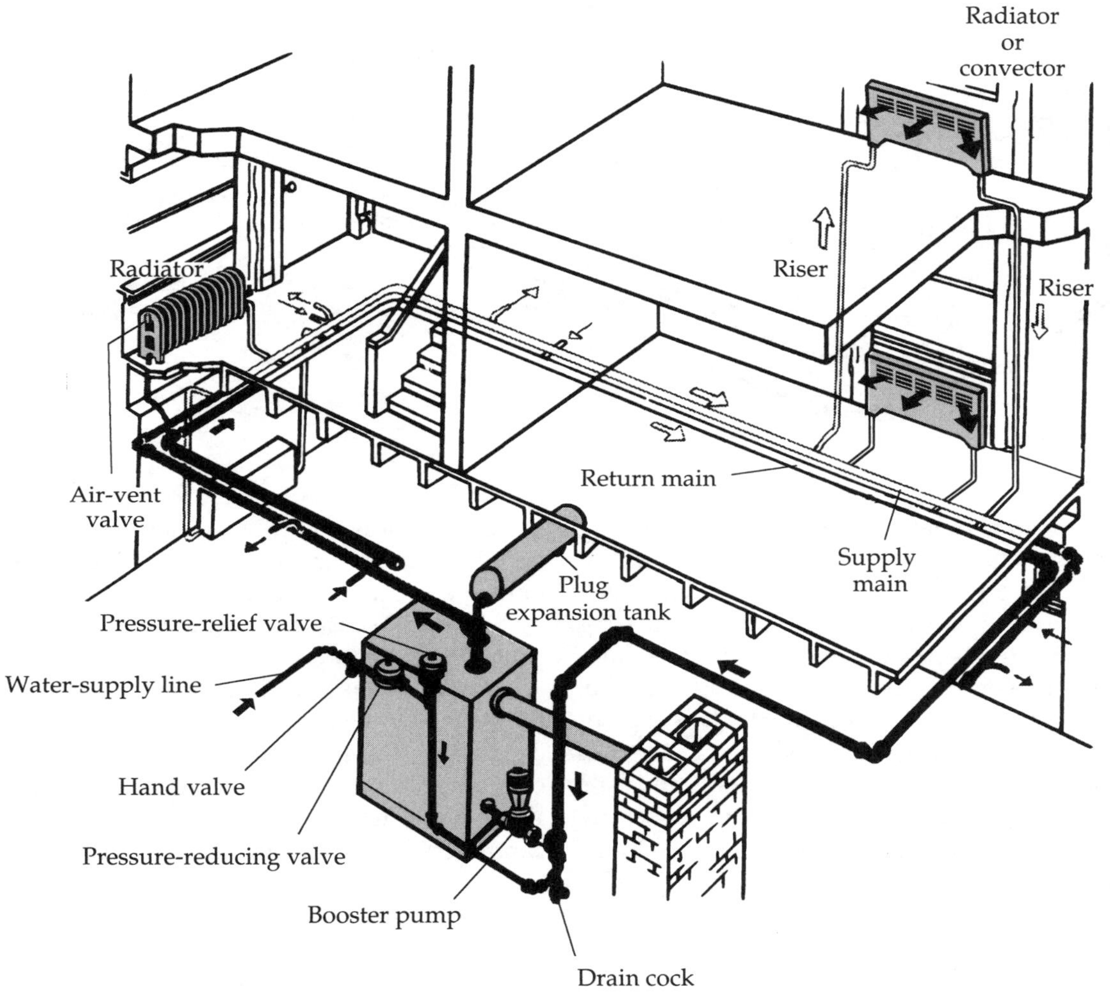

Fig. 14-11. *Two-pipe distribution configuration for forced-hot-water heating system.*

the surface of the panels; which in turn heats the room by radiation and convection. This system produces a very uniform temperature distribution and is particularly effective when the heating panel is the floor slab in a basementless house.

Controls In addition to the thermostat, hot-water systems have safety and operational controls. In a simple forced-hot-water system without individual zones, there is a high-temperature limit control that prevents the boiler water from exceeding a preset temperature and a controller for the circulating pump. Depending on the system design, the circulating pump can operate in any one of three modes: constant-running circulator, aquastat-controlled circulator, and relay-controlled circulator.

In the constant-running circulator, the pump is energized by a manual power switch

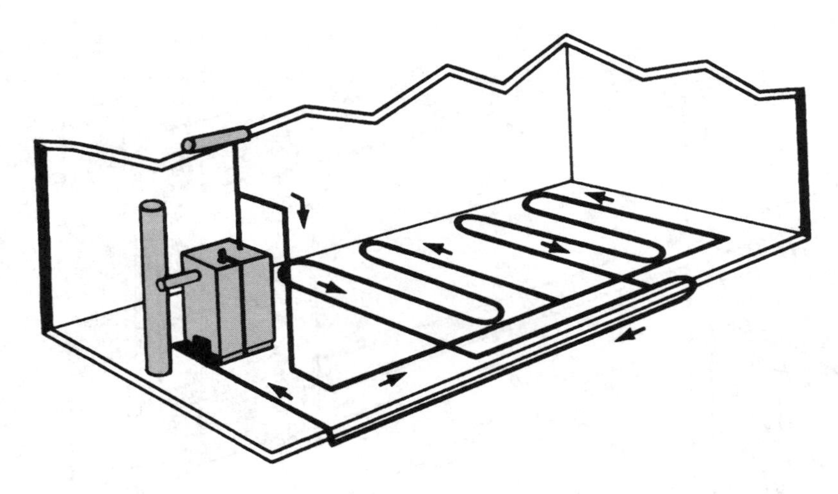

Fig. 14-12. *Forced-hot-water radiant heating panels.*

and operates continuously during the heating season. When the thermostat calls for heat, it fires the burner that heats the boiler. When the heating season is over, the circulating pump must be manually shut off. Sometimes the homeowner forgets to shut the pump, and the pump runs all summer. This type of operation is somewhat wasteful of electrical energy. However, this wastefulness must be weighed against the fact that intermittently operated circulating pumps tend to break down sooner than continuously operated pumps. The constant starting and stopping causes the bearings to wear out sooner. A constant-operating circulating pump is the least expensive to install because there is no relay or temperature controller. It can be converted to either—aquastat-controlled or relay-controlled.

In the aquastat-controlled circulator mode, as with the constant-running circulator, the thermostat will control only the burner. When the boiler water reaches a preset temperature (approximately 120° F), the circulator pump begins to operate. After the thermostat is satisfied and shuts off the burner, the circulator continues to operate until the water temperature drops below the temperature setting of the circulator control. In the relay-controlled mode, the thermostat simultaneously activates the burner and the circulating pump. When the thermostat is satisfied, it will simultaneously shut off the burner and the circulating pump.

Zone control Heating zones in a forced-hot-water system can have a separate circulating pump for each zone or a single circulating pump that services all the zones. When there is a single pump, each zone is controlled by an electrically activated valve that is in turn controlled by a thermostat. When one of the zone thermostats calls for heat, it opens the respective zone valve, fires the burner that heats the boiler water, and starts the circulating pump. After the system is operational, if another zone thermostat calls for heat, it merely opens the zone valve. This allows the hot water to circulate through that zone.

When there are separate circulating pumps for each zone, the first thermostat calling for heat will fire the burner and activate the pump. Thereafter, as long as the system is operational, other thermostats calling for heat will merely activate their respective circulating pumps.

Domestic water heater Most boilers used in hot-water systems can be equipped so that they will also heat the domestic water (water used for washing and bathing). This is discussed in detail in chapter 16. When a heating-system boiler produces domestic hot water, the associated burner must fire all year long and not just for the heating season. In this case, the thermostat does not control the burner. It controls only the circulating pump. The burner is activated by an aquastat that controls the boiler water temperature.

Since the boiler will have hot water during those periods when heat is not required, a flow-control valve must be installed on the supply main. This valve prevents the hot water from rising into the distribution piping and heating the house like a gravity hot-water system. When heat is required, the circulating pump produces sufficient force to lift the flow-control valve and circulate the hot water. In a multizoned system, if zone valves are used, flow-control valves are not needed. When the zone valves are closed, water will not circulate in the distribution piping.

Relief valve Every forced-hot-water heating system must be equipped with an automatic pressure-relief valve as a safety control. To help ensure its effectiveness, the relief valve should be mounted directly on the boiler. Many systems have a relief valve that is mounted in the boiler feed line several feet away from the boiler. This is an improper location because over the years the relief valve can become isolated from the boiler as a result of a lime or scale buildup in the feed line. If this should happen, the valve would be completely ineffective in the event of a pressure buildup. If the heating system does not have a boiler-mounted relief valve, you should have one installed.

Pressure, temperature gauge Although not safety or operational controls, all forced-hot-water systems should have a pressure and a temperature gauge. Newer systems have a combination gauge that measures both pressure and temperature. In older systems, you probably will find a separate pressure gauge and a pencil-type thermometer attached to the boiler. The combination gauge often has two pressure scales, one in pounds per square inch (psi) and the other in altitude (feet of a column of water). (The altitude scale can basically be ignored, although for your information, one psi is equivalent to a column of water measuring 2.31 feet high.) The normal fill pressure of the boiler is 12 psi. This, then, is equivalent to an altitude of 27.7 feet. If the highest radiator in your house is more than 27.7 feet above the boiler, then a higher boiler-water pressure will be needed. Some pressure gauges have two points—one fixed and one movable. The fixed pointer is usually positioned over the normal altitude setting for that house. The movable pointer shows the actual working pressure of the heating system. If this pressure exceeds 30 psi, the relief valve will discharge.

Low water cutoff An automatic water feed valve is normally installed with a hot water boiler. It provides make up water in the event of a low water condition caused by a leak in the system or a discharging relief valve. Over the years these valves can become clogged with sediment and become ineffective. Electronic low water cutoffs are now being installed on new boilers to counter this potential problem. These units have a probe that projects into the water. As long as the water is covering the probe, an electronic circuit is maintained. If the water level drops below the probe, the circuit is broken, which automatically shuts off the burner and prevents damage to the boiler.

Steam heating systems

For the most part, steam heating systems are no longer used in new residential construction, but you will find them in older homes. In fact, in some of these homes, the boiler and its associated controls may be relatively new.

The boiler used in this system is basically the same as that used in a hot-water system. It will be made of either cast iron or steel. (See the discussion on boilers on pages 221–223. You can tell whether a boiler is being used for a steam system or a hot-water system by the type of controls and gauges used. A quick indication is a water-level gauge. (See FIG. 14-15.) If you see such a gauge, the boiler is being used to generate steam.

Unlike the hot-water system that is completely filled with water, the steam boiler is only partially (about three-quarters) filled with water, depending on the size and make. The remaining portion of the boiler, the distribution piping, and the radiators will be filled with air. After the system is fired, the water heats up until it boils, as in a tea kettle. The steam thus formed rises in the pipes of its own accord, without the aid of a fan or pump, pushing the air ahead of it as it moves along. The air from the pipes and radiators is then dissipated into the rooms through air-vent valves that, depending on the piping configuration, are located on the radiator or near the end of the steam main.

When the steam comes into contact with the cool radiator surface, it condenses back into water and in the process gives up its heat. The resulting water then flows back to the boiler for reheating. If because of a blockage in an air vent (often the result of painting the vent) the air cannot be evacuated from a radiator, a pressure will be built up within the radiator that will prevent the steam from entering. In this case, the radiator will be ineffective because it will not heat up. This condition can be easily corrected by replacing the air vent.

Distribution piping

There are two basic distribution piping configurations used with steam heating systems: one pipe and two pipe. You can easily tell which configuration is being used by the number of pipes connected to the radiator. In a one-pipe configuration, there is only one pipe connected to the radiator; in a two-pipe configuration, there are two.

In a one-pipe arrangement, the steam is distributed to the various radiators through the same pipe that carries the condensate back to the boiler. (See FIG. 14-13.) The radiators used in this system must be pitched so that the condensate flows back through the supply valve. Otherwise, the condensate can accumulate and block the steam flow. Each radiator has a manually operated supply valve and an air vent. Some radiators are equipped with an air-vent valve that has an adjustable opening. This opening can be increased or decreased in size, thus allowing the air in the radiators to be evacuated at a faster or slower rate. Adjustable air vents are often found in larger homes and are used as a means of providing a uniform steam supply to all of the radiators. Radiators closest to the boiler will receive steam before those that are farther away. In some cases, depending on the distance apart and the size of the radiator, those closest to the boiler can be fully heated before those farthest away receive any steam. By decreasing the vent opening on those radiators closest to the boiler and increasing the opening on those farthest away, it is possible for all of the radiators to receive steam at about the same time.

In a two-pipe configuration, the steam is supplied to the radiator by one pipe and the condensate returned to the boiler through another. The radiators in this system are not equipped with individual air-vent valves. They have a steam trap on the condensate return pipe. A steam trap allows the air bound in the radiator and the condensate to flow in the return pipe but closes on steam contact and does not allow the passage of steam. The air in the return line is then vented by a main vent. A two-pipe steam system can be converted to a forced-hot-water system. This cannot be done

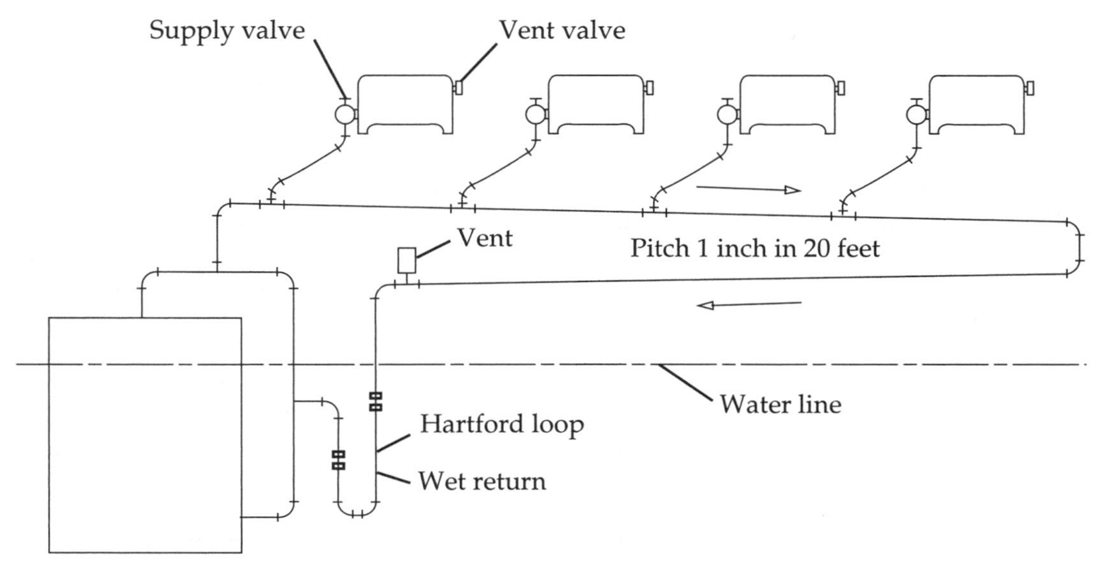

Fig. 14-13. *One-pipe distribution configuration for steam heating system.*

with a one-pipe steam system. In both one- and two-pipe systems, when the condensate is returned to the boiler, if the return line in the boiler room is above the boiler-water level, it is called a *dry return*. If the return line is below the boiler water level, it is called a *wet return*. When the system has a wet return, there should be a special piping arrangement at the boiler, a *Hartford loop*. (See FIG. 14-14.) The purpose of the Hartford loop is to prevent water from draining out of the boiler in the event of a leak in the wet-return piping. If a leak occurs in the return line, boiler water will drain down only until it reaches the top of the Hartford loop. There will still be sufficient water to prevent damage to the boiler if it continues to fire. If the heating system has a wet return, look for a Hartford loop. If you do not see one, you should consider its installation.

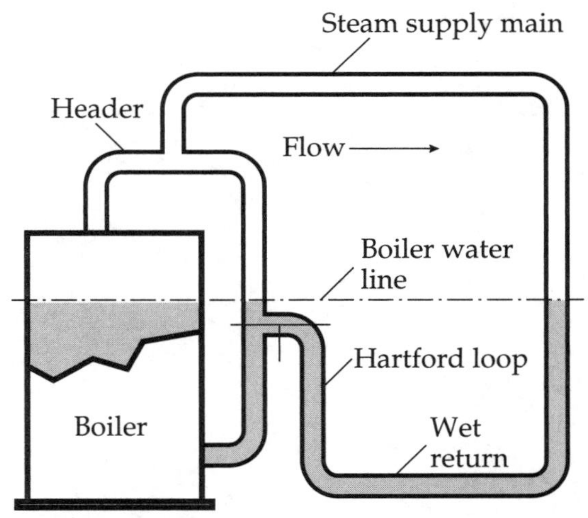

Fig. 14-14. *"Hartford loop" in the condensate return line of a steam heating system. The Hartford loop prevents water from draining out of the boiler in the event of a leak in the wet-return piping.*

Controls

In addition to the thermostat, every steam system should have a high-pressure limit switch, a low-water cutoff, and an automatic pressure-relief valve. Also, to determine whether the boiler is operating properly, there should be a water-level gauge and a pressure gauge. (See FIG. 14-15.) The high-pressure limit switch is connected electrically to the burner control. When the steam pressure exceeds a predetermined setting, the limit switch will shut down the burner, thereby preventing the pressure from building up further. The limit control should be physically connected to the boiler with a pipe that has a curl that looks like a pigtail. The pigtail has water in the bottom of the loop, which prevents the corrosive action of the steam from affecting the control.

The low-water cutoff is a control that shuts down the burner when the level of the water in the boiler drops below the design level. There are two types of low-water controls—one mounted inside of the boiler and one externally mounted. The latter is preferred because it provides a convenient means for testing its operation. This unit has a blow-off valve that when opened drops the water level in the control, causing it to operate. In the past, manufacturers recommended that the low water cutoff be opened and blown down once every month to prevent sludge accumulation, which can affect its operation. They are currently recommending that it should be blown down once a week. All too often, homeowners neglect to perform this simple operation. I have checked many a unit that apparently had not been flushed in years; when I opened the blow-off valve, either nothing flowed out or there was a thick dark-brown sludge oozing out. The low water cutoff units that are mounted in the boiler have an electronic probe that projects into the water. As long as the water covers the probe, an electronic circuit is maintained. If the water level drops below the probe, the circuit is broken, which automatically shuts off the burner and prevents damage to the boiler.

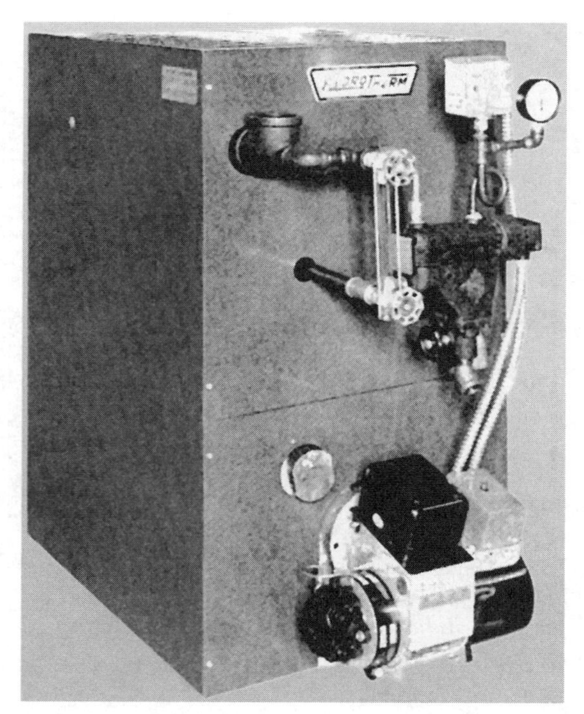

Fig. 14-15. Oil-fired steam boiler, showing the level gauge, low-water cutoff, high-pressure limit switch, and pressure gauge.

The water-level gauge provides a convenient means for determining the level of the water in the boiler. It is usually mounted on the side of the boiler; when there is an exterior-mounted low-water cutoff, it is often part of that assembly. The water level should be at the midpoint or two-thirds up the glass gauge. The exact position is not important. What is important is that you are able to see the water level. If the entire level gauge is filled with water, there is too much water in the system. In fact, the system can be flooded. If too much water is introduced into the boiler, the water level will rise until it fills the entire distribution system and radiators. If any of the radiator valves or fittings are not watertight, and sometimes they are not, the water will leak out all over the room.

When there is no water visible in the gauge, water must be introduced into the boiler. This can be done by manually opening the fill valve in the water-supply line. Some systems have an automatic boiler-water feeder that introduces water to the required level as needed. This is a desirable feature, but occasionally such units malfunction. Often level gauges are coated with sediment so that the water level is not visible. In this case, the glass gauge must be cleaned and all the accumulated sediment removed.

The relief valve is a safety valve that automatically discharges when the operating pressure exceeds the design pressure. For a residential steam system, the relief valve is set to discharge at 15 psi, although the normal operating pressure is considerably less, usually about 1 to 2 psi. Some large systems may even operate at a negative pressure (vacuum).

Domestic water heater

As with hot-water boilers, steam boilers can also be equipped for generating domestic hot water. When it produces domestic hot water, the boiler must be fired all year long. When heat is not required, the boiler-water temperature is controlled by an aquastat. The aquastat activates the burner when the boiler-water temperature drops below a preset figure and shuts off the burner when the boiler-water temperature rises to about 200° F. When steam heat is required, the burner is activated by the thermostat. In this case, the thermostat overrides the aquastat control so that steam is produced.

In a forced-hot-water heating system that also produces domestic hot water, a flow-control valve is needed to prevent the boiler water from circulating as a gravity system during those months when heat is not required. In a comparable steam system, no flow-control valve or equivalent is necessary because the boiler water is not heated sufficiently to produce steam.

Advantages and disadvantages

The advantages of a steam heating system can be appreciated more in a large building than in a residential structure. It is a relatively simple system that does not require a pump or fan for circulation of the steam. Since there is no water in the pipes when the system is not operating, there is no problem of the pipes freezing and bursting. If a repair or replacement is needed to a section of pipe or a fitting, it is not necessary to drain the system. Also, a steam leak would result in very little water accumulation.

On the other hand, a steam system is slow in responding to an initial rapid change in heat demand because the boiler-water temperature must be brought up to 212° F before steam circulation begins. Unless the condensate is returned to the boiler by means of a pump, the boiler must be located below the radiator in the lowest rooms.

Hybrid heating systems

Depending on the extent to which the heating system has been modified or expanded, you may find a hybrid heating system. A hybrid system is composed of two separate heating systems working together.

Steam–hot water

I found the following in an old house. The system functioned as a two-zone system. The main portion of the house was heated by a steam system, and the remaining portion of the house (the lower sections) was heated by a forced-hot-water system. The distribution pipe for the hot-water system was tapped directly off the portion of the steam boiler that was below the water level, and a pump was used to circulate the hot water. The thermostat for this zone controlled the pump and the burner. There was also an aquastat temperature control for the boiler water to keep it from generating steam when the main zone was not calling for heat.

Hydro-air

Hydro-air is basically a hot water–warm air system. It is becoming more popular, especially in larger homes. The system essentially consists of an air handler, which looks like the blower compartment of a furnace, a hot water boiler, and a duct system. Hot water from the boiler is pumped through a series of tubular coils that are mounted above the fan in the blower compartment. As air flows over the coil, it is heated and blown through the duct system to various areas of the house. There are a number of advantages to this system. You can easily add a number of separate zones to the boiler piping; one zone to an indirect fired water heater, another zone to a room that is a new addition and that is heated with baseboard, and a third to another air handler. Another advantage to hydro-air is that the same duct system can be used for the air-conditioning. Also, since no combustion takes place inside the air handler, there is no possibility of combustion gases being distributed throughout the house because of a cracked heat exchanger.

In large homes with more than one air handler, at least one air handler is usually located in the attic. In the event of a prolonged power failure, the water in the coils is subjected to freezing. If the home you are considering has a unit in the attic, as a precaution you should check to determine whether antifreeze has been added to the system either by a tag or a marking on the boiler or by asking the seller.

15

Heating systems II

The boilers and furnaces found in most homes today are heated with oil or gas burners or electrical resistance coils. In many older homes, you will find a boiler or furnace that was originally designed to burn coal but was converted to oil or gas for reasons of economy and convenience. The accessories associated with each heating element are described below.

Oil-fired systems

The most common type of oil burner used in residential heating is the high-pressure or gun-type burner. This unit has a pump that forces oil through a nozzle and produces an oil mist. It also has a fan that mixes the oil mist with a measured amount of air. The resulting flammable mixture is then ignited by an electric spark and burns in a refractory-lined firebox. If the mixture fails to ignite, a safety control will shut off the oil-pump motor. This control is located either in the exhaust stack as a heat-sensitive switch or in the burner as a light-detecting photocell.

The oil burner normally associated with the heating system will generally be either an older conventional unit or a high-efficiency flame-retention burner. Although both burners look somewhat the same to the untrained eye, there is a considerable difference in the flame pattern and overall efficiency. The flammable mixture of oil and air that is produced with a conventional burner is inconsistent. Some parts of the mixture will be fuel-rich and others fuel-lean. Consequently, these burners draw in a large quantity of air so that the fuel-rich portion of the flammable mixture will burn completely and thereby reduce the level of smoke produced.

The large quantity of air and the irregular flame pattern results in inefficient operation of the conventional burner. The flame-retention burner is more efficient. Because of its design, it produces a more uniform mixture of oil and air, requiring a smaller amount of air for complete combustion. This in turn produces a smaller, more compact flame that is hotter. A flame-retention burner gets more Btu out of a gallon of oil than a conventional burner. In fact, a new flame-retention burner is about 16–20 percent more efficient than a new, well-adjusted conventional burner. You can check with the company that services the oil burner to determine whether it is a conventional unit or a flame-retention burner.

The proper draft over the firebox is very important for efficient operation of the oil burner. To ensure the proper draft, most oil burners have a draft regulator mounted in the exhaust stack near the boiler or furnace. The regulator is basically a small swinging damper that can be adjusted to open an inch or two when the burner is firing. (See FIG. 16-7.) I have seen many draft regulators that have been made inoperative by someone cementing shut the opening. This can affect the efficiency of operation and result in greater fuel consumption. Also, to ensure the proper draft, the section of the exhaust stack between the boiler or furnace and the chimney must have an upward pitch.

Another type of oil burner occasionally found in central heating systems is the pot-type vaporizing burner. It contains few moving parts and therefore operates quietly. Basically, the unit consists of a pot containing a pool of oil and a control for regulating the oil flow to the pot. Air needed to produce a flammable mixture is introduced by a small fan. Once the mixture starts burning, the heat produced vaporizes the oil mixed with the air, thus maintaining the burning operation. The initial cost for this burner is less than that for a high-pressure gun burner. However, for efficient operation, it requires a more costly high-grade oil that vaporizes easily.

Oil for the burner is stored in a tank that is either buried in the ground just outside the house or located in the house not too far from the burner. The interior tank should be at least 7 feet from any flame and should have an outside fill connection. These tanks will generally have a 275-gallon capacity; buried tanks generally have a 550-, 1,000-, or 1,500-gallon capacity. Most municipalities allow a maximum of two 275-gallon tanks to be stored in the structure. Oil tanks have a projected life of about thirty years. After that time, pitting and corrosion holes tend to form on the bottom of the tank. This condition is the result of an accumulation of water and corrosive acids that settle on the bottom of the tank. Check to see if the bottom of the tank is at least 6 inches off the ground and whether the tank is at least 2 inches away from the walls. In many municipalities this is a code requirement. Run your hand across the bottom of the tank. If your hand has an oily film, the tank is leaking, even if it is not dripping oil. Recommend that the tank be replaced. Be advised that if there is a buried oil tank that is leaking, it is the homeowner's responsibility to clean it up, including removal of contaminated soil. See "Leaky oil tanks" in chapter 20.

All oil burners require periodic inspection, cleaning, lubrication, and adjustment to ensure an efficient soot- and odor-free operation. During the inspection, which should be performed at least once a year, the exhaust gas should be analyzed to determine whether the burner is operating at maximum efficiency. Preventive maintenance to the oil burner is usually provided by the company that supplies the oil. For an additional fee, oil companies usually issue a maintenance contract in which they agree to maintain and tune the system and provide emergency service. If you enter into a maintenance contract with your oil company, make sure you know exactly what will be maintained. Some companies will inspect and maintain only the oil burner and not any of the peripheral equipment or controls.

I inspected a house one summer that had an oil-fired forced-hot-water heating system that also generated the domestic hot water. The burner, of course, had to fire during the summer to produce the domestic hot water. The circulating pump, however, had not operated for a few months because heat was not needed. When I turned on the heating system (by turning up the thermostat), the circulating pump started to smoke, and I smelled insulation burning. The homeowner, who was with me at the time, said, "How can this be? The oil company just serviced the system two days ago." I called the oil company to find out what kind of maintenance was performed. They informed me that they maintained the oil burner only. They did not even look at the circulating pump. Do not let a maintenance contract lull you into a false sense of security, thinking that your entire heating system is being maintained. It might be. But it must state so in the contract.

Gas-fired systems

If you are given an option and the fixed costs are comparable, a gas burner is preferred to an oil burner. It does not require annual maintenance and is less costly to install. Gas burners have a cleaner, quieter operation, and the supply of gas is not dependent on the weather. In addition, replacement costs are lower.

In this system, gas is supplied at low pressure to the burner through an automatic gas valve. The valve in turn is controlled by the thermostat or aquastat (if the system generates domestic hot water). At the burner, the gas that is mixed with air is ignited by a pilot light. Associated with the pilot light is a safety control, a *thermocouple*. The thermocouple closes the gas valve when the pilot light goes out. Basically, the thermocouple converts heat (from the pilot light) into a small electric current that keeps an electrically operated valve in the main gas line open. When the pilot light goes out, the current ceases, and the valve closes. If the thermocouple is faulty, the pilot will not light. Rather than have a pilot light that is constantly lit, many new boilers use an electronic ignition to fire up the burner.

Some gas-fired systems are equipped with a self-energizing control unit (Powerpile) that enables them to operate in the event of an electrical power outage. This is a particularly desirable feature, especially for those areas prone to power failures. The control unit basically consists of a special pilot-thermocouple assembly, gas valve, and thermostat. It can be used on steam, hot-water, and warm-air systems. With a self-energizing control unit during a power failure, a steam system will operate normally and a forced-hot-water or warm-air system will function as a gravity system.

As with oil burners, gas burners require a small but steady draft. However, rather than using a draft regulator, gas-burning equipment uses a draft diverter-hood. (See FIG. 16-7.) The hood is usually located on the exhaust stack above a boiler and is often built into the sheet-metal casing of a furnace. A draft-diverter hood is completely open on the underside and thus prevents air currents in the chimney (resulting from downdrafts) from blowing out the pilot-light flame.

All gas-burning equipment should have the American Gas Association's seal of approval. This indicates that a similar unit has been tested for safety. Look for the seal as you inspect the equipment. It is usually located on a plate mounted on the front of the boiler or furnace.

What type of pipe is used to supply gas to the furnace or boiler? It should be a rigid black iron pipe rather than a flexible pipe or a copper pipe. Many utility companies buy their gas from the cheapest supplier, and therefore, the gas can contain impurities that will react with copper. Although most municipalities allow flexible gas pipe for movable appliances such as clothes dryers and stoves, they generally require a rigid gas-supply pipe for stationary equipment such as a boiler, furnace, or domestic water heater. If the gas-supply pipe to the heating system is other than a black iron pipe, check with your local utility company for their requirements.

Check the gas meter to see if it is adequately sized to supply gas to the heating system and domestic water heater during peak periods. An inadequately sized gas meter might be found in a house that originally had a gas-fired water heater and an oil-fired heating system. Sometimes when the oil-fired system is converted to or replaced by a gas-fired system, the installer does not notify the utility company to replace the meter, which was sized to supply gas for only the water heater and kitchen stove.

A gas meter is rated in cubic feet per hour, and one cubic foot of natural gas is approximately equal to 1,000 Btu. Therefore, a meter rated at 250 cubic feet per hour can supply gas that is approximately equivalent to 250,000 Btu per hour. You can check to see if the gas meter is adequately sized by adding up the input Btu/hour requirements for the boiler or furnace and the water heater, and comparing the total with the meter. The input Btu will generally be found on the data plate that is mounted on the outer casing of the appliance. (The gas requirement for the kitchen stove is considered negligible for this comparison). The Btu capacity of the meter should be equal to or greater than the sum of the input Btu requirements for the heating system and water heater. If it isn't, make a note on your worksheet to notify the utility company.

Electrical systems

For the most part, electrical resistance heating is used for area heaters such as panel or baseboard heaters. It is also used, although less often, for central heating. The heating mechanism in a boiler or furnace is very simple. Electrical resistance coils are immersed directly into the water (for a boiler) or airstream (for a furnace). As the electrical current passes through the coils, they get hot and directly transmit their heat to the air or water. Unlike oil and gas energy, the conversion of electrical energy to heat is 100 percent. There is no heat lost to the exhaust gas because there are no exhaust gases. There is no fuel combustion with this system. Consequently, there is no need for a chimney.

From an installation point of view, electrical resistance coils are less expensive than oil or gas burners. However, for most parts of the country, they are more expensive to operate. To keep the operating expense from becoming excessive, electrically heated homes must be well insulated. In evaluating the cost of heating a house, do not depend on the owner's assurance ("Oh, it only costs a few hundred dollars to heat the house"). Insist upon seeing the most current year's bills.

Area heaters

These units are self-contained space heaters that are used to provide warmth to a single room. Sometimes they warm up two or three adjacent rooms if the doors between the rooms are kept open. The heat distribution is not very desirable from a comfort point of view. Other than a fireplace or stove, most area heaters of today will be gas-fired or electric.

Gas-fired units

Gas-fired space heaters are basically small warm-air furnaces and can operate either as gravity or forced-air units. Some of the smaller gravity heaters have an open combustion area exposing the flames. This is a potential fire hazard and is particularly undesirable when there are small children in the house. The exhaust gases from all gas-fired space heaters must be vented to the outside. In addition, the units should have the American Gas Association seal of approval. These units are controlled either automatically by a wall-mounted thermostat or manually by a calibrated knob on the gas valve. They are inexpensive to install, relatively maintenance-free, and require only periodic cleaning and adjustment.

Electrical units

Electrical space heaters have an advantage over central heating systems in that each room is usually individually wired, so there are as many independent heating zones as there are rooms. There are three general types of electrical area heaters: panel, baseboard, and wall. *Panel* heating, sometimes called *radiant heating*, is similar in concept to forced-hot-water panel (radiant) heating, as described in chapter 14. However, instead of hot-water pipes being embedded in the walls or ceiling, electrical resistance cables are installed with each room having a separate control. Similarly, *baseboard* heaters are like baseboard convectors (see chapter 14) but they contain electrical resistance coils rather than circulating hot water. They are very popular because of the ease of installation and low initial cost. They also produce a uniform heat distribution. Some baseboard heaters have a thermostat mounted directly on the unit rather than on the wall. This is not a desirable thermostat location because in many cases it will be necessary to move furniture every time a temperature change is desired. *Wall heaters* are compact units mounted on or recessed into a wall. Most heaters operate in conjunction with a fan that blows the warm air into the surrounding area. They are usually used to provide heat in nonhabitable areas such as a garage or workshop and are sometimes used to supplement a central heating system. These heaters do not produce a uniform heat distribution but are very responsive to a call for heat. They are usually manually controlled rather than automatically controlled.

Heating system inspection procedure

A full heating system inspection consists of an evaluation of the operation of the boiler or furnace, the burner, the condition of the distribution system (wherever visible), and heat outlets—radiators or registers. During your interior inspection, each room should be checked to determine whether there is a heat outlet and if there is, whether it is properly located for maximum effectiveness. The area below the radiators should be checked for signs of leakage, and the dampers in registers should be checked for ease of operation. Distribution piping or ducts, which are often visible in the basement, crawl space, attic, and garage, should be checked for leaky joints and the need for insulation. Also, the thermostats should be checked for location (should not be in a draft), condition, and type. An automatic clock-type thermostat is more convenient and if used properly, will result in fuel savings. A broken thermostat must be replaced, and one that is loosely mounted must be resecured.

The boiler or furnace and the associated burner are inspected after completing the interior inspection. By the time you are ready to perform this inspection, you should know whether the house is heated by warm air, hot water, or steam. You can tell by the type of heat outlets—registers for warm air or radiators for hot water and steam. A radiator with one pipe attached is a steam radiator. When there are two pipes, it might be difficult to tell whether the radiator is used for hot water or steam. In this case, you should look at the boiler.

At first glance, two or more thermostats can lead to a false conclusion—that the house has a multizoned heating system. A house with a multizoned heating system will have more than one thermostat. However, a house with more than one thermostat need not have a multi-zoned heating system. The number of heating zones should be verified when you inspect the boiler or furnace. In a hot-water system, the zones are controlled by electrically operated valves, circulating pumps, or a combination of the two. For a warm-air system, the zones are controlled by electrically operated damper motors. Usually the number of thermostats indicates the number of independent heating zones. However, you might find more thermostats than zone controllers. I have inspected homes that had a one-zone heating system that was activated by two separate thermostats. When either thermostat called for heat, the entire house would be heated. Also, in older homes you might find an old nonfunctioning thermostat on a wall.

When you are ready to inspect the boiler or furnace, stand where you can see the burner, about 3 feet away. Have someone turn up the thermostat and activate the burner. It is important to be near the unit when the burner ignites to determine if there is a problem condition. A puffback with an oil burner, or flames licking back under the cover plate with a gas burner, is an abnormal condition and a potential hazard that must be corrected. Record it on your worksheet. If the unit does not fire, check the master switch. If the switch is on and there is an oil burner, push the *reset* button once. If the burner does not fire, there is a problem condition that must be corrected. If the gas burner does not fire and the pilot is lit, there is also a problem.

Once the unit is firing, check the overall condition. Is the boiler or furnace an aging unit? Do the burner and boiler or furnace appear to have been neglected? Are there signs of excessive corrosion (rust), dust, and flakes? Are there mineral deposits, indicating a past or current water leakage? If there is more than one heating zone, check each zone independently. This can be done by waiting until the burner or circulating pump (if a forced-hot-water system) shuts down and then activating the thermostat that controls the zone being checked. The thermostat should activate the burner or circulating pump, depending on the system (as described in chapter 14). Sometimes there is a time delay of about one minute or so between engaging the thermostat and activating the burner or circulating pump. This occurs because of the time that it takes to open the zone valve or damper physically. Also, some gas valves have a built-in time delay. However, if after five minutes nothing happens, there is a problem with the controls.

Check the condition of the smoke pipe. This is the horizontal section of sheet-metal pipe that connects the boiler or furnace to the chimney. In some cases, the boiler or furnace is connected directly to a prefabricated metal chimney, and there is no horizontal run. When there is, however, the pipe must have a slight upward pitch from the boiler or furnace to the chimney. If the pipe is long, it should be supported to prevent sagging sections. The pipe should not have corrosion holes, and the joints between sections should be tight. This pipe gets very hot and should not be within several inches of combustible material.

In recent years, *direct-vent appliances* (appliances that vent exhaust gas through a sidewall) have become more common. During your exterior inspection, if you see a direct-vent terminal below a window, check the distance between the top of the vent and the bottom of the window. If the terminal is too close and the window is open, flue gases could enter the house. The recommended distance for an appliance with an input of 50,000 Btu per hour is a minimum of 9 inches, and for an appliance with an input of over 50,000 Btu per hour, the vent terminal clearance is at least 12 inches. Also the bottom of the vent terminal should be at least 12 inches above the ground.

The various types of heating systems have specific items that should be checked during an inspection. Each is discussed below. At the conclusion of the heating system inspection, remember to turn the thermostat back to its original setting.

Of particular concern is whether there is an adequate supply of air for ventilation and combustion for the heating equipment. When the heating system is located in an unconfined space such as an unfinished basement, there is generally sufficient infiltration so that the air supply is usually not a problem. However, if the system is located in a confined space such as a small furnace room, provisions must be made for an adequate air supply.

To provide the needed air supply, openings must be made within 12 inches of the top and 12 inches of the bottom in one of the walls of the enclosure. Each opening should have a *net free area* of at least 1 square inch per 1,000 Btu of total input rating of the equipment in the enclosed space. The size of the opening can be reduced if the air supplied to the furnace room comes from the outdoors rather than from inside the building. If the adequacy of the air supply is a concern, you should have it checked by a professional.

Warm-air systems

This system will be either gravity or forced. You can determine which by whether there is a fan. In some of the old "octopus"-type furnaces, the fan might be difficult to locate. It might be on top of the unit 5½ feet off the ground, or there might be fans in return ducts.

After the burner has fired, wait until the fan begins to operate. It should begin before the burner is shut down by the high-temperature limit control. If it begins after the burners are shut off, the fan controller is either faulty or out of adjustment. Also, if there is a canvas connection between the furnace and the main supply duct, check its condition. Torn or open sections should be patched. The canvas connection is used to isolate the supply ducts from the vibrations and sounds that develop within the furnace.

Be advised that with some older systems, the fabric connector may contain asbestos. Since warm air is constantly flowing across the fabric when the heating system is operating, the presence of asbestos would be considered a potential health hazard. See "Asbestos" in chapter 20.

When the fan is operating, is there excessive noise or vibration? There should be none. If there is a power humidifier, it is usually wired so that it will operate only when the fan is running. Turn the humidifier on by turning the humidity control up. If there are portholes in the unit, see if the fan or drum is rotating. If there are no portholes, listen to hear if the motor goes on and shuts off when the humidity control is lowered. Very often the humidifier is in need of a tune-up. This is usually indicated by excessive mineral deposits around the humidifier and the duct to which it is attached.

When the system is operational, check the flow and temperature of the air discharging from the various heat-supply registers. Put your hand in front of the register. A weak flow can be the result of air leakage in the distribution ducts, a dirty air filter, a loose fan belt, or the need for rebalancing. When the temperature of the discharging air in one room is lower than in other rooms, it is probably caused by a heat loss in the branch duct leading to that room. The return register can be checked by placing a tissue on the grille. Because the return duct draws air in, the tissue should stay in place without falling down.

The item of main concern is the heat exchanger. When this portion of the heating system breaks down, the furnace or its heat exchanger must be replaced. When the furnace is aging or excessive corrosion is noted, the condition of the heat exchanger should be suspect. Also, if the flames in a gas-fired furnace are unstable and appear to be dancing, this often indicates a crack in the heat exchanger. It is recommended that prior to the contract closing, the furnace and associated equipment be tuned by a competent service organization. At that time, the serviceman can check the condition of the heat exchanger. Checking the heat exchanger usually requires the use of sprays, propane torches, and mirrors or disassembling a portion of the furnace and should be performed by a professional.

If the furnace is the high-efficiency condensing type, check the unit and associated piping for corrosion. Corrosion is a major concern in these furnaces because of the acidic nature of the condensate. Sulfur in the fuel oil or natural gas can combine with the flue-gas condensate to form sulfuric acid. Also, when indoor air is used for combustion, if the air contains chlorides, the chlorides can combine with the water vapor in the flue gas to form a corrosive condensate containing hydrochloric acid. Chlorides in the air can come from laundry bleach, household cleaners, even chlorinated tap water. To counter these problems, most condensing furnaces use outdoor air for combustion. They also generally use a plastic pipe for the exhaust stack and condensate drain.

Check the condensate drainpipe for blockage and loose sections. The condensate pipe should discharge into a floor drain that is connected to the house sewer. If the floor drain is not lower than the condensate drain or there is no floor drain, a condensate pump should be used. It is not good practice to have the condensate discharge into a hole in the floor slab.

The air supply should also be checked for blockage. If the supply pipe is run through the roof, the exposed end should be shaped like an inverted U with the open end facing down. This keeps the rain out, and since the supply and exhaust pipes are generally near each other, it prevents exhaust gases from being drawn into the supply pipe.

With a condensing furnace, the exhaust vent should not be combined with the exhaust from another appliance such as a water heater. The exhaust gases can be vented through a side wall, the roof, or a masonry chimney. However, if connected to a chimney, the plastic pipe should run up to the top of the chimney where the exhaust gas can vent to the atmosphere. If the exhaust pipe terminates at the base of the chimney, the corrosive condensate that forms will attack the masonry.

If the exhaust pipe has horizontal sections, see if any low spots or sagging sections could trap condensate and thereby block the exhaust gas from venting. Horizontal sections should have a slight pitch back to the furnace. This is especially important if the section is near the termination by a side wall or attic. If the exhaust pipe is sloped away from the furnace, the condensing water vapor in the exhaust gas could freeze and eventually block the opening in the pipe.

Condensing furnaces are noisier than conventional furnaces. They have a higher airflow, which is needed to absorb the additional heat resulting from the added heat exchanger. The higher airflow produces more sound and vibration. There is also a draft-inducing blower that is needed to overcome the increased resistance to the flowing exhaust gases. Although the operation of a condensing furnace is noisier than that of a conventional furnace, it is generally not considered annoying. If you find it bothersome, record that fact on your worksheet. The system is in need of a tune-up and adjustment by a competent heating contractor.

Condensation of water vapor in the exhaust gases also occurs in a pulse-combustion furnace. Consequently, the overall concerns of exhaust-gas venting and condensate removal, as discussed previously for condensing furnaces, apply. Because of the nature of the combustion process in a pulse furnace, the sound of sixty to seventy tiny explosions per second can be heard outdoors at the side-wall terminations of the air-supply and exhaust-vent piping. In those cases, where the sound is considered objectionable, mufflers can be installed on the supply and exhaust pipes to reduce the sound level.

Hot-water systems

When inspecting a hot-water heating system, you should first determine whether it is a gravity or forced system. Sometimes an open expansion tank is found in the attic during the interior inspection. This would normally indicate a gravity system. Often, however, the expansion tank is no longer functional and was replaced by a closed expansion tank when the system was converted to forced circulation. You can easily tell whether a system has forced or gravity circulation by checking the equipment associated with the boiler. A forced system will have a circulating pump in the return line near the boiler. Record the type of system on your worksheet.

Next, look at the pressure gauge and record the operating pressure. The normal operating pressure should be between 12 and 22 psi. Since the water-makeup valve (pressure-reducing valve) is set to introduce water into the system when the pressure drops below 12 psi, any pressure below that value indicates either a malfunctioning valve or the need for adjustment. A low-pressure reading is fairly common.

If the pressure gauge indicates an operating pressure of 30 psi, the relief valve should be discharging. Sometimes the relief valve has already discharged and lowered the pressure to about 28 psi. In this case, the exposed end of the relief valve will have water dripping from it, and there will be a pool of water on the floor. A high-pressure condition that results in the relief valve discharging will usually be caused by a waterlogged or undersized expansion tank. If domestic hot water is generated through the heating system, another possible cause for a high-pressure condition is pinholes in the coils of the water heater. Regardless of the cause of the high pressure, the condition must be corrected.

Locate the relief valve. From a safety point of view, it should be boiler-mounted. If there is no boiler-mounted relief valve (and often there is not), the installation of one should be considered. Do not check the valve to determine whether it is operational. Bits of corroded material, sediment, or mineral deposits may prevent the valve from reseating properly and shutting off. Your best bet is to check the valve after you move into the house. At that time, if it does not reseat properly, it should be replaced. A relief valve is inexpensive. These valves should be checked at least once a year by pulling the lever and allowing a small quantity of water to flow. This action flushes any sediment that tends to build up before it has a chance to clog the valve. Some relief valves do not have levers for testing the unit. If you have such a relief valve on your boiler, you should consider replacing it.

As the various zones are being checked, inspect the zone valves for dripping water and deposits. Even though the valves might be working during your inspection, if water is dripping from the valves, they should be replaced. The dripping water results in a mineral-deposit buildup that can eventually cause the valve to "freeze" in an open or closed position. This might happen, and usually does, on the coldest night of the year. Zone valves are perhaps the weakest link in the heating system, and their occasional replacement should be anticipated.

Is the circulating pump operating properly? Just because you can hear the pump operating or feel its vibration does not mean that it is functioning properly. The coupler between the pump and the motor may be broken, and the noise that you hear or the vibration that you feel may be only the motor turning, not the pump. You can be sure the pump is operating by putting your hand on the return pipe associated with that pump. After a while, the pipe will get hot if the pump is operational. Check the gasket between the pump and the motor for signs of current or past leakage. This is a vulnerable joint for leakage.

Is the circulating pump emitting any unusual or loud sounds? If it is, the noises might indicate faulty bearings, and the condition should be checked by a service organization. If you see smoke coming from the circulating pump or smell electrical insulation burning, the entire system must be shut down immediately and the condition corrected. One last check of the circulating pump is to determine whether it operates continuously or intermittently. If the circulating pump continues to run long after the thermostat is not calling for heat, it is in continuous-mode operation. This mode, discussed in chapter 14, can be modified to an intermittent operation.

If the heating system is used for generating domestic hot water, the thermostat will not control the burner, only the circulating pump. In this case, check the temperature of the boiler water (look at the temperature gauge). If the temperature drops below 150° F and the burner does not fire, it will indicate a faulty or improperly set aquastat. In most cases, the burner will be activated when the temperature drops below 180° F.

In addition to the above, with high-efficiency condensing and pulse boilers, look at the piping associated with the inlet air supply, exhaust gas vent, and condensate drain. Check to see that there are no loose sections, that there is no blockage at the pipe terminations, that the exhaust pipe is not combined with the exhaust from the water heater, and that the horizontal portions of the exhaust pipe slope back to the boiler. Also check that the condensate pipe discharges into a floor drain or condensate pump rather than a hole in the floor slab.

Steam systems

In this system, a water line showing the level of water in the boiler should be visible in the level gauge. Look for the gauge. If it is dirty, it must be cleaned. A gauge that is completely filled with water indicates too much water in the system, and an empty gauge indicates an insufficient amount of water in the boiler. When the water level in the gauge is unsteady (rising and dropping), it indicates a problem condition that might reflect an excessive grease and dirt buildup in the boiler or that the boiler is operating at an excessive output. In either case, it should be corrected.

Check the heating unit to determine whether it has all the required safety controls. A steam system should have a low-water cutoff, a high-pressure limit switch, and a relief valve. If any of these items are missing, record the fact on your worksheet. Look for the low-water cutoff. Is it the built-in type or externally mounted? If it is externally mounted, you should test its operation. This is a normal procedure that the manufacturer recommends the homeowner perform periodically. If the relief valve is old, then as a precautionary measure it should be replaced after you take possession of the house.

Look for the condensate return line to the boiler. Is it a wet or dry return? If it is a wet return, it should be connected to the boiler by means of a special piping arrangement, a *Hartford loop*. Is there one? If not, the installation of a loop should be considered.

Oil burners

When the heating system is oil-fired, the first thing you should note is whether the top portion of the chimney (as seen during your exterior inspection) is covered with soot. This is an indication that the oil burner has not been operating efficiently. When you are inspecting the heating unit, look for a card or tag near or on the unit that contains the maintenance service record. Whenever a serviceperson performs any type of maintenance, even a tune-up, he or she records the date on this card. This will at least give you an idea of whether the burner has been maintained or neglected. The burner should be tuned up and checked for combustion efficiency at least once a year.

Other telltale signs indicate the need for oil-burner maintenance. Is there a smoke odor in the furnace room? When the oil burner first fired, was there a puffback of smoke and fire? At startup, while running, or at shutdown, do you feel pulsations in the boiler room? Are there particles of soot on the furnace or around the burner? Is the oil burner emitting an abnormal noise during operation? If any of the above questions are answered yes, oil-burner maintenance is needed. There are also tests that can be made, such as sampling the exhaust gas for combustion efficiency. This analysis, however, should be performed by a competent serviceperson.

The proper draft over the firebox for an oil-fired system is critical to efficient operation. Therefore, there must not be any cracks or open joints that will allow excess air to infiltrate into the combustion chamber or smoke pipe. Look for cracks and open joints at the front of the boiler near the burner, especially around the air tube, mounting plate, and front boiler doors. If you can see the flame through the cracks or open joints, the open areas should be sealed with a refractory cement. A vulnerable joint for air leakage is the joint between the smoke pipe and the chimney. If this joint is cracked, it must be sealed. Also, to ensure a proper draft, there should be a functioning draft regulator on or near the smoke pipe. Look for one. If it is missing or inoperative, record the fact on your worksheet.

If you see a small hole in the flue stack about ¼ inch in diameter, don't be concerned. It is a test hole that is used by the oil-burner serviceperson for measuring operational parameters such as draft, percentage of carbon dioxide, and stack temperature.

Next, look inside the firebox. The refractory lining of the fire chamber should be intact. There should not be any broken sections or holes in the lining. If there are, maintenance is needed. While looking inside the firebox, if the heating unit is a boiler rather than a furnace, look for water dripping or seeping around the top or sides. As discussed earlier, this condition might be caused by joint movement as a result of thermal expansion. Nevertheless, as a precautionary measure, since it can also be caused by a cracked boiler, it should be checked by a professional.

The oil burner receives its oil through a feed line that runs to the oil storage tank. The feed line is usually a small-diameter copper tube that can easily be damaged and must be protected. Look for the line. It should not run exposed over the floor. (See FIG. 15-1.) Someone could accidentally step on and damage the feed line, restricting the

Fig. 15-1. Oil-fired forced-hot-water boiler with an exposed oil-feed line. The feed line is a small-diameter copper tube that can easily be damaged. It should be protected. Usually the feed line is located below the floor slab and emerges near the oil burner.

flow to the burner and causing oil to leak on the floor. If the storage tank is the interior type, the underside of the bottom section should be checked for oil leakage. Run your hand along the underside of the tank. If your hand has a slight or heavy film of oil, the unit may be deteriorating, in which case patching repairs would be in order. The average projected life for these tanks is about twenty years.

Gas burners

Gas burners do not require as much preventive maintenance as oil burners. All that is normally needed is periodic cleaning and adjustment. Look at the gas burners. In neglected units, they are covered with excessive corrosion (rust) dust. Quite often some of the gas ports are clogged by the rust particles and are no longer effective. This condition requires a professional cleaning. The quality of the flames on the gas burner is important. The flame should be primarily bluish in color, with little or no yellow. If the flame is lifting off the burner head, it is an indication that too much air is being introduced into the mixture.

A yellow tip in the flame will result in a soot layer building up in the flue passage of a boiler. After some years, the soot buildup can completely block the flue passage. I found this condition in a boiler that was only seven years old. A blocked flue passage is a potentially dangerous condition because it prevents the exhaust gas from rising up to the chimney. Instead, the gas, which contains poisonous carbon monoxide, seeps into the house. Fortunately, this condition can be easily corrected. Clean the boiler flue passages with a long stiff brush to remove the soot buildup and adjust the air intake to the burner.

Another safety check that should be made is to test the draft in the chimney. After the heating unit has been operational for a few minutes, put your hand near the draft diverter (located on the exhaust stack above a boiler and built into the sheet-metal casing of a furnace). There should be no outward flow of hot exhaust gas toward your hand. If you do feel a flow on your hand, a potentially dangerous condition exists that should be checked by a professional.

Checkpoint summary

General considerations

- ❍ How old is the heating system?
- ❍ Is the boiler or furnace an obsolete unit (i.e., a converted coal-fired boiler or an "octopus" warm-air furnace)?

- ❍ Check each room for heat-supply registers or radiators.
- ❍ Are supply registers or radiators located on or near exterior walls (preferably under the windows or next to exterior door openings)?
- ❍ Check supply-register dampers for ease of operation.
- ❍ Are there separate return grilles for each room or centrally located returns for each floor?
- ❍ Check radiators for broken shutoff valves and blocked air vents (steam systems).
- ❍ Inspect area below radiators for signs of leakage.
- ❍ Check visible portions of ductwork and piping (basement, attic, garage, etc.) for open or leaking joints and uninsulated sections.
- ❍ Are the thermostats properly located (no drafts)?
- ❍ Note whether the thermostats are manual or automatic clock units.
- ❍ Is the clock functional?
- ❍ Check each heating zone and thermostat control independently.
- ❍ Is there an adequate air supply for the heating system?
- ❍ If the heating unit has a direct vent, is it properly located?

Boiler or furnace

- ❍ Check boiler or furnace during startup. Note any puffback with an oil burner or a licking back of flames on a gas burner.
- ❍ Is the boiler a steel or cast-iron unit? Is it aging? Is it neglected?
- ❍ Check for signs of excessive corrosion (rust), dust, flaking metal, and mineral deposits indicating past or current leakage.
- ❍ Inspect firebox, if accessible, for water seepage through joints and dripping water. Have this condition checked by a professional.
- ❍ Check smoke pipe for corrosion holes, open joints, sagging sections, and adequate clearance.
- ❍ If furnace or boiler is the condensing type, check for corrosion, signs of condensate leakage, blockage in the supply and vent pipes, and horizontal vent pipe sections sloping back to the furnace or boiler.
- ❍ Check master shutoff switches for proper operation.

Warm-air systems

- ❍ Check furnace, burner area, and sheet-metal casing for excessive corrosion, rust, and flaking metal.
- ❍ Check for unstable or dancing flames (gas-fired furnace). If these conditions are noted, have the heat exchanger checked by a professional.
- ❍ If there is a canvas connection between furnace and main supply duct, check it for torn and open sections. This material may contain asbestos.
- ❍ Check fan controller during furnace operation. If fan starts after the burners shut off, controller is either faulty or out of adjustment.
- ❍ Check fan operation for excessive noise or vibrations.
- ❍ If system contains a power humidifier, check for operation, signs of leakage, and excessive mineral deposits.
- ❍ During furnace operation, check airflow and temperature in room supply registers.
- ❍ Check return grilles for proper air circulation (tissue test).

Hot-water systems

- ❍ Check whether system is forced or gravity circulation.
- ❍ During operation, check pressure gauge. System pressure should generally be between 12 and 22 psi.

- ❍ Inspect for relief valve, which should be boiler-mounted. Do not check valve operation.
- ❍ Check circulating pumps for proper operation (motor and pump both functioning). Note any unusual loud sounds.
- ❍ Inspect pumps for leaking fittings or gaskets.
- ❍ Check each zone valve independently. Inspect valves for dripping water and deposits.
- ❍ If boiler also generates domestic hot water, check temperature gauge to determine whether aquastat is faulty or improperly set.

Steam systems

- ❍ Check water-level gauge for proper level, accumulated sediment, and fluctuations (rising and dropping).
- ❍ Is the water supply to the boiler a manual or an automatic feed?
- ❍ Check boiler for required safety controls: a low-water cutoff, a high-pressure limit switch, and an automatic relief valve.
- ❍ If low-water cutoff is an external type, check operation by flushing the unit.
- ❍ Inspect boiler relief valve. *Do not operate*. If valve is old, consider replacement.
- ❍ Inspect the condensate return line. Is it a wet or dry return? If a wet return, is there a Hartford loop?

Oil burners

- ❍ During exterior inspection, check top portion of the chimney for accumulated soot.
- ❍ Has oil burner been maintained or neglected? Look for a dated maintenance service card.
- ❍ During operation, note any smoke odors or soot particles in the furnace room.
- ❍ Are any abnormal noises or pulsations emitted from the oil burner during operation?
- ❍ Check for cracks and open joints near the burner, particularly around the air tube, mounting plate, and front boiler doors.
- ❍ Check for air leakage at the joint between the smoke pipe and the chimney.
- ❍ Check exhaust stack for an operational draft control.
- ❍ Check refractory lining of the fire chamber for cracked and deteriorating sections.
- ❍ Check oil-burner feed lines (copper tubing). Are there kinked or exposed sections vulnerable to damage?
- ❍ Check underside of oil storage tank (interior type) for leakage.

Gas burners

- ❍ Check gas burners for corrosion dust and clogged gas ports.
- ❍ Check flame pattern for safe and efficient operation.
- ❍ Check draft diverter on the exhaust stack for backflow of exhaust gases.
- ❍ Is the gas meter adequately sized?

16

Domestic water heaters

Hot water for bathing and washing (domestic hot water) is produced through a separate tank-type water heater, through the heating-system boiler (steam or hot water) by a tankless coil, or through an indirect-fired storage water heater. Domestic hot water is considered by manufacturers as cold water whose temperature has been raised 100° F. An important consideration in the design of water heaters is the ability of the unit to raise the temperature of 1 gallon of water 100° F. This criterion is used in evaluating the recovery rate for tank heaters and the water-flow rate for tankless coil heaters. These items will be discussed in the sections that follow.

The temperature of domestic hot water is important for energy conservation and safety. The operating temperature range for most water heaters is between 120° and 160° F, although in some cases it is possible to achieve temperatures of 180° F. Normally, a thermostat setting that produces a water temperature of 140° F will be adequate for household chores, including the use of automatic clotheswashers and dishwashers. However, for washing dishes by hand or bathing, temperatures of 120° F are more than adequate. In fact, 120° F water is above the tolerance level of most people and will have to be mixed with cold water. Setting the thermostat so that the water temperature exceeds 140° F is wasteful of energy and will shorten the life of the water heater. In addition, a water temperature in excess of 160° F is a potential hazard because of the possibility of being scalded while showering (if the mixing valve should be faulty).

Tank water heaters

Most tank water heaters are located near the heating system or in a utility room, sometimes in the kitchen or a closet. These heaters basically consist of a lined steel tank that is covered with an insulated sheet-metal jacket to reduce heat loss. The lining in the tank is usually vitreous enamel (glass) but can be concrete (stone) or copper. To minimize tank corrosion, most heaters are equipped with a "sacrificial" magnesium anode rod that is suspended inside the tank. The electrochemical action causing corrosion takes place between the water and the anode rather than between the water and the tank, thereby extending the life of the tank. Some tanks are constructed so that the magnesium anode can be inspected and replaced if necessary. Tank water heaters are gas- or oil-fired or electric. (See FIG. 16-1.)

Reversed connection

Many water heaters have at the top of the tank hot-water outlet and cold-water inlet fittings that are clearly marked HOT and COLD, respectively. Yet it is surprising how many times I have found heaters with a reversed water-line installation. You can tell whether the installation is reversed by touching a pipe about 4 feet away from the unit and comparing it with the HOT outlet fitting on top of the tank. A reversed connection results in an inefficient operation of the heater. Even though there is a cold-water inlet tap at the top, the cold water does not enter the tank at the top. There is a pipe (dip tube) inside the tank connected to the inlet fitting that carries the cold water to the bottom of the tank. Because the hot water tends to rise above the cold water, the hot water is normally at the top of the tank where it flows through the outlet fitting. When the inlet-outlet

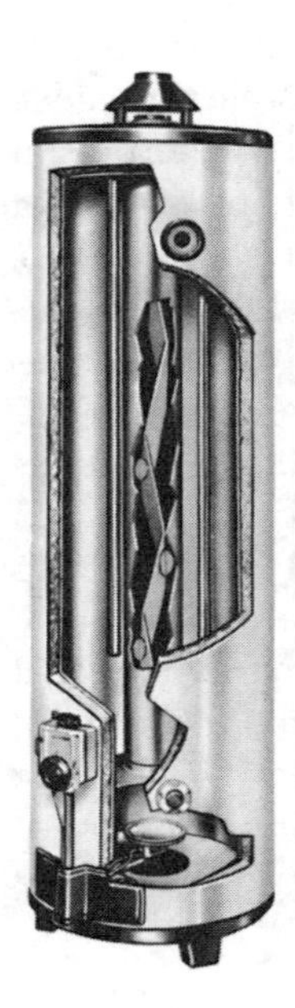

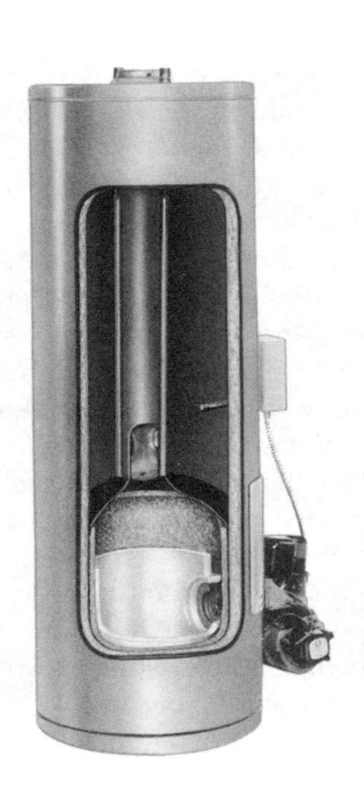

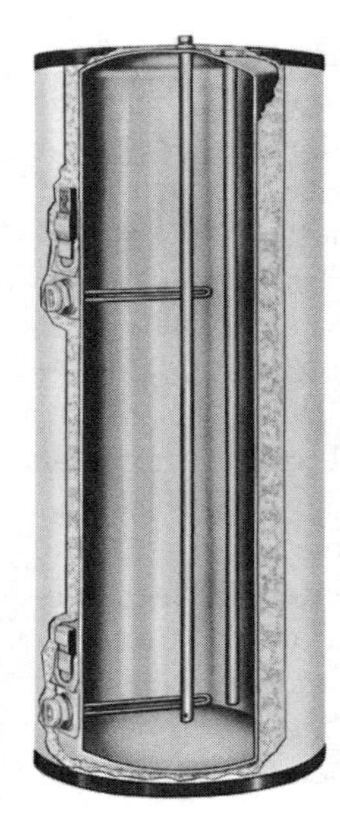

Fig. 16-1. *Tank water heaters. Left—gas-fired. Center—oil-fired. Right—electric. Rheem Mfg. Corp.*

connection is reversed, the cold water enters the tank at the top and mixes with the hot water as it settles to the bottom near the outlet fitting. Consequently, for the same thermostat setting, the temperature of the hot water is lower than it would otherwise have been. To compensate for the lower temperature, homeowners usually turn the thermostat to the high setting, thereby increasing the fuel cost and decreasing the life of the water heater. Some heaters have interchangeable cold- and hot-water fittings. In these units, the dip tube can be shifted from one side to the other.

Fig. 16-2. *If this relief valve discharges, someone standing nearby could be scalded. The relief valve should have an extension that discharges into a bucket located on the floor.*

Relief valve

All water heaters must have a relief valve that is both temperature- and pressure-sensitive. A relief valve that responds to pressure only will not offer adequate protection for the homeowner. An excessively high pressure alone can cause the tank to rupture. Though this will cause water damage, it is not necessarily dangerous. However, if the tank should rupture as a result of a high-temperature failure, the superheated water contained therein will flash into steam, instantaneously releasing high energy in an explosion. A relief valve that is sensitive to both temperature and pressure can avoid this problem.

The relief valve should have an extension that will allow it to discharge into a bucket on the floor. Sometimes, a relief valve is found without this extension. (See FIG. 16-2.) This is a potential hazard because a person standing near the water heater when the relief valve discharges could be scalded. I have also found an extension on the relief valve that leads to a sink or floor drain located near the water heater or in the next room. (See FIG. 16-3.) This type of piping arrangement is usually installed to eliminate water damage if the relief valve discharges. Water discharging from the valve will flow directly down the drain rather than

Fig. 16-3. The relief valve has an extension that leads to a sink in the adjacent room. This is undesirable because the effluent will go down the drain, and the homeowner will not be aware of a discharge condition.

accumulate on the floor. The disadvantage of this installation is that the homeowner might never know a discharge condition had occurred. A discharging relief valve indicates an abnormal condition of some type that should be checked out. Obviously, if the homeowner does not know that there is a problem, he or she will not have it corrected. By having the relief valve discharge into a bucket, water damage can be eliminated, and the presence of a problem condition will be indicated.

If you notice that the relief valve is dripping and the temperature of the hot water is not excessive, the problem will be either a defective relief valve or the result of thermal expansion. Many municipalities require a backflow preventer or a nonbypass pressure regulator on the inlet water service. This creates a closed system when a faucet is not drawing water. When the water is heated and expands, the resulting pressure buildup causes the relief valve to drip. Installing an expansion tank on the cold-water inlet to the water heater can control this condition. (See FIG. 16-4.) Record a dripping relief valve on your worksheet for later correction.

It's also important that the Btu capacity of the relief valve exceed the Btu input of the water heater. If the capacity is less, the relief valve will not be able to discharge the overheated water at the same rate that it's produced, resulting in an unsafe condition or possible explosion. Check the capacity of the relief valve by looking at the nameplate mounted on the valve. Specifically look for the AGA (American Gas Association) rating given in Btu/hour (see FIG. 16-5). This rating must exceed the Btu input to the water heater listed on the data plate mounted on the side or top of the tank casing. You can also check the Btu rating of electric water heaters even though they are rated in watts rather than Btu/hour. You can determine the electrical energy input to the heater by converting the total wattage of the heater, which is listed on the data plate, to Btu/hour. 1 watt = 3.413 Btu/hour.

To be effective, the relief valve should be located so that its sensing element is immersed

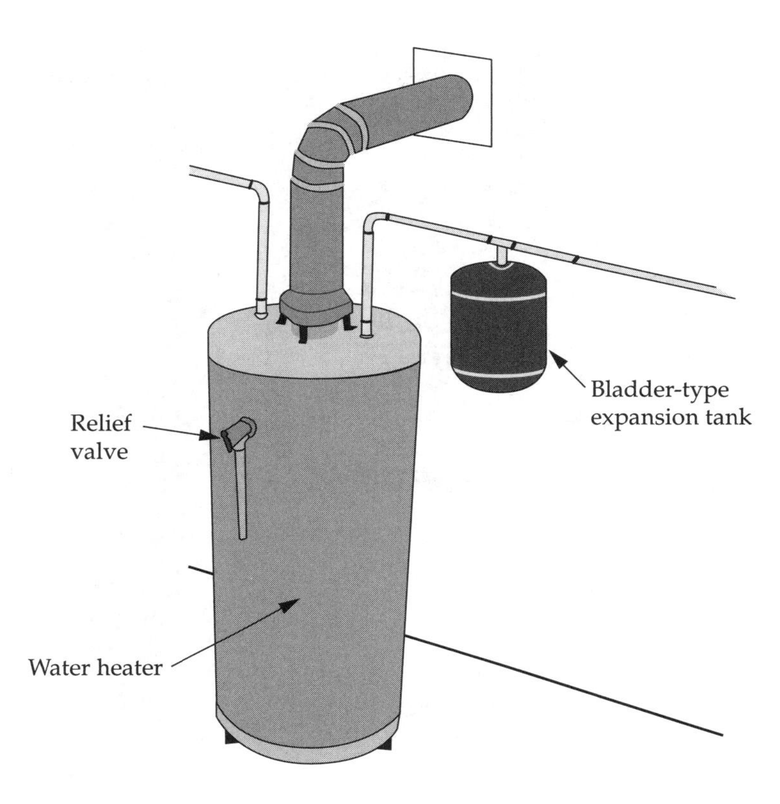

Fig. 16-4. Bladder-type expansion tank installed to prevent relief valve from leaking due to thermal expansion of the water in the domestic water heater tank.

in the top 6 inches of water in the tank rather than in the hot-water outlet pipe. This is important because there is a temperature difference between the tank and the hot-water outlet pipe. For instance, when the relief valve is installed in the hot-water outlet pipe 5 inches away from the tank, the water in the tank could be raised to 250° F before the temperature at the valve reaches 210° F.

During my inspections, I have often found the relief valve located on the cold-water inlet pipe. This location negates the effectiveness of the valve. If you find a relief valve on the cold-water inlet pipe, you should notify the homeowner of the potential hazard and have it relocated after you take possession of the house.

It is important that the diameter of the relief-valve extension pipe be the same as the outlet opening of the relief valve. If the extension pipe has a smaller diameter, it reduces the discharge capability of the valve and reduces its effectiveness as a safety device.

Rumbling noise

Over the years, sediment scale and mineral deposits tend to build up at the base of the tank. Manufacturers suggest that a few quarts of water be periodically drained from the water heater to help remove these deposits. However, this practice is not always effective, and if sufficient deposits accumulate, a rumbling or pounding sound can be heard when the unit is firing. If you hear a rumbling noise while inspecting the water heater, do not be alarmed. It is not a dangerous condition, although the noise can be annoying. In addition to their annoyance value, accumulated deposits at the

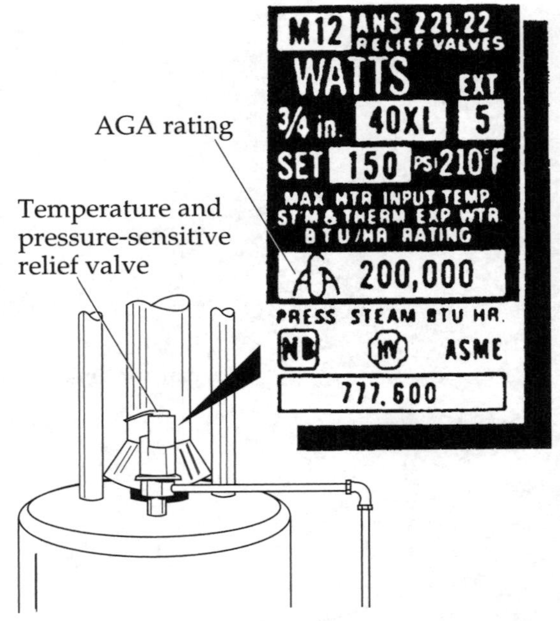

Fig. 16-5. Look for the AGA rating on the plate hanging from the relief valve. The rating must exceed the Btu input to the water heater.

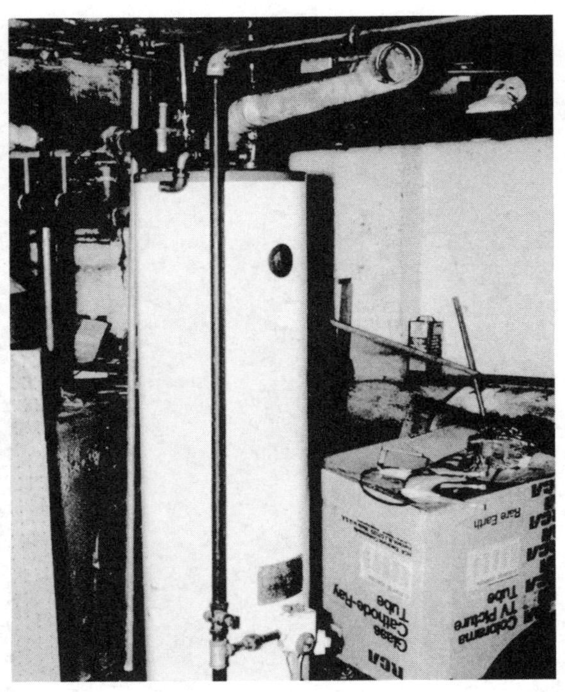

Fig. 16-6. Disconnected exhaust stack. Exhaust gases from this gas-fired water heater discharge into the basement rather than a chimney—a potentially dangerous condition.

base of the tank can act as an insulator between the water and the flame, decreasing the overall efficiency of operation.

Exhaust stack

All oil- and gas-fired water heaters must have an exhaust stack to vent the products of combustion to the outside. If you find an exhaust stack that is loose or broken, so that the exhaust gases are discharging directly into the house (see FIG. 16-6), you should notify the homeowner of this potentially dangerous condition. As with a heating system, the exhaust pipe should have an upward pitch from the heater to the chimney connection. The exhaust pipe gets quite hot and therefore should not be in contact with combustible material.

Operational inspection

The water heater should be inspected while the burner is operational. This can be achieved by opening the hot-water faucet at a nearby sink. Within a few minutes, depending on the size of the tank and the water flow at the sink, the burner should fire. (This assumes that the pilot is lit for a gas burner or that the electronic ignition is operational for an oil burner, and that there is oil in the storage tank.) If the burner fails to fire, there is a problem condition that should be checked. Oil and gas burners and their controls are discussed in detail in chapter 15.

To provide an adequate draft for combustion gases, oil-fired units should have a draft regulator, and gas-fired units should have a draft diverter. (See FIG. 16-7.) After the unit has been firing for several minutes, check the draft in a gas-fired unit. Put your hand near the draft diverter. If you feel a flow of hot gases on your hand, there is either a blockage in

Fig. 16-7. Draft regulator (left) for an oil-fired water heater; draft diverter (right) for a gas-fired water heater.

the flue passage or there is a negative pressure in the house. This is a condition that must be corrected because the exhaust gas contains carbon monoxide. This colorless, odorless gas is poisonous and can cause asphyxiation.

Another check for faulty venting of the exhaust gases is to hold a lit match at the draft diverter opening. If the exhaust gases spill out into the room, they will blow out the flame. If the venting system is functioning properly, the flame will be sucked into the draft diverter hood. As a safety feature, some water heaters have a spill switch sensor mounted on the draft diverter. (See FIG. 16-8.) When combustion gases spill out of the draft diverter, the spill switches sense the heat from the exhaust gases. This causes the spillage sensor circuit to open, which shuts the pilot, causing the burner to shut down.

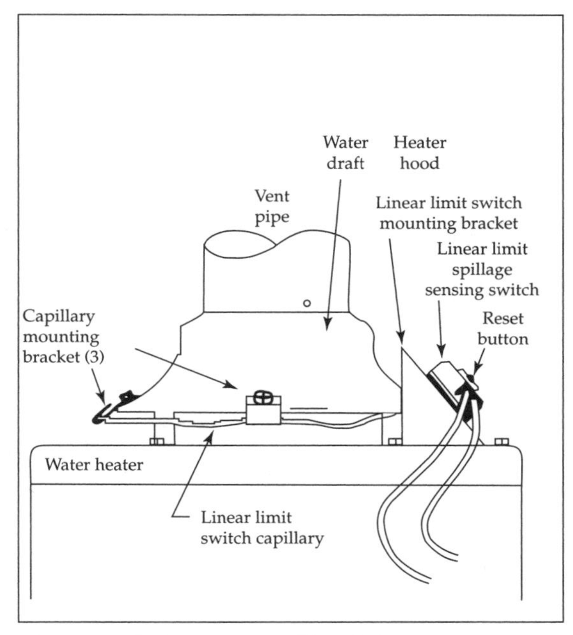

Fig. 16-8. Exhaust gas spill switch sensor. Courtesy of: Tjernlund Product, Inc.

Whenever there is a heavy draw of water, if the temperature of the inlet water is low, condensation can form on the tank. If this should occur, you might see some water beneath the heater; in a gas-fired unit, you may hear sizzling or pinging sounds caused by water droplets falling on the burner. Consequently, if you see a little water on the floor or hear the pinging noise of droplets, do not assume that the water heater is leaking; it may be condensation. If it is, the condition will disappear after the water in the heater becomes heated. If the condition does not disappear, a leak should be suspected, and the unit may require replacement.

If you notice a water pipe connected to the drain fitting at the bottom of the tank, it is probably a hot-water return line. (See FIG. 16-9.) This is the typical connection for providing "instant" hot water at the bathroom fixture that is farthest from the water heater. It is more often found in large pricey houses than in more modest homes. You can check out the effectiveness of the instant hot-water system during the interior inspection of the bathrooms.

As with a gas-fired heating system, the pipe supplying gas to a domestic water heater should be rigid black iron and not a copper pipe or a flexible pipe. If the pipe is not black iron, check with the utility company for its requirements.

Some gas-fired water heaters use bottled gas (propane) because natural gas is not available. In that case, a copper gas-supply pipe is usually acceptable because the supplier provides a quality control over the impurities in the gas. All impurities that would react with the copper are removed.

Water heater replacement

A water heater requires replacement only when the tank is leaking. If there are problems with any of the controls, they can be replaced individually as needed. You cannot tell by looking at a unit when it will start leaking. The projected life for an oil- or gas-fired water heater is from seven to ten years, although units have been known to last over fifteen years. Electrical units have a longer projected life than oil- and

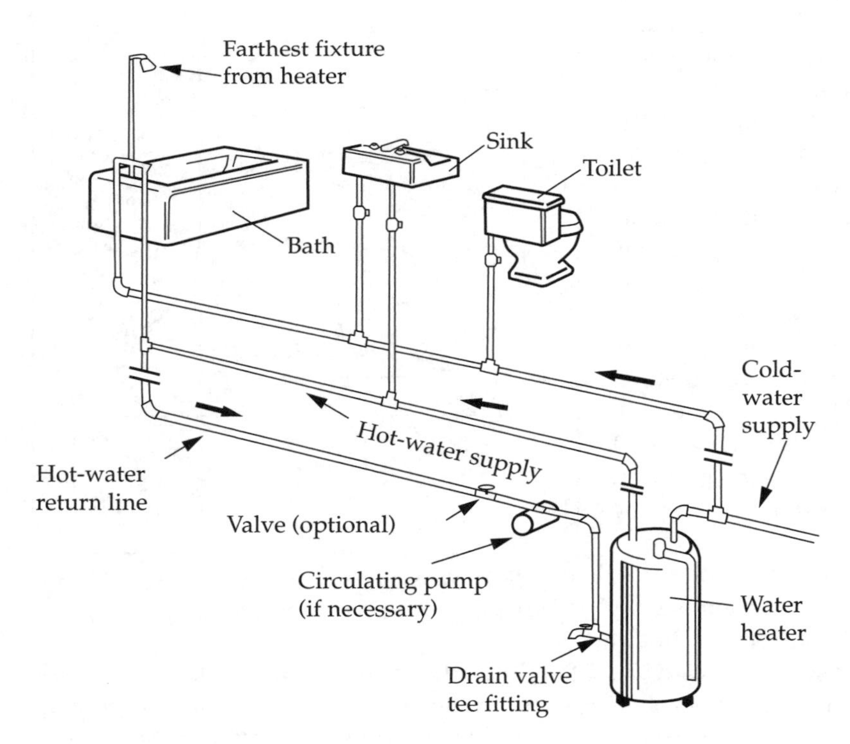

Fig. 16-9. *Hot-water return line supplies "instant" hot water to a fixture. This results from the constant recirculation of the water back to the water heater.*

gas-fired heaters. This is because the electrical heating element is immersed in the water, imparting its heat directly to the water and not to the tank, whereas in an oil- or gas-fired heater the flame imparts its heat directly to the base of the tank. The intense heat results in expansion and contraction stresses that effectively shorten the life of the tank. Look at the temperature control knob (thermostat) on the tank. If it is at the maximum setting, the unit has been operating at a high temperature, and a shorter life span should be anticipated.

The age of a tank water heater can generally be determined from the serial number found on the tank's data plate. The numbering system, however, varies by manufacturer. For example, for Rheem or Rudd water heaters, the first four numbers of the serial number indicate the month and year of production (thus 0286 would indicate that the heater was manufactured in February 1986). Usually a water heater is installed within several months of its manufacture.

With newer A. O. Smith water heaters, the date of manufacture is listed on the data plate. However, with older A. O. Smith water heaters, the date can be found by checking the serial number. For example, in serial number 800-A-89-12345, the letter stands for the month of manufacture (January in this case); and the middle number (89) stands for the year of manufacture, 1989. The letter code for month of manufacture is

A—January	G—July
B—February	H—August
C—March	J—September
D—April	K—October
E—May	L—November
F—June	M—December

If you determine that the water heater is over ten years old, you should not anticipate an extended life for the unit, even though it might look relatively new (shiny and clean on the outside).

I have been asked by clients whether it is necessary to replace a gas-fired water heater that has been in a flooded basement. It is not necessary to replace the unit. The tank is designed to hold water and therefore cannot be damaged by it. However, the burner assembly and controls (pilot, thermocouple, and thermostat) can be damaged and should be checked by a competent serviceman. If any of the controls are malfunctioning, they should be replaced as needed. Also, when the insulation between the tank and the jacket gets wet to the extent that it becomes thoroughly soaked, it can sag and be less effective as an insulator. If this should occur, you will feel "hot spots" on the jacket. This condition can easily be remedied. Insulation jackets are available that fit over water heaters to reduce heat loss.

Water heater capacity and recovery

The amount of hot water that your family requires depends on the number of people, their bathing and washing habits, and the number of tubs and showers available for simultaneous use. Do not judge whether a water heater is adequate by the capacity of the tank alone. Equally as important is its recovery rate—the volume of water that will have its temperature raised by 100° F in one hour. The capacity of residential water heaters typically varies between 30 and 82 gallons. The following example, although somewhat exaggerated, illustrates the point.

Let us take an average home. Assuming normal water pressure and flow, the typical faucet delivers about 5 gallons of water per minute. Let us further assume that there is approximately a sixty-minute demand for hot water flowing at the rate of 2 gallons per minute. This hot water will usually be mixed with cold water for washing and bathing. If the home is equipped with an electric water heater that has a 60-gallon capacity and a 10-gallon recovery rate, after about thirty minutes there would be no more hot water. The low recovery rate of 0.17 gallons per minute is less than 10 percent of that required and would be totally ineffective in satisfying the hot-water demand. On the other hand, if the home had been equipped with an oil-fired water heater that had a capacity of only 30 gallons (half the capacity of the electric heater) and a 120-gallon recovery rate, the hot water would never run out, and the demand would be completely satisfied.

What is the capacity and the recovery rate of the water heater that you are inspecting? The capacity of all water heaters will usually be stamped on the data plate. Gas-fired units might also have the recovery rate on the plate. If yours doesn't, you can get an approximate value by dividing the number of input Btu (on the data plate) by 1,000 and multiplying the results by 0.85. Oil-fired water heaters might not indicate their recovery rate. This should be of little concern because oil-fired units have a high recovery rate, often on the order of 120 gallons per hour. Electric units will not have their recovery rate on the data plate. However, the wattage of the upper heating element, the lower heating element, and the total wattage will usually be stamped on the plate. You can easily compute the approximate recovery rate for an electric water heater by using the following formula: Each 250 watts heats about 1 gallon of water 100° F in one hour. Thus a water heater rated at 4,500 watts has a recovery rate of only 18 gallons per hour. This is a typical recovery rate for an electric water heater and is quite low in comparison to a gas- or oil-fired unit. To compensate for the low recovery rate, electric units should, and often do, have large-capacity tanks.

I have found the following rule of thumb effective in determining the adequacy of a water heater for a house, regardless of whether the unit is gas- or oil-fired or electric: For a house with one full bathroom (sink, bowl, and shower-tub), the sum of the capacity plus the recovery rate should total about 70. For two full bathrooms, the total should be about 90; for three full bathrooms, about 105; and for four full bathrooms, about 115. This also assumes that there is a washing machine in the house that might be operating while all the showers-tubs are being used. Of course, your requirements might be different. If you buy a house with three full bathrooms and there are only two people in your family, you obviously do not need a water heater whose capacity and recovery rate total 105.

From an energy-conservation point of view, a small tank with a high recovery rate would be better than a large tank with a low recovery rate. In addition, it is more costly to maintain a large tank of hot water, especially during periods when there is no hot-water demand.

Tankless (coil) water heaters

Tankless water heaters, sometimes called "instantaneous water heaters," should not be confused with tankless (wall-mounted) water heaters, which will be discussed in the next section. Tankless coil-type water heaters are used in conjunction with steam and hot-water boilers. Most tankless heaters consist of a small-diameter pipe shaped in the form of a coil that is located inside of the boiler (internal generator). (See FIG. 16-10.) The coil might also be located in a casing outside, but connected to, the boiler (external generator). The coil is surrounded by hot boiler water that gives up its heat to the water flowing in the coil. Cold water enters the coil and leaves as hot water with its temperature increased by 100° F. These units are designed for a specific

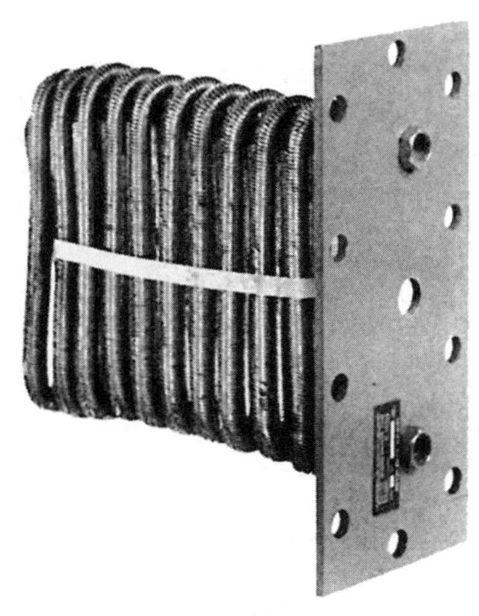

Fig. 16-10. Tankless water heater. Unit is mounted inside a boiler. Triangle Tube and Specialty Co., Inc.

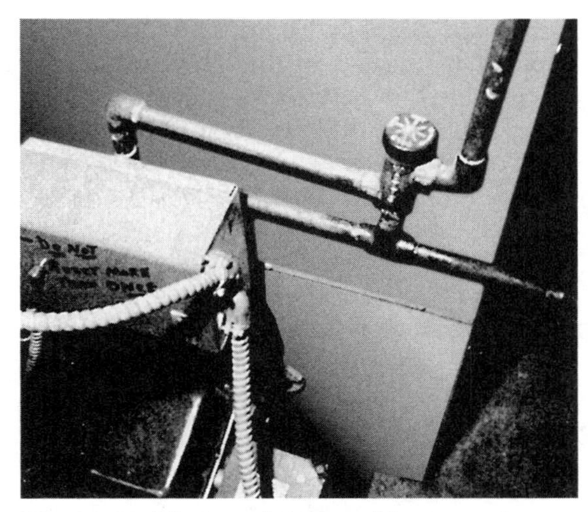

Fig. 16-11. Mixing valve for tankless water heater.

water-flow rate (gallons/minute) through the coil to achieve the desired increase in temperature.

If water flows through the coil at a greater rate, less heat will be transferred to each gallon, resulting in a lower outlet temperature. Consequently, many units have a flow-regulating valve installed in the cold-water supply pipe to the tankless heater to limit the water flow to the capacity of the heater. The design flow through these heaters is usually about 3 to 4 gallons per minute. Assuming a normal water supply to the house, the typical cold-water flow will be between 4 to 8 gallons per minute.

To regulate the temperature of the hot water being distributed throughout the house, there should be a mixing valve between the cold-water inlet and the hot-water outlet pipe. (See FIG. 16-11.) A mixing valve is considered a necessary feature to eliminate the possibility of scalding. Over the years, mineral deposits tend to build up inside the heater coil, especially in hard-water areas. This in turn further reduces the water flow through the heater and results in a higher water temperature at the outlet. The high-temperature water, which is a potential hazard, can be cooled to the design temperature by mixing it with cold water in the mixing valve. This water can then be distributed to the plumbing fixtures throughout the house.

As mineral deposits form inside the heater coil, they restrict the water flow. It is possible for the deposits to build up to such a point that the hot water flow becomes a trickle when two or more faucets are turned on at the same time. The hot-water flow should be checked during your interior inspection, as discussed in chapter 10. A low hot-water flow that is caused by a blockage in the tankless coil can often be corrected by chemically flushing the unit and dissolving the deposits. However, when the flow becomes a trickle, the heater coil should be replaced.

A tankless (coil) has virtually no storage capacity. Consequently, if there is a hot-water demand that exceeds the design flow (and it often does), there will not be enough hot water. To provide additional hot water, some installations are equipped with a storage tank. Water heated in the tankless coil will be circulated to the tank by gravity flow or a pump. The pipe that distributes hot water to the plumbing fixtures will be connected to the storage tank rather than the tankless heater. When the hot-water flow between the tankless heater and the storage tank is by gravity, the storage tank must be located above the boiler. When the flow is induced by a circulating pump, the tank can be located alongside the boiler. (See FIG. 16-12.)

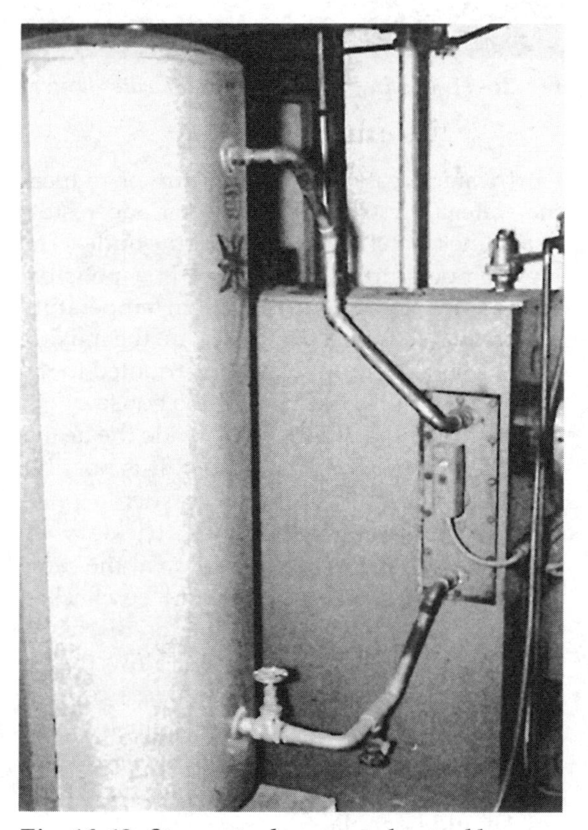

Fig. 16-12. Storage tank connected to tankless water heater. The hot water circulates between the tank and the coil to maintain high temperature. The tank increases the overall volume of available hot water.

From an energy-conservation point of view, a tankless (coil-type) water heater is not desirable. It is inefficient and wasteful of energy. To produce domestic hot water, there must be a double heat transfer. The boiler water will first be heated with oil or gas, and it will then transfer its heat to the water in the tankless coil. To produce domestic hot water, the boiler must be heated all year long, winter and summer. As discussed in chapter 14, controls associated with the heating system prevent steam or hot water from being distributed throughout the house when heat is not required. Nevertheless, the heated boiler water during those months does represent wasted energy. It also introduces an additional heat load on the house at a time when heat is not desired. If you should decide to replace a tankless coil with a tank-type water heater, it can be done at a reasonable cost. An oil-fired water heater, however, is more expensive than a gas-fired or an electric unit.

Inspection procedure

The first portion of the tankless water heater inspection is performed in the bathroom. Turn on the hot water at the sink and the tub simultaneously and look at the flow. If the flow appears to be low, mineral deposits might have built up in the tankless coil, and maintenance might be needed. On the other hand, the low flow might reflect a constriction in the distribution piping. Record the fact that there is a low flow on your worksheet as a reminder that the condition must be corrected regardless of the cause.

At the boiler, look at the joints around the tankless heater mounting plate. This is a vulnerable area for leakage. If there are water stains and deposits around the joint, even if you do not see water dripping, maintenance is needed. The deposits tend to self-seal the leak, but this is not a permanent fix. Any movement or vibration of the boiler can cause the deposits to loosen, reactivating the leak. Often, all that is needed is tightening of the mounting bolts around the joint or replacing the gasket around the joint. Look at the inlet and outlet pipes to the heater and the mixing valve (if there is one) for signs of past or current leakage.

The temperature of the boiler water that heats the domestic hot water is controlled by an aquastat. You can check the operation of the aquastat by opening one of the hot-water faucets. After a while, the temperature of the boiler water will drop to a point where the aquastat will cause the oil or gas burner to fire. Tankless coils are designed to operate with a boiler-water temperature of about 200° F. Look at the boiler-mounted thermometer when the burner fires. If the boiler-water temperature is less than 180° F, the aquastat may be in need of adjustment. In some homes, the aquastat is intentionally set lower. This, of course, will result in a lower hot-water temperature.

Look for a temperature-pressure relief valve on the hot-water outlet line of the tankless coil. If there is a storage tank associated with the tankless coil, the relief valve might be located on the tank. A temperature-sensitive relief valve is a necessary safety control and is normally factory-set to discharge when the domestic hot water reaches a scalding temperature. If you do not find a relief valve, record the fact on your worksheet as a reminder to have one installed after you move in.

Tankless (wall-mounted) water heaters

A tankless (wall-mounted) water heater, also sometimes called an "instantaneous water heater," is an appliance that heats water as the water is used. It's about the size of a medium-capacity suitcase. Operating a hot-water faucet activates either a burner (natural gas or propane) or an electrical resistance heater that transfers thermal energy to a heat exchanger. As cold water flows through the heat exchanger, it is heated to a preset temperature, which is normally 120° F. However, for most units there is a temperature controller within the unit, which is accessible by removing the cover, that can adjust the temperature setting from 100° to 140° F. Some units have a remote control option for the homeowner to control the temperature. Tankless water heaters are designed for a maximum flow rate through the heat exchanger; exceeding the flow rate results in a cooler water outlet temperature. However, for high-demand applications, two or more tankless units can be connected together to operate as one. Closing the hot water faucet deactivates the heating mechanism.

One advantage of this water heater is that because of its compact size it can be placed close to the point of use so that very little of the water's heat is lost between where the water is heated and where it is used. However, if a gas-fired unit is installed as a retrofit, as opposed to an electric unit, it must be located in an area where the exhaust gases can be vented to the outside. One item of concern with a retro fit is the size of the gas line. It's important that the gas pipe being used meets the requirements of the new gas-fired tankless water heater. The gas requirement of the tankless unit often exceeds that of the existing tank-style water heater. If the water heater is a retrofit, you should check with the owner to determine whether the installation was approved by the local municipal building department. If the house is located in an area where there are exceptionally hard-water conditions, the heat exchanger inside the unit may require periodic flushing by a plumber to remove built-up scale. To minimize this problem, a water softener or a descaling device should be installed in the water line before the heater.

Indirect-fired storage water heaters

In the mid-1970s, a new type of water heater, an *indirect-fired storage water heater*, was developed. The unit is called indirect-fired because the medium used for heating the domestic water is boiler water, which itself is heated in a separate boiler used for either central heating or space heating. The boiler water circulates through a heat exchanger that is located either within the heater tank or in an external compartment. Figure 16-13 shows one type of indirect-fired storage water heater. This unit has a specially designed heat-exchanger coil mounted in the base of the tank. Hot boiler water flows through the coil, which is surrounded by domestic water. The design of the heat exchanger, which consists of an inner core surrounded by an outer helical coil, is such that it results in a relatively high recovery rate.

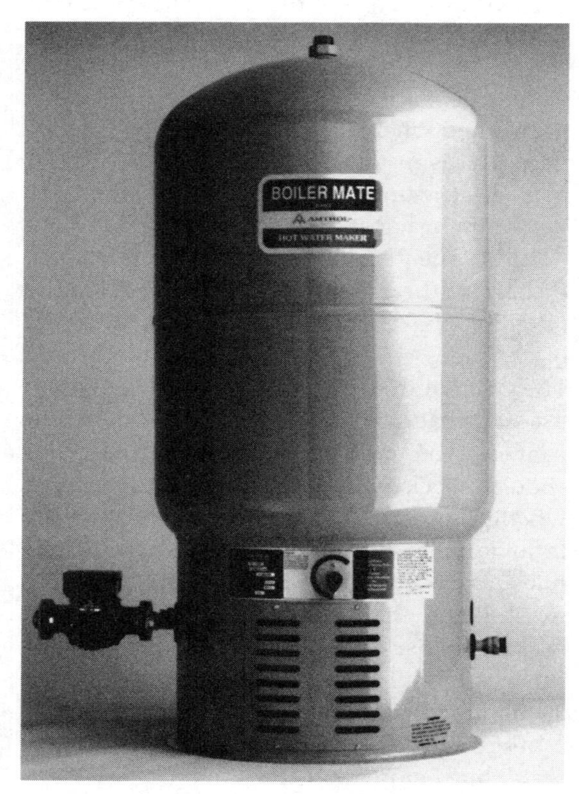

Fig. 16-13. *Indirect-fired storage water heater.*

The typical capacity of this residential indirect-fired water heater is about 40 gallons. With water heaters, whether there is an adequate supply will depend not only on the capacity but also on the recovery rate of the unit. The actual rate will depend on the Btu rating of the boiler that feeds the heat exchanger. For a typical house with a boiler that has an output rating of about 100,000 Btu per hour, the recovery rate is about 90 gallons per hour.

The indirect-fired storage water heater works off the heating system as a separate zone, utilizing its own circulator pump. According to manufacturers, 15–20 gallons of domestic hot water is removed before the zone aquastat calls for hot water from the boiler. Consequently, with the high recovery rate, the boiler firings are kept to a minimum.

In checking the unit, either raise the thermostat setting to HOT or let the hot water run until the boiler fires. You can tell if the pump is operating by touching the water supply to the heater before and after the boiler fires. The pipe should get hot. Look for signs of leakage around the base of the unit. Also, make sure that there is a pressure-temperature relief valve mounted directly on the tank casing. If you adjust the thermostat to activate the boiler and circulator pump, don't forget to reset it.

In hard-water areas, it is possible that even though the boiler is functioning properly, the domestic water does not get very hot. This is caused by a lime buildup on the heat exchanger, which acts as an insulator and inhibits the heat transfer. In this case, the coil is in need of a cleaning, although it may require replacement.

Checkpoint summary

- ❍ Is hot water supplied by a separate-tank type of water heater, a tankless coil, or by a wall-mounted tankless heater?
- ❍ If you have a tank-type heater, record the capacity (gallons), the recovery rate (gallons per hour), and the age (from serial number).
- ❍ Are exhaust gases spilling out of the draft diverter?
- ❍ Is tank-type heater adequate for house? (See table.)

Full bathrooms	Sum of tank capacity plus recovery rate
1	70
2	90
3	105
4	115

- ❍ Is hot-water flow adequate when two fixtures are turned on?
- ❍ Are hot and cold supply lines incorrectly installed (reversed)?
- ❍ Does water heater contain a properly installed temperature-pressure relief valve?
- ❍ Is the relief valve dripping?
- ❍ Is there an expansion tank on the cold water line?
- ❍ Look for signs of corrosion or past leakage:
  - — Tank type (gas-fired unit): corrosion dust and flaking metal in burner area.
  - — Tankless coil: rust and deposits around fittings and gasket.
- ❍ Is there an "instant" hot-water return line?

17
Air-conditioning

Before outlining the procedure for inspecting a central air-conditioning system, let me discuss the components of a typical residential system and their functions. Central air-conditioning systems provide comfort cooling by lowering the air temperature and removing excess moisture. This is achieved by recirculating air from the house across a cooling coil. As the air flows across the coil, its temperature drops, causing some of the moisture in the air to condense out on the coil. The cool, dehumidified air is then distributed throughout the house, and the moisture is disposed of through a drain.

The basic components of an air-conditioning system are the compressor, condenser, expansion device, and evaporator. The evaporator is the cooling coil mentioned above. The other components simply provide the means for the evaporator coil to cool the circulating air. This is done using a cooling medium, a *refrigerant*, that cycles between the components. (See FIG. 17-1.) The refrigerant is normally a gas at atmospheric pressure and temperature. By (1) applying pressure to the refrigerant and (2) removing absorbed heat, the refrigerant will change from a gas to a liquid. At this point, the refrigerant exists as a high-pressure liquid. When the pressure on the liquid refrigerant is released, the refrigerant (3) expands and changes back to a gas. In the process of changing from a liquid to a gas, the refrigerant (4) absorbs heat from its surroundings, thus cooling the air passing over the coil.

An air-conditioning system is a closed system, and theoretically there should never be a need for additional refrigerant. However, in practice the various fittings on the connecting pipes can loosen or develop hairline cracks that can allow some of the refrigerant gas to escape. Usually, when a system is low in refrigerant, the practice is to introduce additional refrigerant and not to look for the openings in the system that allowed the refrigerant to escape. However, if the air-conditioning system cannot hold a refrigerant charge for at least one season, the

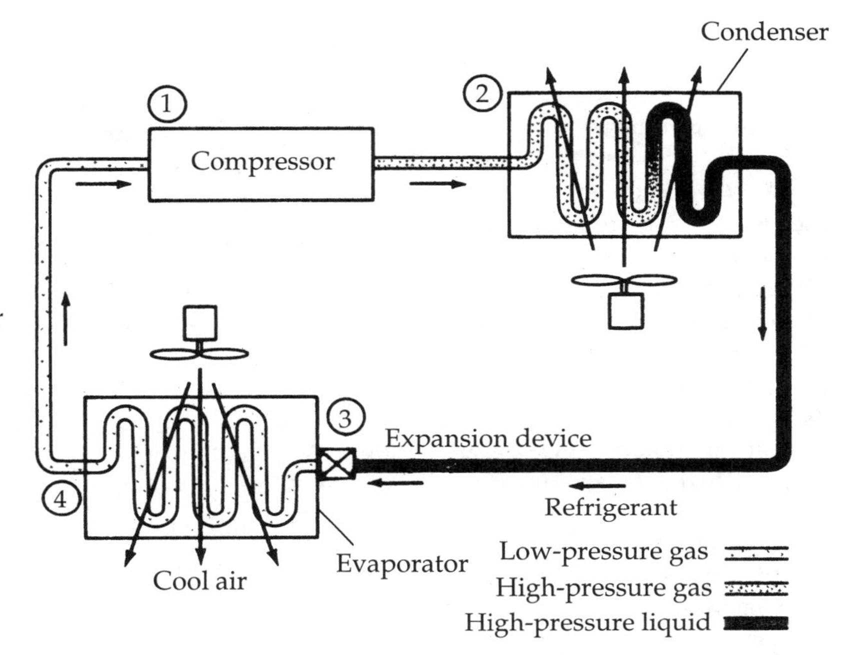

Fig. 17-1. *Basic air-conditioning cycle. Normally the compressor and condenser are contained in a single housing.*

leaks in the pipes or fittings should be located and corrected.

Air-conditioning capacity

The overall cooling capacity of an air-conditioning system is usually measured in Btus (British thermal units) or tonnage. One Btu is the heat required to raise the temperature of 1 pound of water 1° F. A rating in Btus per hour indicates the amount of heat that the unit will remove in that time. A *ton* of air-conditioning, historically, represented the cooling effect achieved by melting 1 ton of ice in twenty-four hours. An air-conditioning system that is rated at 1 ton will provide 12,000 Btus of cooling in an hour.

The cooling needed for a house depends on several variables, such as size of the rooms, whether there are cathedral ceilings, the number and type of windows, the amount of insulation in the walls and ceilings, and whether shading is provided by trees. The following rule of thumb can be used to determine the size of the air-conditioning system needed to cool the house that you are inspecting. One ton of cooling capacity (12,000 Btu) is needed for every 550 square feet in the structure. Therefore, if you are inspecting a house with 1,800 square feet, you will require an air-conditioning system rated at 3 tons. More recently, because newer homes are tighter and are better insulated, a figure of 700 square feet per ton has been used. The exact requirement can be determined by a professional. However, the rule of thumb is quite effective in determining whether the unit servicing the house is grossly oversized or undersized. Calculate your cooling requirements and then ask the homeowner the capacity of the air-conditioning system.

You can check the capacity yourself by looking at the data plate on the compressor-condenser. Even though most manufacturers donot put the system's capacity on the data plate, it can be approximated. The model number of the compressor on most air-conditioning systems contains digits that represent the approximate number of thousands of Btu of cooling capacity. See TABLE 17-1.

Another figure on the data plate that can be used for determining the cooling capacity is the full load amperage (FLA) or rated load amps (RLA). There are approximately 7 amps per ton of cooling. The FLA for the Tappan compressor in TABLE 17-1 is 27.1. When this figure is divided by 7, you get approximately 4 tons, which verifies the figure indicated by the model number.

An air-conditioning system should be properly sized or slightly undersized, but it should not be oversized. A unit that is too large operates intermittently. It quickly chills the air and then shuts down. When the system is shut off, moisture in the air does not condense, and the system does not dehumidify the air. For the system to remove enough moisture to make the air comfortably dry, the evaporator coil must be kept cold. This means that the compressor should run almost continuously for maximum comfort. Even though a system that is too small will run continuously, its cooling capacity is not adequate to remove enough heat to cool the room to a comfortable level.

Air-conditioning systems

There are basically two types of central air-conditioning systems used in residential structures—the *integral* system and the *split* system.

Integral system

This system, sometimes referred to as a *single-package unit*, is self-contained. That is, all the components—compressor, condenser, expansion device, and evaporator, plus the electrical controls and fans—are contained in a single housing. This system, which must be vibration-mounted, is often installed in the attic or a crawl space with ducts projecting through the exterior wall or roof to provide air for cooling the condenser. The integral system is less expensive than the split system. However, the noise level from the compressor makes an interior installation less desirable.

Split system

In this system, the compressor-condenser is physically apart from the evaporator coil. To eliminate the interior noise and provide outside air for cooling the condenser, the compressor and condenser are housed in a unit located outside the structure, usually at the rear or side. The evaporator coil, on the other hand, is located in the house, either in the attic or inside the heating system. The specific location often depends on the type of heating system. When the house is heated by forced warm air, the evaporator coil is usually located in the furnace plenum and utilizes the furnace blower to move the air through the coil. When the house is heated by steam or hot water, the evaporator coil is usually located in the attic with a separate blower (fan) to move the air through the coil and distribute it throughout the house. When there

Table 17-1. Air conditioner capacity.

Manufacturer	Model number	Btu/hour	Tons
General Electric	BTB390A	30,000	2½
Bryant	567CO36RCU	36,000	3
Carrier	38CC042-1	42,000	3½
Tappan Company	CM48-42C	48,000	4

is no attic, the evaporator unit can be located in a closet or the basement.

The compressor-condenser is connected to the evaporator coil by two copper pipes that contain the refrigerant that cycles between the two units. (See FIG. 17-2.) The diameters of the two pipes are different—one pipe is the size of a pencil, the other the size of a broom handle. The small pipe (pencil size) is the liquid line; it carries the high-pressure liquid refrigerant from the condenser to the expansion valve. The larger pipe (broom-handle size) is the suction line; it carries low-pressure refrigerant gas from the evaporator coil to the compressor. The suction line should be insulated. Usually it is covered with black foam-rubber type of insulation.

Inspection procedure Whether the air-conditioning system can be checked operationally depends on the outside air temperature. Most manufacturers do not recommend turning on the system at temperatures below 60° F because of the possibility of damage to the compressor. If the outside temperature is below 60° F during your inspection, do not start up the air conditioner. If the system cannot be checked prior to purchasing the house, the seller should provide you with a guarantee of its operational integrity. If the temperature is above 60° F during your inspection, walk over to the compressor and have someone turn down the thermostat that controls the air conditioner so that the system will begin to operate.

Compressor The compressor is the most important part of any air-conditioning system and the most costly to replace. Its projected life is about eight to ten years, although units have been known to last over fifteen years. In areas of the sunbelt with long air-conditioning seasons,

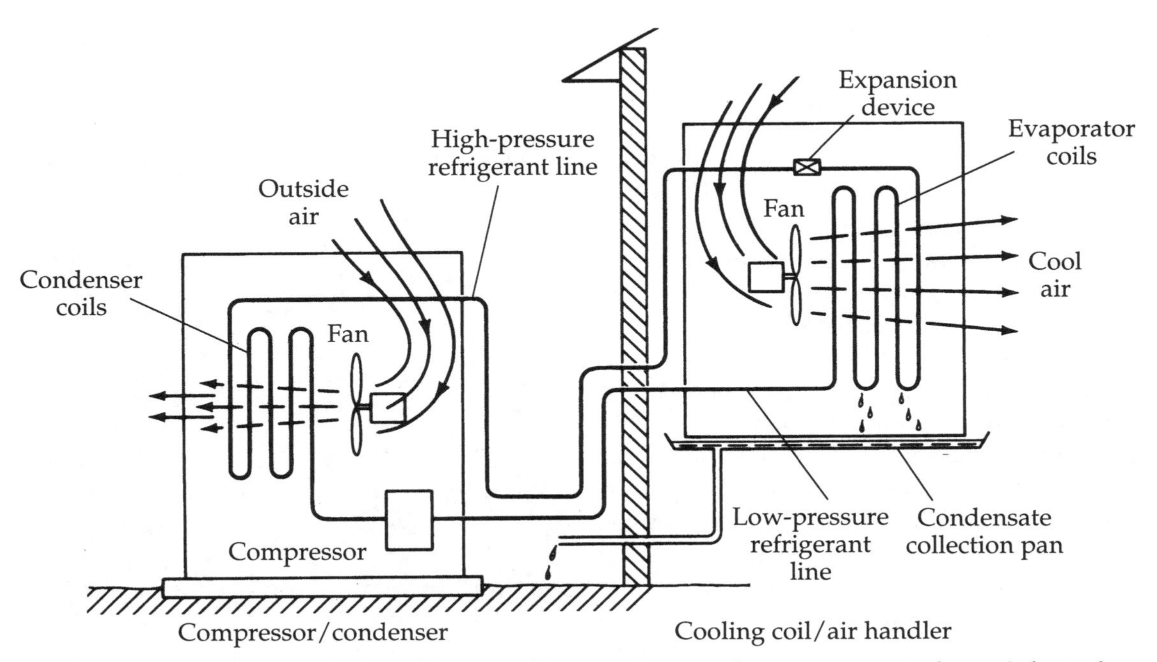

Fig. 17-2. Schematic diagram of a split air-conditioning system. The compressor-condenser is located outside the house, and the cooling coil is inside the house (usually in the furnace plenum or attic).

the projected life will be somewhat less. After the system has been turned on, listen for any unusual sounds. The compressor should start up smoothly. A straining, grunting, groaning, or squealing noise indicates a problem condition that should be checked and corrected by a competent service organization. Shut the unit off and note the condition on your worksheet. Once the compressor starts up smoothly, it should then operate continuously without any noise (except a low hum) or squeaks. The compressor should not operate in short cycles (on and off repeatedly). If this occurs, there is a problem condition and the unit should be shut down.

When the air-conditioning system is turned on, the fan associated with the condenser should begin to operate at the same time as the compressor. Look inside the unit to see if the fan is turning. If the interior portion is not visible and you cannot see whether the fan is operating, place your hand over the unit. Air rushing over your hand is an indication that it is operating. After the system has been operational for about fifteen minutes, the air being discharged through the condensing coil should be warm. This air is removing the heat that has been generated during the compression of the refrigerant. Air that is not warm is usually an indication that the compressor is not operating properly, a condition that should also be checked by a competent service company.

After the system has been operational for about fifteen minutes, look at the low-pressure refrigerant line (the pipe about the size of a broom handle). Usually the pipe is covered with insulation. If a section of the pipe is exposed, grab it with your hand. If the compressor is working properly and there is an adequate refrigerant charge, the pipe will be quite cool to the touch. On many occasions, the pipe and end fittings will be "sweating" as a result of condensation. This is a normal operating condition. However, a low-pressure line covered with frost usually indicates a deficiency in refrigerant. Even though cooling can be obtained with an air-conditioning system that is deficient in refrigerant, the efficiency of the system is greatly reduced.

Occasionally you will find a sight glass on the small-diameter refrigerant line. The sight glass is a small device installed directly into the high-pressure liquid refrigerant line that allows the homeowner to see whether there is a problem with the flow of the liquid refrigerant. The refrigerant is colorless, and if the system is operating properly, nothing unusual will be noted when looking at the sight glass. However, when the system is low in refrigerant, bubbles will show up in the liquid as it passes under the glass. Record it on your worksheet.

The location of the compressor-condenser is important for efficient operation. (See FIG. 17-3.) The compressor should be located where it will receive a minimum of direct sunlight, since the cooler the air flowing across the condenser, the more efficient the cycle. See if the compressor is positioned so that the condenser air intake is at least 12 inches away from any obstruction

Fig. 17-3. *Typical compressor-condenser for a split air-conditioning system. Airflow through unit must be unobstructed.*

or dense shrubbery. If it isn't, there will not be an adequate airflow for condenser cooling.

The compressor-condenser can be a noisy piece of equipment and should be vibration-mounted on a concrete slab or precast concrete blocks that will not settle. The unit should be level. Excessive uneven settlement can cause fractures in refrigerant-line fittings and thereby allow the refrigerant to escape.

Look for an electrical disconnect switch on the exterior wall near the compressor. The purpose of this switch is to allow the maintenance man to disconnect the unit so that if someone in the house unknowingly turns the thermostat down, the unit will not be activated while he is making repairs. The absence of a disconnect switch should be recorded on your worksheet. The overall compressor-condenser unit should be checked to see whether it is in need of a cleaning. These units require periodic cleaning because leaves, seed pods, twigs, and dust tend to clog the condenser, thus restricting the airflow. The condenser has fins like an automobile radiator's, which can clog easily.

The condenser discussed above is air-cooled. On occasion, you might see a condenser that is water-cooled. A water-cooled unit can be located inside the structure since it does not require outside air. These condensers are generally not used for residential structures because they are quite wasteful and costly to operate. To cool the condenser adequately, cold water flows through a jacket around the coils absorbing heat and is then directed into a sink or floor drain. If the central air-conditioning system operates for twelve hours a day, this type of system can waste several thousand gallons of water a day. If you find this type of condenser in the air-conditioning system, you should consider its replacement.

In large air-conditioning systems such as those found in apartment buildings, water-cooled condensers are not wasteful or costly to operate because the cooling water is recirculated. After the water absorbs heat in the condenser, it is pumped up to a cooling tower, often located on the roof, where it loses its heat and is then recirculated to the condenser.

Evaporator After checking the compressor-condenser, you should inspect the evaporator unit, commonly called the cooling coil. The evaporator will usually be located in the attic or the basement. When the unit is located in the attic, you will often find the refrigerant lines from the compressor running up along the outside of the structure and entering the building at the attic level. When the evaporator coil is located in the basement, either as a separate unit or in the furnace plenum, the refrigerant lines from the compressor are short and run directly into the structure. If possible, the evaporator coil should be observed after the unit has been operational for about thirty minutes. The coil might not always be accessible because it may be covered with a sheet-metal casing that cannot be easily disassembled. If the evaporator is accessible, look at the coil and the associated refrigerant tubing. The evaporator is used to cool and dehumidify the circulating air. If you notice a frosting condition (a buildup of ice) on portions of the coil and refrigerant tubing rather than dripping water, the system is not operating properly. The frosting is usually the result of an insufficient airflow through the evaporator coil or an inadequate amount of refrigerant in the system. This condition should be indicated on your worksheet.

Furnace-mounted evaporator When the house is heated by forced warm air, the most common location for the evaporator is in the *furnace* plenum. A furnace-mounted unit takes advantage of the ducts that have been installed for heating the house and also uses the heating-system blower to circulate the cool air.

You can tell if the evaporator coil is located inside the furnace plenum by whether there are refrigerant lines entering the sheet-metal casing of the plenum. The refrigerant lines inside the casing are connected directly to the evaporator coil. The most common type of coil that is found in a furnace plenum is a two-section design, an *A-coil* (because of its shape). Sometimes an inclined or horizontal coil is used. (See FIG. 17-4.)

Below the evaporator coil is a pan that collects the water condensing out of the circulating air. The water is then removed by means of a plastic drain line that will be visible when looking at the furnace plenum. Look for it. Depending on the location of the furnace, the condensate drain line will run to a nearby sink where the condensate drips down the drain or will run through the foundation wall, where the condensate drips on the outside. When the condensate drain line is extended through the foundation wall, there should be a splash plate below the end of the pipe so that the dripping water can be directed away from the foundation.

Fig. 17-4. *Evaporator (cooling) coil mounted in furnace plenum.*

Sometimes the condensate drain line runs from the furnace down to a small hole in the floor slab. The condensate trickling out of the drain line accumulates below the slab. This method of removing the condensate is not very desirable in those areas where the water will not readily drain because of a high water table or a high clay content in the soil. Even though the amount of water discharging from the condensate drain is small, the introduction of additional water could aggravate a condition that makes the lower level of the structure vulnerable to water seepage.

The condensate drain line should terminate at the plenum with a U-shaped trap. Since the condensate drain line is an open pipe leading directly into the cooling coil, the trap, which is on the negative side of the blower, prevents the blower from sucking air up through the drain line. This would inhibit the condensate from flowing out and could cause a flooding problem. Look for a trap. If it is missing, record that on your worksheet.

Occasionally the condensate drain discharges into a small rectangular box located near the furnace. This box is the reservoir for a lift pump. The purpose of the pump is to lift the condensate to a level where it can then flow to any desired location. Without a pump, it is often necessary to position the drain line so that it blocks a portion of the room or interferes with foot traffic. Check the pump's operation. These pumps have a float control that is activated when the water reaches a preset level. If there is only a small amount of water in the reservoir, the pump can be checked by pouring water from a glass into the reservoir. A malfunctioning pump should be recorded on your worksheet.

Now look at the overall furnace plenum around the evaporator coil. Rust and mineral deposits indicate a past or present problem in condensate removal. Water overflowing the condensate drain pan can damage the heat exchanger below. If you see this condition, you should have the furnace heat exchanger checked by a heating contractor for signs of deterioration. (See chapter 14.)

When the evaporator coil is located inside the furnace plenum, the blower for the heating system is also used as the blower for circulating the cool air. Because cool air is heavier than warm air, when the blower is used for air-conditioning, it should operate at a higher speed. Most often, however, the furnace is equipped with a one-speed motor. Consequently, the air-conditioning system is often not as effective as it might be. A pair of double- or triple-sheaved hubs can be installed to allow multispeed operation. When the blower is turned off by the master switch, check the tension in the fan belt. There should be no excessive slack. Press the belt midway between the pulleys. If the belt gives more than ¾ inch to an inch, it is too loose, and adjustment is needed. When the blower is operating, listen for any unusual noises or vibrations. They should be recorded on your worksheet. You might also ask the owner when the unit was last serviced—there is no substitute for periodic maintenance.

Blower coil When the evaporator coil is housed in a separate casing that contains a blower for circulating the cool air, the coil is commonly referred to as a *blower coil*. Most often, the blower coil is located in the attic. However, it can be located in a closet or in the basement. The blower coil should be vibration-mounted to prevent the noise of the blower unit from being transmitted into the living area. Vibration mounting can be achieved by placing the unit on rubber, cork, or styrofoam pads. (See FIG. 17-5.) The vibrations might also be isolated in the attic by suspending the unit from the roof rafters.

Fig. 17-5. *Attic-mounted blower coil. Unit is resting on a styrofoam pad to minimize vibrations. Below the unit is an auxiliary condensate drain pan and associated drain line. Note that the main condensate drain line does not have a U-shaped trap.*

The base of the blower coil is basically a condensate collection pan. The accumulated condensate is removed by means of a drain line that will extend through the exterior wall, terminating on the outside, or extend through the lower portion of the roof, terminating in the gutter. Sometimes the condensate drain line terminates in the plumbing vent stack located in the attic. (See FIG. 17-6.) In many communities, this type of termination is not permitted because it is not in compliance with the plumbing code. If the drain terminates in the vent stack, record that on your worksheet. The legality of this type of connection should then be verified with the local building department.

The purpose of the vent stack is to channel sewer gases in the plumbing system to the outside. If the condensate drain line is connected to the vent stack and there is no trap on the drain line, the sewer gases may back up into the condensate drain line, enter the blower coil, and be circulated throughout the house. An additional concern is bacterial growth developing in the air handler. The condensate drain line should have a U-shaped trap near

Fig. 17-6. Air-conditioning condensate drain line terminating in plumbing vent stack. In many communities, this type of termination does not comply with the plumbing code.

its connection to the blower-coil housing. On many installations, this trap is omitted. Look for it. If it is missing, one should be installed.

When the blower coil is located in the attic, certain steps must be taken to prevent cosmetic damage to the ceiling below in the event of a blockage in the main condensate drain line. Some blower-coil housings have a fitting for an auxiliary drain line that is located just above the main condensate drain fitting. If the main drain becomes clogged, the level of the condensate will rise and be drawn off by the auxiliary drain.

For those blower coils that do not have a fitting in the housing for an auxiliary drain line, there should be an auxiliary drain pan below the unit. The auxiliary pan will collect any condensate that overflows from the main pan when there is a blockage in the main drain line. Look for an auxiliary drain pan. In some parts of the country auxiliary drain pans are installed when the blower coil is located over any furred space, even when the blower housing has an auxiliary drain fitting. Record the absence of one on your worksheet. Unfortunately, many air-conditioning contractors do not install the auxiliary drain or drain pan. Because rising costs make it difficult to remain competitive, they cut costs wherever they can.

The auxiliary drain pan must have a separate drain line that discharges to the outside. It should not be connected to the main drain line. (See FIG. 17-7.) If it is, it reflects poor-quality workmanship; if the main drain line becomes clogged near the discharge end, the auxiliary drain line will also not function.

If the evaporator coil is accessible, it should be inspected for frost buildup. From an efficiency point of view, the attic is the least desirable area for locating the blower coil because of the high temperatures, easily reaching the 140° to 150° F that normally occur during the summer. Even though the blower coil is insulated, there will be a heat gain because of this high temperature. The overall attic temperature, however, can be lowered by increasing the number or size of the attic vent openings. A ridge vent is quite effective, as is a thermostatically controlled power ventilator.

Ducts After the air-conditioning system has been operational for about fifteen minutes, the air discharging from the registers should be felt to determine whether it is relatively cool. The temperature of the air discharging from the supply registers should be about 15 degrees lower than the temperature in the room. If the air does not have a slight chill, it might be because there is a heat gain along the duct leading to that register as a result of inadequate insulation, or the system may be undersized or low in refrigerant.

While checking the temperature of the air leaving the supply registers, also check the airflow. If the air discharging from the registers has a low flow and appears to be

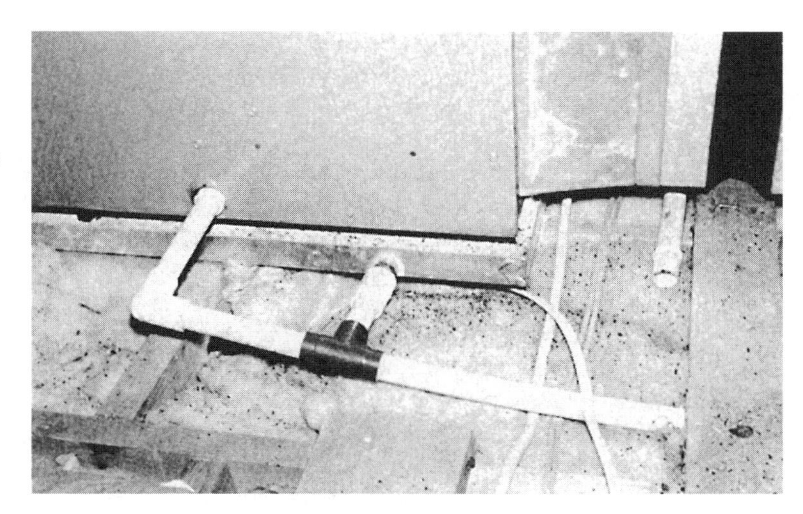

Fig. 17-7. *Auxiliary condensate drain line connected to main condensate line. This negates the use of an auxiliary drain and reflects poor-quality workmanship. If the main drain becomes clogged near the discharge end, the auxiliary drain will not function. Note that the U-shaped trap is missing.*

sluggish, it might indicate that there is an obstruction within the system caused by dirty filters or icing on the evaporator coils. Sometimes the condition is caused by an undersized fan or the need for balancing the airflow between the registers. In any case, the condition is abnormal and should be recorded on your worksheet.

As with a heating system, the location of the supply registers is important for effective air-conditioning. Since cool air is heavier than warm air, the cool air will tend to accumulate near the lower portion of the room and the warm air near the top. As a result, there is usually a temperature difference between the ceiling and floor. This stratification of heat layers can be minimized by adequate circulation in the room. Adequate circulation can be achieved by locating the supply registers on the opposite side of the room from the return grille. When the return grille is near the supply register, the air discharging from the supply is drawn in by the return grille and does not have a chance to circulate adequately around the room. In many houses, the rooms do not have individual return grilles. Instead there might be a large central return located in the hall. In these cases, the supply registers should be located on a wall that will allow the supply air to circulate completely prior to being drawn off and returned to the central grille. Also, the doors to the individual rooms must be undercut so that when they are closed, the supply air will be able to flow to the return grille.

Ideally, air-conditioning supply registers should be located in or near the ceiling. To minimize air-conditioning installation costs, rather than install new ducts many new homes use the ducts and registers provided with a forced-air heating system. These registers are usually located near or at the floor level and are quite effective for heating purposes. However, when they are used for air-conditioning, they are less effective and tend to increase the stratification effect. Some houses that have forced-warm-air heating have what are called *high-low registers.* The duct supplying the heat register is extended vertically to a point near the ceiling level where it terminates at another register. When the system is used for heating, the damper controlling the upper register is manually closed, and the lower register is opened. When the system is used for cooling, the damper controlling the lower register is closed, and the upper register is opened. This type of arrangement is very desirable and is often found in high-quality construction.

The presence of a central air-conditioning system does not mean that the entire house is air-conditioned. Look specifically for registers as you walk through the house. In many raised ranches, I have found that the upper level is air-conditioned and the lower level is not. If you have any questions, check each room while the air-conditioning system is operating.

There are two basic types of ducts used in residential structures—sheet metal and glass fiber. While checking the distribution portion of the air-conditioning system, look for exposed ducts. The glass-fiber type of duct is by its very nature insulated. However, the metal duct may or may not be insulated. The fact that there is exposed metal on the outside does not mean that the duct is not insulated. The insulation might be located inside the duct. Whether the metal duct is insulated can be determined by feeling the duct when the system is operating (if no insulation, the duct will be quite cool) or by striking the duct with your fingernail. If there is no insulation, you will hear a ringing sound, and if there is insulation, you will hear a dull thud. All ducts that lead through unfinished areas such as crawl spaces and attics must be insulated so that the cool air flowing through the ducts will not absorb heat from its surroundings. Pay particular attention to the joints for indications of air leakage. Any open joints should be sealed with inexpensive duct tape. Also, whether the evaporator coil is located inside the furnace plenum or in the attic, check the joints around the housing for air leakage. Very often, there are open joints that must be sealed.

Heat pump

A heat pump is a year-round air-conditioning system that provides warm air during the winter and cool air during the summer. It is basically a compressor-cycle air-conditioning system (similar to the one described previously) that can operate in reverse. During the reverse operation, the condenser functions as an evaporator, and the evaporator functions as the condenser. The overall refrigerant cycle, however, remains the same. (See page 262 and FIG. 17-1 for the operational flow details.)

When the system is operating, the condenser (which is located in the house) is cooled by the air that is circulated around the house. As this air passes over the condenser, it absorbs heat that is used for heating the house. For the cycle to operate properly, the evaporator (which is located on the outside) must absorb heat from the outside air. Even when the outside temperature drops to as low as 20° F. the evaporator can absorb heat because the refrigerant within the evaporator is at a lower temperature. However, as the temperature of the outside air drops, the ability of the evaporator to absorb heat decreases, decreasing the effectiveness of the heat pump. Even though a heat pump may be operational at temperatures as low as 20° F, the Btu output is sufficiently reduced so that auxiliary heaters are usually required.

A heat pump should be sized for the air-conditioning load and not the heating load. Otherwise, the air conditioner will be oversized. Except for a small section of the Deep South, the heating load on a house will always be greater than the cooling load. A heat pump will produce approximately 20 percent more Btus per hour for heating than it does for cooling. Consequently, in most parts of the country, when a house is heated by a heat pump, auxiliary heat will also be needed. For example, in the New York area, a typical eight-room house would require a furnace that could produce about 100,000 Btus per hour for winter heating and also a 3½ ton (42,000 Btu/hour) air-conditioning unit for summer cooling. If a properly sized heat pump was used for heating, it would produce only 50,400 Btu/hour. The remaining 49,600 Btu/hour would have to be provided by auxiliary heaters, which are usually electrical resistance heaters.

The auxiliary heaters in heat pumps are automatically activated when the pump cannot supply sufficient heat to keep up with the heat loss of the structure during the winter. In northern communities where there is a considerable amount of moisture and low temperatures during the winter, there is a tendency toward an ice buildup on the metal fins of the outdoor evaporator coil. An excessive ice buildup could cut off air circulation across the coil and result in a loss of heating capacity. With some heat pumps, this icing condition is automatically controlled by a defrost cycle that reverses the flow of the refrigerant for a short time. The hot refrigerant heats the outdoor coil and melts the ice. During the defrost cycle, the auxiliary heaters are usually energized to offset the cycle's cooling effect on the indoor circulating air.

A heat pump can operate in either the heating or air-conditioning mode. Most manufacturers suggest that the unit be operated in the air-conditioning mode when the outdoor air temperature is above 65° F (unlike a regular air conditioner, where the recommended temperature is 60° F) and in the heating mode when the temperature is below 65° F. Operating a heat pump in the wrong mode can result in damage to the compressor.

The components and problems of heat pumps are basically similar to those of air-conditioning systems. Consequently, the overall inspection procedure outlined earlier in this chapter for air conditioners should be used when inspecting heat pumps. However, you should not check both modes of operation. As long as the unit is functioning properly in the mode tested, it is an indication that the major and most costly components (compressor, fans, and coils) are operational. The system should then also function in the opposite mode. If it doesn't, a faulty reversing valve is usually the cause.

Evaporative cooler

One of the benefits of air conditioning is the dehumidification of the circulating air. This benefit is not without cost. Cooling and dehumidifying the air is more costly than cooling alone. In the southwestern part of the United States, the outdoor air is relatively dry, and dehumidification is not necessary. In this area, cooling can be achieved by means of an evaporative cooler. Because of the low humidity, water readily evaporates. In the evaporation process, the water absorbs heat from its surroundings and lowers the temperature.

The typical evaporative cooler consists of a sheet-metal and plastic casing containing a fan, pads, filter, and a water source. The pads, which hold the water, can be wetted by a spray, by a trickling stream, or by passing through a reservoir on a rotating drum. In some units, the wetted pads also function as air filters. As the air passes over or through the pads, it is cooled by the evaporating water and then distributed throughout the house.

When inspecting an evaporative cooler, turn the unit on and listen for any unusual sounds or vibrations in the blower compartment. Also look for signs of water leaks and check the pads for deposits and crusting. For efficient operation, the pads may require cleaning or replacement.

Checkpoint summary

General considerations

- ❍ How old is the air-conditioning system?
- ❍ When was the unit last serviced?
- ❍ Do not turn system on if the outside air temperature is below 60° F.

Compressor-condenser

- ❍ Check compressor during startup. Listen for any unusual sounds, such as straining, groaning, or squealing.
- ❍ Does compressor operate smoothly without short-cycling (repeated startup and shutdown)?
- ❍ Is condenser fan operating properly?
- ❍ Does compressor appear to be functioning properly? (Warm air should be discharging from unit.)
- ❍ Are there indications that the system is low in refrigerant (frosting on low-pressure refrigerant line or air bubbles in sight glass)?
- ❍ Is compressor-condenser located properly for maximum effectiveness (minimum sun exposure and unrestricted airflow)?
- ❍ Is unit in need of a cleaning (clogged with leaves, twigs, dust, etc.)?
- ❍ Check that unit is level and adequately supported by a concrete pad or blocks.
- ❍ For safety and maintenance, check for a main electrical disconnect for the compressor located near the unit.

Evaporator (cooling coil)

- ❍ During operation if possible, check cooling coil for frosting (ice buildup), usually the result of an insufficient airflow or lack of refrigerant.

Furnace-mounted evaporators (installed in the furnace plenum)

- ❍ Check for signs of leakage, mineral deposits, and areas of rust and corrosion.
- ❍ Note method of condensate discharge. Is it
  —to a nearby sink?
  —to the exterior?
  —to a floor drain?
  —to a hole in the floor slab (less desirable)?
- ❍ If condensate discharges into a reservoir lift pump (small rectangular box), check operation of pump.
- ❍ Check blower motor for unusual noises and vibrations. Is blower a single- or two-speed unit?

Blower coil (housed in a separate casing and most often located in the attic)

- ❍ Is unit vibration-mounted?
- ❍ If access is available, inspect evaporator coil for frost buildup.
- ❍ Check for a condensate drain line.
- ❍ Does this drain line discharge the condensate to a nearby roof gutter or directly into a plumbing vent stack? (The latter type of connection is usually not permitted and should be verified with the local building department.)
- ❍ Check blower-coil casing for auxiliary drain-line fitting; if not present, check for auxiliary drain pan below the unit. The auxiliary pan should contain an independent drain line that is *not* connected to the main drain line.
- ❍ If blower coil is located in attic, is the attic adequately ventilated?
- ❍ Is ventilation provided by ridge vent or thermostatically controlled power ventilator?

Ducts or registers

- ❍ Check airflow and temperature after fifteen minutes of operation.
- ❍ Note type of supply registers—ceiling units, combined type (heating and air conditioning), or high-low registers.
- ❍ Check whether all rooms are air-conditioned.
- ❍ Are supply registers and return grilles efficiently located?
- ❍ If rooms do not have individual return grilles, check for a large central return grille (often located in the hall).
- ❍ Are doors to the rooms undercut to permit proper air circulation?
- ❍ Check ductwork for open joints, signs of air leakage, and uninsulated ducts (particularly in the attic and crawl spaces).

18

Swimming pools

Swimming pools, like central air-conditioning systems, are no longer considered a luxury that only a privileged few can afford. New construction methods and materials have considerably lowered the cost of construction. Swimming pools can be installed either in-ground or aboveground. You will find them with various sizes and shapes that can fit into almost every budget. Because the shape of an in-ground pool can be freeform, the shapes available are limitless. On the other hand, the shape of aboveground pools is basically limited to circular, rectangular, or oval. There are three types of construction for the shell of an in-ground pool: concrete, vinyl lined with sidewall supports, and preformed fiberglass. These items as well as the pool's associated equipment and accessories will be discussed below.

Concrete pools

Concrete pools can be constructed using any of four methods: shotcrete, gunite, poured concrete, and concrete block. Both shotcrete and gunite are applied and sprayed from a hose that is directed behind, over, and above previously installed reinforcing rods (re-bars). (See FIG. 18-1.) Shotcrete is premixed wet concrete, and gunite is a dry mix, which combines with water as it discharges from the hose. The spraying of the concrete mix allows complete freedom of size and shape because it can follow the contours of any excavated shape. Poured concrete and concrete block walls are more restrictive with regard to the shape of the pool. Whichever method is used to construct the shell, an interior finish must be applied to provide a waterproof surface.

The most common finish used on a concrete pool is plaster. It gives a smooth finish to the pool and also provides a nonskid walking surface. (See FIG. 18-2.) The projected life of a plaster finish is about 7 to 10 years depending on how well the pool and water quality has been maintained over the years. In many homes where the pool and the chemistry of the water have been diligently maintained, the plaster finish has lasted considerably longer.

Fig. 18-1. Gunite being sprayed around reinforcing rods for a concrete swimming pool.

Fig. 18-2. Plaster finish being applied to a concrete pool.

Other finishes are paint, fiberglass coating, and ceramic tile. Tile is by far the most expensive; so most homeowners opt for a decorative ribbon of tile above the water line. The top edge of the pool shell is covered with coping, which prevents water from getting behind the shell. It can also serve as a handhold for swimmers and as a shove-off point into the water. The coping can be in the form of precast coping stones, flagstones, or brick, or it can even be the extension of a concrete deck over the edge of the pool shell.

Vinyl-lined pools

The advent of vinyl-lined pools brought the ownership of swimming pools to within the reach of many families. Construction of the pool shell consists of two phases: building of the perimeter frame or sidewalls and installing the liner. The most common sidewall panels are galvanized steel or aluminum. The frame must be structurally adequate to support the various forces exerted on the pool. After the sidewalls are bolted together, the bracing is installed. (See FIG. 18-3.) Then the area around the bracing is backfilled so that it is flush with the surrounding terrain. The sidewalls must have a smooth surface so as not to abrade the vinyl. They also don't have to be waterproof since the vinyl lining is. Prior to installing the liner, the base must be prepared. The bottom of the liner generally rests on a 2- to 3-inch sand base, and the top is secured with a special coping. The life of a vinyl liner is affected by the sun's ultraviolet light and the chemistry of the water. Manufacturers generally guarantee a liner against defective workmanship for 10 years. As vinyl pool liners age, the material becomes less supple and more prone to leaks from cracks or damage. It is possible to patch a small hole in a vinyl pool with a similar vinyl sheet material and a waterproof adhesive. Patching, if done correctly, can be effective, but once a liner has reached the age when several patches are needed, it is worth considering replacing the liner.

Aboveground pools are also vinyl lined. They are considered portable pools in that they can be dismantled and moved to a new location; however, many pools are partially or fully surrounded by a deck and are more permanent. The walls of the pool must be self-supporting and capable of withstanding the pressure being exerted on them by the water. The walls are usually galvanized steel, plastic, or aluminum and are 48 or 52 inches high. Because of the height of the walls, most municipalities do not require a safety fence surrounding the pool.

Preformed fiberglass pools

There are a number of advantages to a preformed fiberglass pool. The major advantage is its durability; the pool can flex and not get damaged. Also time is not lost in constructing the pool shell. After the site is excavated and prepared, the pool shell, which is brought to the site by truck, is lifted by a crane and placed into the excavation. (See FIG. 18-4.) Another advantage is the low maintenance. It is difficult for algae to cling to the sides of the pool because of its very smooth surface. As a result, the walls are very easy to clean. The main drawback of preformed pools is that

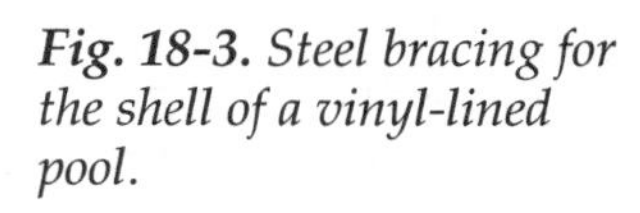

Fig. 18-3. *Steel bracing for the shell of a vinyl-lined pool.*

Fig. 18-4. *Preformed fiberglass pool shell being installed in excavated area.*

they are limited in size and shape. However, once the installation of the pool is complete, you cannot readily tell the difference between it and an on-site constructed pool.

Pool equipment

The standard equipment associated with circulating the water in a swimming pool are the pump and motor, filter, heater, and surface skimmer. The water circulation system is a closed system. The pump, which is driven by the motor, is normally located on the intake side of the filter. It draws water from the pool and forces it through the filter where dirt particles are removed. The water then flows to the heater and then back to the pool. To prevent the pump from getting clogged, there is a strainer basket on the water inlet side of the pump, which catches hair, lint, and other debris.

There are three basic types of filters available for swimming pools: sand filters, cartridge filters, and diatomaceous earth (DE) filters. The filters have no moving parts, and any one of the three will provide effective filtration. No matter which filter is used, the filtering medium will require periodic cleaning. The process used to clean DE and sand filters is backwashing, which is sending water backward through the filter, thereby flushing the debris onto the lawn, into the street, or to an approved drain line or sewer. Cartridge filters are not cleaned by backwashing; the cartridge is simply removed from the casing and washed. Most filters have a pressure gauge that is mounted on the top of the casing. The normal operating pressure is usually around 10 to 12 psi, although for some of the newer filters, it may be between 5 and 7 psi. As the filtering medium becomes dirty and clogged, it takes more pressure for the water to flow. It is generally time to clean the filter when the filter pressure increases approximately 10 psi above the normal operating range.

A swimming pool heater is not mandatory, but most houses with pools have one for comfort and because the heater can extend the swimming season. Most swimming pools are heated with a gas-fired heater, although they can also be heated with an electric heater and to a limited extent with solar heating or a heat pump. The ignition system in gas-fired heaters will be either a pilot light or electronic spark ignition. Either natural gas or propane gas (LP) can be used as the fuel for firing the heater. However, a heater that is designed for natural gas should not be used with propane gas and vice versa, because it will not operate properly. Keeping the pool water in the proper chemical balance is very important not only for health reasons and maintaining the quality of the water but also to extend the life of the pool heater. Water that is out of balance will result in a scale buildup within the heater, which if not corrected will cause a blockage of water circulation.

Most swimming pool shells have at least one to two built-in surface skimmers with skimmer baskets that are tied into the water's circulation system. When the pump is operating, dirt, leaves, oils from lotions, algae, and other debris that float on the water's surface are drawn into the skimmer by floating over the entrance weir. (See FIG. 18-5.) The baskets trap and collect the larger debris, and the oils and dirt particles circulate back to the filter where they are removed from the water. Some pools have skimmers that are also connected to a pipe located below the water level. This pipe prevents air from being sucked into the circulation system when the water level is down. Otherwise, when the pool water level is below the bottom of the weir, air will be drawn into the circulation system, which could damage the pump. The pump is not designed to run dry. Running a pump with air in the system or running it dry can

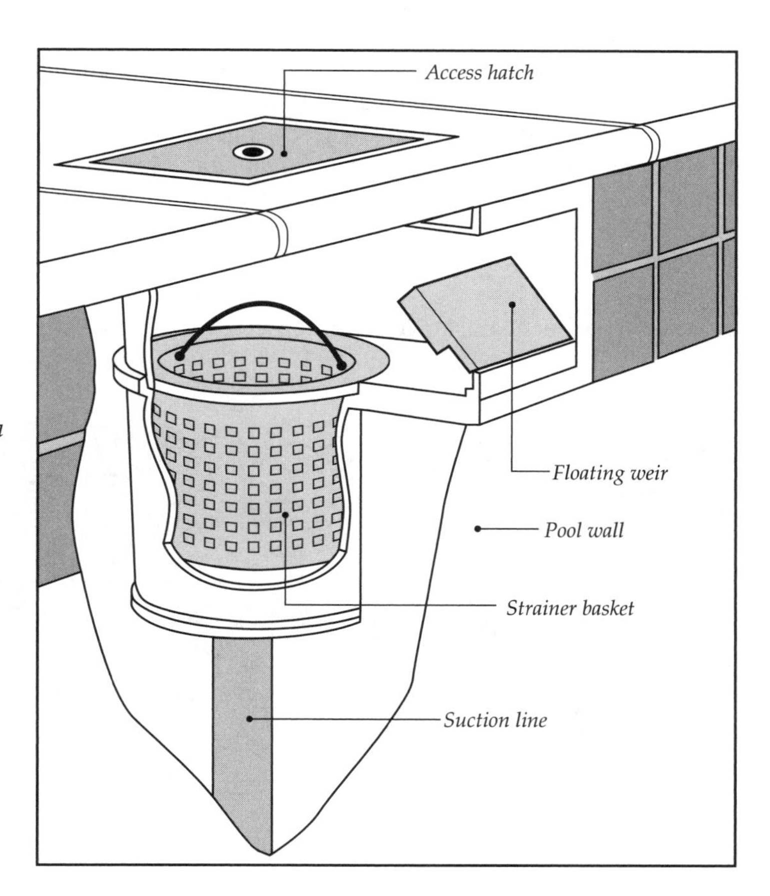

Fig. 18-5. *Cutaway view of a typical surface skimmer for a concrete pool.*

cause overheating and seriously damage both the pump and the motor.

Pool accessories

There are a number of accessories associated with a swimming pool, but the ones of concern from an inspection point of view are the pool cover, diving board, ladder, slide, and grab rails.

The main use for a pool cover is to keep debris out of the pool. It is mostly used when the pool will not be used for an extended period of time, such as during the winter months. (See FIG. 18-6.) However, depending on the material and design, the covers are also used by some homeowners during the swimming season to minimize heat loss and maximize heat gain. There are also safety covers that prevent small children or animals from falling into the pool. These covers are reinforced mesh with spring-loaded straps that hook onto the deck. There is also a "thermal or solar" cover available for a pool. It is essentially a sheet of vinyl bubble wrap that floats on the water. One puts it over the pool during swimming season because it allows the sun to heat the water and because it provides a barrier against evaporation.

Diving boards are available in spring-assisted and simple platform models. They are covered with fiberglass to make them waterproof and topped with a nonslip coating. Because of the danger of injury, some homes have jump boards rather than diving boards. Jump boards are shorter and are considered safer than diving boards. The concern about ladders, slides, and grab rails is whether they are adequately anchored to the deck.

Inspection procedure

In most municipalities a swimming pool is considered to be an auxiliary structure, and, as such, a building permit is required prior to construction. After completion, depending on the municipality, a Final Inspection Certificate, a Certificate of Compliance, or a Certificate of Occupancy is normally issued. If the house that you are inspecting has a swimming pool, check to see if there is a completion certificate for the pool.

There are certain limitations in a home inspection with regard to a swimming pool that you should be aware of. With no splashing, the average water loss in one week from evaporation is less than 1 inch. There

Fig. 18-6. *Pool cover keeps debris out when pool is not in use for extended period of time.*

could be a leak in the underground piping that causes the water level to drop more than what would be expected over a 24-hour period. This is a major problem to correct, but it will never be picked up during a home inspection. Pools lose water continuously through evaporation, splashing, and on the bodies of swimmers as they get out of the water. This is normal. However, a steady drop in the water line even when the pool is not being used may indicate a leak. This item should be discussed with the seller. In addition, the chemical composition of the water is generally not checked as part of a home inspection because it is considered to be normal ongoing maintenance.

As you walk around the house during your exterior inspection, check to see if there is a fence that encloses the swimming pool. This is important from a safety point of view. The fence should provide limited access to the pool. The gate(s) must be self-closing and self-latching. Check the condition of the fence and the operation of the gate. If maintenance is needed, record it on your worksheet for later correction.

The swimming pool and equipment can be inspected either before or after the inspection of the house, whichever is convenient. Check the condition of the deck that surrounds the pool. With an in-ground pool the deck is often concrete, although it could be stone, tile, or brick. Specifically look for cracked, chipped, or settled sections. Uneven joints between the sections are considered tripping hazards. If there are any open joints, particularly between the deck and the coping, they must be sealed. While on the deck if there is a grab rail, ladder, and slide, check to see if they are adequately anchored to the deck. If there is a diving board, look for cracks. If there are any, the board should be replaced. Is the board topped with a nonslip coating? Many older pools with diving boards do not meet new diving board standards. The safety upgrades outlined in the standards are mandatory. Check with a local pool company for the latest standards. With an aboveground pool the deck is generally wood constructed. Check the decking and railing for cracked, broken, or rotting sections. If there is access to the area below the deck, inspect the support framing.

After the deck inspection check the condition of the pool's sidewalls. With a vinyl-lined pool, look for staining, discolorations, or tears in the lining. Faded liners are more likely to tear. Has any portion of the lining pulled out of the edge retainer? If there are stains in the liner at the bottom of the pool that don't continue up the sides, it may indicate a fungal growth in the sand below the liner. With tile-lined concrete walls, look for cracked, loose, chipped, or missing tiles. If the walls are finished with plaster, check for spalling (flaking or chipped) or cracked sections. With painted walls, look for flaking or faded paint and cracks. As you walk around the pool looking at the sidewalls, check the skimmer. Does the weir move freely? Is it broken or missing? Is the strainer basket in place? Is it damaged?

Before you inspect the pump, filter, and heater, look closely at the pool water. If it is turbid, it would indicate that the filter is in need of a cleaning. If there are tiny bubbles discharging at the water-supply outlet, it would indicate that air is getting into the system. This condition must be corrected to prevent possible damage to the pump. Check the cover to the pump's hair and lint strainer. Most covers are transparent. If bubbles are visible, it would normally indicate that the cover is not tightly secured or that the gasket needs replacement. However, if there are no bubbles, you should suspect a leak in the water circulation system. In this case, further investigation is required by a pool service company.

When inspecting the pump, listen for any abnormal sounds. If it is noisy, there may be impeller damage or worn bearings. A pump that is very hot to the touch is also a problem condition. It could indicate that the pump is running dry. If there is water dripping around the pump, it may indicate leakage around the shaft because of worn seals or because the hair and lint strainer cap is not tightly secured. Check the pressure gauge on the filter. Is it inoperative? Is the pressure excessively high? If the heater is not on, turn it on. If it doesn't fire up, record it on your worksheet. If the heater cycles on and off, it may indicate low water flow, a condition that must be corrected. To whatever extent the burner area is visible, look for a scale buildup with rust flakes and dust. Is water dripping in that area? If it is, you should suspect pin-hole leaks. If this condition exists, have the heater inspected by a pool service company. If there is an underwater pool light, check to see if it is operational. Also if there are electrical outlets nearby, check to see if they are GFI protected.

If the pool is not being used and is covered over, check the condition of the cover. Is it worn, are there any torn or damaged sections? If the cover is held in place with straps, are any straps missing? Since the operation of the pool equipment and the condition of the pool shell cannot be checked at this time, you should contact the company that has been servicing the pool and ask them for a copy of their service record. If the record is not available, have the seller include a clause in the contract to guarantee that the pool shell and equipment are in good operating condition.

Checkpoint summary

General considerations

- ❍ Has a Certificate of Occupancy been issued for the pool?
- ❍ Is there a fence that encloses the swimming pool?
- ❍ Are fence gate(s) self-closing and self-latching?

Deck

- ❍ Are there any cracked, chipped, or settled sections?
- ❍ Look for uneven joints between sections.
- ❍ Cracked or open joints should be sealed.
- ❍ Are grab rails, ladder, and slide adequately anchored to deck?
- ❍ Is diving board cracked or warped?

Vinyl-lined pool

- ❍ Is lining stained, discolored, or torn?
- ❍ Has lining pulled out of edge retainer?
- ❍ Is bottom of liner stained?

Concrete pool

- ❍ Are there cracked, chipped, loose, or missing tiles?
- ❍ Does the plaster finish have flaking, chipped, or discolored sections?
- ❍ Are any cracks visible in the sidewalls?
- ❍ Is the painted surface flaking or faded?
- ❍ Check the skimmer weir and strainer basket.

Pool equipment

- ❍ Is pool water turbid?
- ❍ Are tiny bubbles discharging into pool?
- ❍ Is the pump noisy? Is it very hot to the touch?
- ❍ Check for water dripping around the pump.
- ❍ Is the pressure gauge on the filter operative?
- ❍ Does heater cycle on and off?
- ❍ Look for scale buildup with rust flakes and dust around heater.
- ❍ Is water dripping under or around the heater?
- ❍ Does pool cover have torn or damaged sections?
- ❍ Check the operation of underwater pool light(s).
- ❍ Are electrical outlets GFI protected?

19

Energy considerations

Many states have enacted an energy conservation construction code to supplement their building codes. The purpose of the energy conservation code is to provide a construction standard that will minimize energy consumption in a building while maintaining the necessary comfort factors. The conservation code applies to new construction, renovations, and additions to buildings. The code is not retroactive and therefore does not apply to buildings constructed prior to its enactment. You can check with the local building department to determine whether there is an energy conservation construction code in your state and if so, whether it was in effect prior to the construction of the house you are planning to buy.

If you are considering the purchase of an existing building, there is a high probability that the house is not as energy efficient as it could be. Although the energy-deficient items are usually found during a prepurchase home inspection, they are often not upgraded until the buyer takes possession of the house. Consequently, after you move into the house, you should perform an energy audit to bring back into focus those items that are needed for conservation improvements.

Having an energy-efficient house is not only good citizenship; it is also good for your pocketbook, resulting in reduced utility bills. The actual dollar savings will of course depend on the gas, oil, and electricity rates in your area; the climate; and the extent to which your house is already energy efficient. You can often determine the projected savings and payback period for the costs involved in making your home energy efficient by contacting your local utility company. For a nominal fee, many utility companies will analyze your energy-conserving improvements, taking into account current and projected energy costs, and will estimate your dollar savings per year.

Energy audit

An energy audit is an inspection of the house to determine the extent of deficiencies that result in energy being wasted. The decision whether to upgrade deficiencies is usually based on economics. Are the dollars spent in making energy-conservation improvements a wise investment? Will the improvements save you enough money on heating and cooling to pay for themselves? For most homes, the answer is yes, especially with ever-increasing costs for fuel.

One cause for wasted energy in a house, especially an older one, is the lack of adequate insulation. Insulation is a basic energy saver and should be used in all houses regardless of location. In colder climates, it reduces heat loss and thereby reduces fuel costs. In warmer climates, the insulation reduces heat gain and consequently reduces cooling costs (electricity is used for running most residential air-conditioning units). During your energy audit, you should determine whether your house is adequately insulated.

Insulation

Insulation is available in a variety of forms and materials. The three most common forms are flexible insulation, loose-fill insulation, and rigid insulation. *Flexible* insulation is manufactured in two types, batts and blankets. Both are made of fibrous materials such as glass fibers, rock wool, wood fibers, or cotton. Organic fibers are treated chemically to make them resistant to fire and decay. Batts are precut in 4- or 8-foot lengths and are available in thicknesses between 2 and 6 inches. Blankets are furnished in continuous rolls and are available in thicknesses between 1½ and 3 inches. Both batts and blankets are manufactured in 15- and 23-inch widths so that they can be readily used in homes that have been constructed with joist and stud spacing of 16 or 24 inches.

Loose-fill insulation is generally made from rock wool, glass fibers, vermiculite, pearlite, cellulose, granulated cork, shredded redwood bark, sawdust, or wood shavings. It is normally supplied in bags or bales and can be poured, blown, or placed by hand. Loose-fill insulation is suited for use in the sidewalls of homes that were not insulated during construction or between the floor joists of unheated attics. However, if there is no floor covering, it is not recommended for use between the floor joists when an attic fan could blow the loose material around.

Rigid insulation is generally made from extruded polystyrene, polystyrene bead board, urethane, fiberglass, or wood fiberboard. It is often used to insulate masonry walls and comes in widths of 24 and 48 inches. Most rigid insulation boards are not fire resistant and should be covered with at least ½-inch gypsum wallboard to ensure fire safety. Rigid insulation boards are also used as backer boards for aluminum and vinyl exterior siding.

Another type of insulation that was popular in the late 1970s was *foamed-in-place* insulation. It was made from urea formaldehyde. In many homes, the foam ingredients were improperly mixed and installed. This resulted in excessive formaldehyde vapor being released into the house, causing adverse health effects. Urea formaldehyde foam insulation (UFFI) is no longer being installed; however, many homes have UFFI in their walls. This is no longer considered a problem. See "Formaldehyde" in chapter 20.

One measure of the effectiveness of insulation is its resistance to heat flow, the *R-number*. The higher the R-number, the greater the resistance to winter heat loss or summer heat gain. Table 19-1 shows typical R-numbers for various types and thicknesses of insulation.

R-numbers are additive. You can add an insulation rated R-11 to one that is rated R-19 to achieve a resistance value of R-30. The thermal resistance of an area covered with loose-fill or flexible insulation can change over the years. The insulation value depends not only on the

Table 19-1. Insulation ratings.

Insulation type	R-number 11	13	19	22	30	38
Batts/blankets						
Fiberglass	3½″	4″	6″	7″	9½″	12″
Rock wool	3″	4″	5½″	6″	8½″	11″
Loose-fill						
Fiberglass	5″	5½″	8½″	10″	13½″	17″
Rock wool	4″	4½″	6½″	8″	10½″	13″
Cellulose	3″	3½″	5½″	6″	8½″	11″
Vermiculite	5″	6″	9″	10″	14″	18″
Rigid board						
Polystyrene (extruded)	3″	3½″	5″	5½″	7½″	9½″
Polystyrene (bead board)	3″	3½″	5½″	6″	8½″	10½″
Urethane	2″	2″	3″	3½″	5″	6″
Fiberglass	3″	3½″	5″	5½″	7½″	9½″

material but also on the amount of trapped air contained within the material. If the loose fill is disturbed or the flexible insulation crushed (because of items being stored on top of it), it will no longer be as thick as when it was installed. Consequently, its effective R-number will be reduced. To determine the current R-number of the insulation in your home, you should measure its thickness.

The insulation recommended for your house can be determined from the map in FIG. 19-1. You might be surprised to learn how much insulation is recommended. The R-numbers, however, are based on current and projected fuel costs. If your house is already insulated, once you determine the amount of existing insulation, you can add the difference. Remember, the R-numbers are additive. In some homes, it might not be economically justifiable to increase the insulation to the recommended value.

In determining whether your house is adequately insulated, you should check the exterior walls and the ceilings and floors that face unheated areas, such as the attic and crawl space. (See FIG. 19-2.) In unfinished areas where the insulation is exposed (often in the floor of an attic or ceiling of a crawl space), the thickness can easily be measured. If the attic floor is covered, you can pry up one board and look for insulation. Determining the amount of insulation in a finished exterior wall is more difficult. However, you can make a quick determination whether the wall is inadequately insulated or has no insulation by feeling the inside surface during the heating season. If the wall feels cold to the touch, insulation is needed.

Sometimes you can determine the amount of insulation by removing the cover to a light switch and peering into the wall space, using a flashlight. Caution should be observed, since the switch is electrically hot. Because of the small amount of open space between the wall and switch box, this method is usually not effective. Also, it is possible that the electrician who installed the wiring might have pulled the insulation away from the switch and outlet boxes to facilitate installation. In this case, you might think that there is no insulation in the wall. The only way to determine positively how

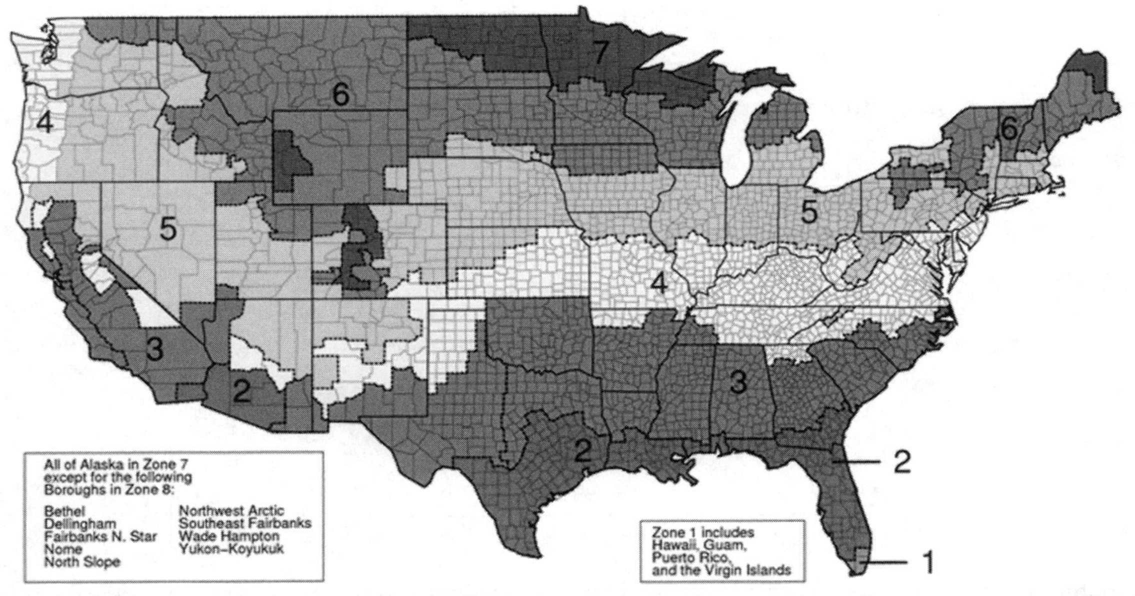

Zone	Add Insulation to Attic		Floor
	Uninsulated Attic	Existing 3–4 Inches of Insulation	
1	R30 to R49	R25 to R30	R13
2	R30 to R60	R25 to R38	R13 to R19
3	R30 to R60	R25 to R38	R19 to R25
4	R38 to R60	R38	R25 to R30
5 to 8	R49 to R60	R38 to R49	R25 to R30

Fig. 19-1. *Recommended insulation levels for retrofitting existing wood-framed buildings. Courtesy of: Energy Star (U.S. Dept. of Energy and EPA)*

much insulation there is in a finished exterior wall is to make a small hole in the wall (in a nonobvious location such as a closet) and measure it. The hole can then be patched.

While you are determining the amount of insulation, you should also determine whether there is a *vapor barrier* associated with the insulation. A vapor barrier is a thin sheet material such as polyethylene film, aluminum foil, or an asphalt-impregnated kraft paper through which water vapor cannot readily pass. Many insulation materials are produced with a vapor barrier applied on one side. If the insulation does not have a vapor barrier, a separate one can be installed. The purpose of a vapor barrier is to prevent moisture problems in exterior walls and ceilings, and floors that face unheated areas, due to condensation of water vapor (normal in a house) that passes through those surfaces. To be effective, the vapor barrier must be facing the heated room rather than the cool, unheated area. (See FIG. 19-3.)

In addition to the need for insulation of the building shell (exterior walls, ceilings, and floors), all hot-water pipes and heating and cooling ducts that pass through unheated

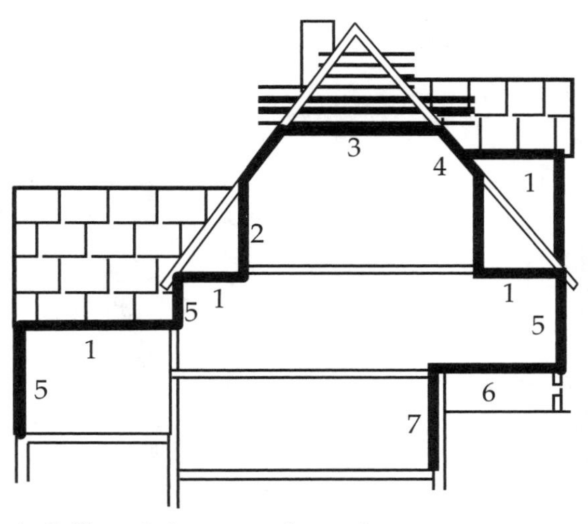

1. Ceilings below an unheated area.
2. "Knee" walls of a finished attic level room.
3. Floor of a crawl attic.
4. The sloping portion of the roof in a finished attic. Leave an airspace between insulation and roof.
5. Exterior walls.
6. Floors above cold crawl spaces. Floors above a porch or an unheated garage.
7. Walls of a heated basement.

Fig. 19-2. *Where to insulate.*

portions of the house (such as a crawl space, garage, or unfinished attic) must be insulated. Most houses usually have no more than 1 or 2 inches of insulation wrapped around ducts in unheated areas. Because of increasing fuel costs, this is considered minimal for most areas, and additional insulation can usually be justified. Check the condition of the insulation. Are there any loose, torn, or missing sections? Also, if there are any exposed duct joints, check them to see if they are sealed tightly. When the ducts are used exclusively for air conditioning or serve a dual function (such as heating and air-conditioning), the outside of the insulation should be covered with a vapor barrier to prevent condensation. A vapor barrier, however, is not needed on ducts used only for heating. If there is a vapor barrier on the ducts, check its condition. Look for torn and missing sections. All vapor-barrier joints must be tightly sealed.

If the domestic hot water is produced in a tank-type water heater located in an unheated area, the tank should be covered with an insulation jacket. These jackets can be purchased in most building-supply or hardware stores. Although tank-type water heaters are normally insulated by the manufacturer, by installing an outer insulation jacket, you will further reduce heat loss and thereby minimize the energy needed to maintain the desired water temperature. The temperature of the hot water should not exceed 140° F. (See chapter 16.) Temperatures in excess of 140° F are not only wasteful of energy but will also shorten the life of the water heater.

Attic ventilation

While in the attic checking the insulation, see if the area is adequately ventilated. As discussed in chapter 9, attic ventilation is necessary not only to prevent condensation problems during the winter months but also from an energy-conservation point of view to reduce the heat load on the structure during the summer. Because of trapped air, attic areas can become excessively hot during the summer, reaching temperatures of about 150° F. If there is an air-conditioning blower coil located in the attic, the high temperature will tend to lessen its efficiency of operation.

Ventilation in an attic should be provided by at least two vent openings located so that air can flow in one opening and out of the other. Vents in the eaves and at gable ends are better than gable vents alone. One of the most effective methods for ventilating the attic is a combination of vents in the eaves and a continuous ridge vent. Remember, the attic must also be adequately

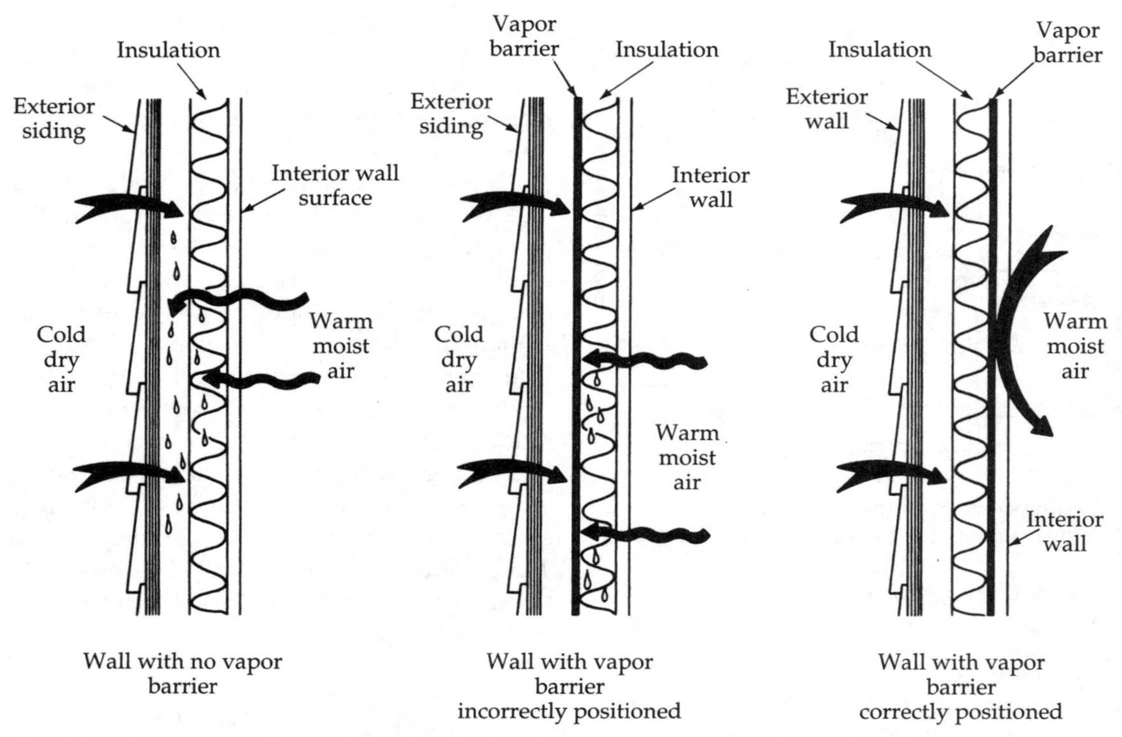

Fig. 19-3. *How a correctly positioned vapor barrier prevents condensation in the exterior wall.*

ventilated during the winter. As long as the attic is adequately insulated, the benefits of ventilation greatly exceed any fuel savings that might result from blocking the vent openings.

Storm windows

Most homes with single-pane windows will benefit by the installation of storm windows. (See chapter 5.) A storm window, whether a storm sash, panel, or combination unit, reduces the heat loss through a single-glazed window by about 50 percent. In cold climates, it also adds to physical comfort by reducing an apparent draft. Body heat radiates toward a cold surface. Also, warm air hitting a cold surface loses its heat, becomes dense, and falls to the floor. This combined effect creates what appears to be a draft to someone sitting or standing near the cold surface (window). If the storm window is a sash or combination unit (both of which cover the window frame), it also helps reduce cold-air infiltration through the movable and fixed joints around the window.

Basically, a storm window is effective because it traps a layer of air between itself and the window. This dead air space acts as an insulator and thus reduces heat loss. If installing storm windows is not economically justified because you are planning to move or your budget does not presently permit the purchase of storm windows, you can still make your windows energy efficient. Simply cover them with plastic sheets and secure the edges with tacks, molding strips, and caulking. Although temporary, these inexpensive homemade storm windows are an effective approach to reducing heat loss. They will more than pay for themselves in fuel savings in the first year.

Caulking and weatherstripping

In a well-insulated house with thermal-pane or storm windows, air leakage is the greatest source of heat loss. To conserve energy further by reducing the cold-air infiltration during the winter and the loss of air-conditioned air in the summer, the movable joints (such as those around windows and doors) should be weatherstripped, and the fixed exterior joints should be caulked. (See chapter 5 for types of caulking compounds.)

Check windows and doors for weatherstripping and tightness of fit. Loose-fitting windows and doors not only lose heat but result in uncomfortable drafts. All window sashes, exterior doors, and interior doors or hatches leading to unheated areas (such as an attic, basement, or crawl space) should be weatherstripped on their sides, tops, and bottoms. Periodically check the condition of the weatherstripping. Over the years, some types of weatherstripping will wear, tear, crack, and generally deteriorate so that replacement is required.

The condition of the caulking on the exterior joints should be checked during your energy audit. Specifically, look at the joints (1) between the exterior siding and the window and door frames, (2) at the inside and outside corners formed by the exterior siding, (3) just under the bottom side of the exterior siding and the foundation wall, (4) between dissimilar siding materials (such as a masonry-and-shingle wall), (5) where the chimney meets the siding, and (6) where storm windows meet the window frame (except for drain holes at the windowsill). If the caulking is old, brittle, broken, or missing, the joints should be recaulked. Good home maintenance includes an annual check of all exterior joints, particularly in view of rising fuel costs.

Fireplaces and wood-burning stoves

Most fireplaces are used for creating a relaxed, cozy atmosphere rather than heating. Usually their use as an auxiliary source of heat cannot be economically justified, since they are very inefficient. A fireplace requires a large volume of air for combustion. Normally this air is drawn from the heated air in the house. Since the combustion air flows up the chimney, all the fuel that had been used in heating the air is wasted. To minimize this problem, a fireplace should have dampered air vents that provide combustion air from the outdoors. Although newer fireplace installations have provisions for outside air, most existing fireplaces do not have this feature and therefore waste more heat than they generate. In fact, in terms of the total heating value of the wood being burned, a fireplace does not generate much heat at all. Only about 10 to 15 percent of the available Btus actually find their way out into the room, while much more warm air disappears up the chimney.

Even when a fireplace is not being used, it is a potential source of heat loss. If the flue is not blocked, heat in the house will be drawn up the chimney. This condition can normally be minimized by closing the damper. Check the fireplace to see whether there is a damper and if so, whether it properly seals the flue opening. Fireplaces in many older houses do not have dampers. If a movable damper is too costly to install, you can always block the flue with a piece of sheet metal supported by guides on the sidewalls. It can be inserted and removed on an as-needed basis.

Some fireplaces have built-in convection ducts around the firebox that function as a heat exchanger. Cool air entering the bottom of the ducts is heated as it circulates around the firebox. The warm air rises and flows out into the room through openings in the upper section of the ducts. This type of fireplace has a greater heating efficiency than the conventional open-faced masonry fireplace. Nevertheless, a much more efficient way to heat the house is by using a wood-burning stove. A good-quality stove will allow only the air needed for combustion into the unit. Consequently, heated room air is not lost up the chimney. If you control the airflow through the stove, a log will produce more heat by burning for a longer period of time. In addition, since the stove is physically located in the room that it is intended to heat, the entire surface of the unit will heat up and warm the room by convection and radiation.

The cost-effectiveness of a wood-burning stove for heating will depend on the cost of the wood being burned. In those areas where wood is inexpensive, a stove is an economically viable method for providing auxiliary heat and reducing your overall heating bill. Even in those areas where wood is more costly, as utility rates increase, the use of a wood-burning stove becomes more attractive. Different woods have different burning characteristics. For example, wood from conifer trees (softwood) such as pine, spruce, and fir burns more quickly and gives less heat than wood from deciduous trees (hardwood) such as maple, oak, and beech. To aid you in making a proper selection for your fireplace or stove, TABLE 19-2, prepared by the Maine Bureau of Forestry, shows the various characteristics of wood.

Heating and air-conditioning systems

During your energy audit, you have been mainly concerned with those items needed to reduce heat loss. Another item of concern is maximizing the Btu output of your heating system. As you walk around the house, check the radiators or heat registers to see that they are unobstructed. If your house is heated with a warm-air furnace, check the filter to see if it needs replacement. If the oil or gas burner for the heating system was not cleaned and tuned up prior to your energy audit, it should be done as soon as possible. A heating system out of adjustment results in

Table 19-2. Characteristics of woods for use in a fireplace or stove.

Species	Ease of starting	Ember generation	Sparks	Fragrance	Heating value
Apple	Poor	Excellent	Few	Excellent	Good
Ash	Fair	Good	Few	Slight	Good
Beech	Poor	Good	Few	Slight	Excellent
Birch (white)	Good	Good	Moderate	Slight	Good
Cherry	Poor	Excellent	Few	Excellent	Good
Cedar	Excellent	Poor	Many	Good	Fair
Elm	Fair	Good	Very few	Fair	Good
Hemlock	Good	Low	Many	Good	Fair
Hickory	Fair	Excellent	Moderate	Slight	Excellent
Locust (black)	Poor	Excellent	Very few	Slight	Excellent
Maple (sugar)	Poor	Excellent	Few	Good	Excellent
Oak (red)	Poor	Excellent	Few	Fair	Excellent
Pine (white)	Excellent	Poor	Moderate	Good	Fair

greater fuel consumption, which can be quite costly. Heating systems require a periodic tune-up for efficient operation.

Various devices on the market to be fitted on or around the chimney flue are supposed to cut fuel consumption and conserve heat. Many of these furnace attachments, however, do not live up to the manufacturers' claims of saving significant energy.

One device of concern from a safety point of view is the *automatic flue damper*. This device is designed to close the furnace chimney after the burner turns off. The intention is to conserve heat by preventing room air from being drawn up the chimney. However, if the automatic flue damper fails to open when the burner fires, poisonous carbon monoxide fumes will vent into the house. If your furnace has such a device, you should find out whether it has been approved by a nationally recognized testing agency and by your local utility company. If it has not been approved, you should consider its removal or replacement as a precautionary measure.

Table 19-3. Energy savings for lowering thermostat.

	Percent of energy savings for an 8-hour temperature setback	
Degree days	**5° F**	**10° F**
5,000	8.1	12.1
6,000	7.2	10.8
7,000	6.1	9.6
8,000	5.2	8.5
8,500	4.6	7.8

By lowering the thermostat setting 5 to 10 degrees each night before going to bed and raising it in the morning, you will reduce fuel consumption and save dollars. The amount of the savings will depend on the duration of the setback, the climate, and the fuel costs in your area. See TABLE 19-3.

The *degree day* is a unit that expresses the severity of the climate in an area. The reference temperature for evaluating degree days is 65° F. Degree days are the number of degrees that the average (of the high and low temperatures for a 24-hour period) is less than 65° F. For example, if for a twenty-four-hour period (during the heating season), the high and low temperatures are 50° F and 30° F, respectively, the average temperature would be 40° F. The degree-day number for that day is then 65–40, or 25. The total number of degree days for an area, therefore, is simply the sum of the degree-day numbers during the heating season. You can check with the local utility company to find the total number of degree days for your area.

From a convenience point of view, if your heating system is controlled by a manual thermostat, you should consider replacing it with an automatic-lock thermostat. With this type of thermostat, you can regulate the amount of the setback and its duration. Depending on your requirements, thermostats with double setbacks are also available.

If there is a central air-conditioning system in your house, it must also be cleaned and tuned up to maximize the efficiency of operation. Prior to the cooling season, have the system checked to see if it needs a refrigerant charge. The system will cool even if it is low in refrigerant; however, it will operate inefficiently. Check the location of the compressor to see if it is in the shade and whether the airflow into and out of the unit is unobstructed. If the compressor is in the midday and afternoon sun, it will not perform efficiently. You should build a sunscreen to shade the unit if necessary. Be careful, however, not to obstruct the airflow.

20

Environmental concerns

In the past, whenever people bought a home, their main concern was the physical house; that is, the structural integrity of the building; the condition of the mechanical equipment such as the heating system, plumbing, water heater; the adequacy of the electrical system; the condition of the roof; whether the basement was dry; and whether there was a termite condition. In recent years, another factor has entered into the decision process—environmental problems. Some of these problems, such as a high radon concentration or deteriorating asbestos insulation, are potential health hazards. Others, like a leaky buried oil tank, can contaminate the soil and eventually the aquifer (water table). In all cases, it costs money to correct the problems. This cost should be added to the overall sale price of the house to determine the true cost of purchasing the house.

In this chapter, I will discuss environmental problems that are or should be of concern to the home buyer or homeowner. There is no doubt that in the future, as technology improves and more statistical health information becomes available, additional items will be added to the list of environmental problems.

Radon

Although the health risks associated with exposure to high concentrations of radon have been known for decades because of experience with uranium miners, it wasn't until December 1984 that it was realized that people in homes can also be exposed to high concentrations of radon resulting from uranium deposits in the soil on which the houses are built. A worker in a nuclear generating plant passed through a radiation detection monitor as he entered the plant. It turned out that his home had twenty times more radiation than is allowed in a uranium mine.

Radon is a gas present in varying quantities in the atmosphere and soils around the world. It is colorless, odorless, and tasteless, and is produced by the natural radioactive decay of uranium deposits in the earth. Prolonged exposure to high concentrations of radon can cause cancer. According to the U.S. Environmental Protection Agency (EPA), scientists estimate that between 5,000 and 20,000 lung-cancer deaths a year in the United States can be attributed to radon.

The concentration of radon in the air is measured in units of *picocuries* per liter of air (pCi/l). Based on currently available information, the EPA has set guidelines for radon levels in residential structures. (See TABLE 20-1.) Their "action level" is 4 pCi/l; that is, no action is needed if the radon concentration is below 4 pCi/l. The EPA acknowledges that even the action level is not risk free because exposure to an annual radon level of 4 pCi/l is equivalent to the risk of smoking ten cigarettes a day or having two-hundred chest X-rays a year. The action level was set at 4 pCi/l because in some cases reduction below that level might be difficult or impossible to achieve.

Table 20-1.
U.S. EPA guidelines for residential radon.

Annual average Radon level	Recommended action
Over 200 pCi/l	Take action to reduce levels within several weeks.
Between 20 and 200 pCi/l	Take action to reduce levels within several months
Between 4 and 20 pCi/1	Exposures in this range are considered above average You should take action to reduce levels to about 4 pCi/l or below within a few years (sooner if levels are at the upper end of this range).
Below 4 pCi/l	Exposures in this range are considered average or slightly above average. Although exposures in this range do present some increased risk of lung cancer, reductions of levels this low may be difficult, and sometimes impossible, to achieve.
Below 1.0 pCi/l	These are average first-floor residential levels.

Although radon gas is present in varying quantities in soils around the world, not every house has a problem with high radon levels. It is possible for one house to have a very low radon concentration while an adjacent house has a very high level. It depends on the construction of the house, the uranium-radium content in the soil, and geological formation below the house. The only way to know if a house has a radon problem is to test it. The radon concentration in a house varies with time. It is affected by a number of environmental factors such as rain, snow, barometric pressure, and direction of wind relative to open windows, and by induced negative pressures caused by periodic use of exhaust fans, attic fans, fireplaces, and heating systems. Consequently, the most accurate method of determining the average annual radon concentration is a long-term test. However, since time is usually limited in real estate transactions, a long-term test is not practical, and consequently a charcoal canister with a test period of three to seven days is generally used.

Canisters can be purchased in hardware stores and home centers. Also, in many areas professional home-inspection companies offer radon-testing services. If you intend to place the radon-detection canister yourself, make sure that you follow the directions carefully; otherwise, the results might be inaccurate.

Since radon is a gas, it can seep into the lower level of a house through dirt floors, cracks in concrete floors and walls, floor drains, sump pits, open joints and tiny cracks or pores in hollow block walls. (See FIG. 20-1.) If after a house is tested it is determined that it has a high radon concentration, don't panic. The condition can be corrected at a reasonable cost.

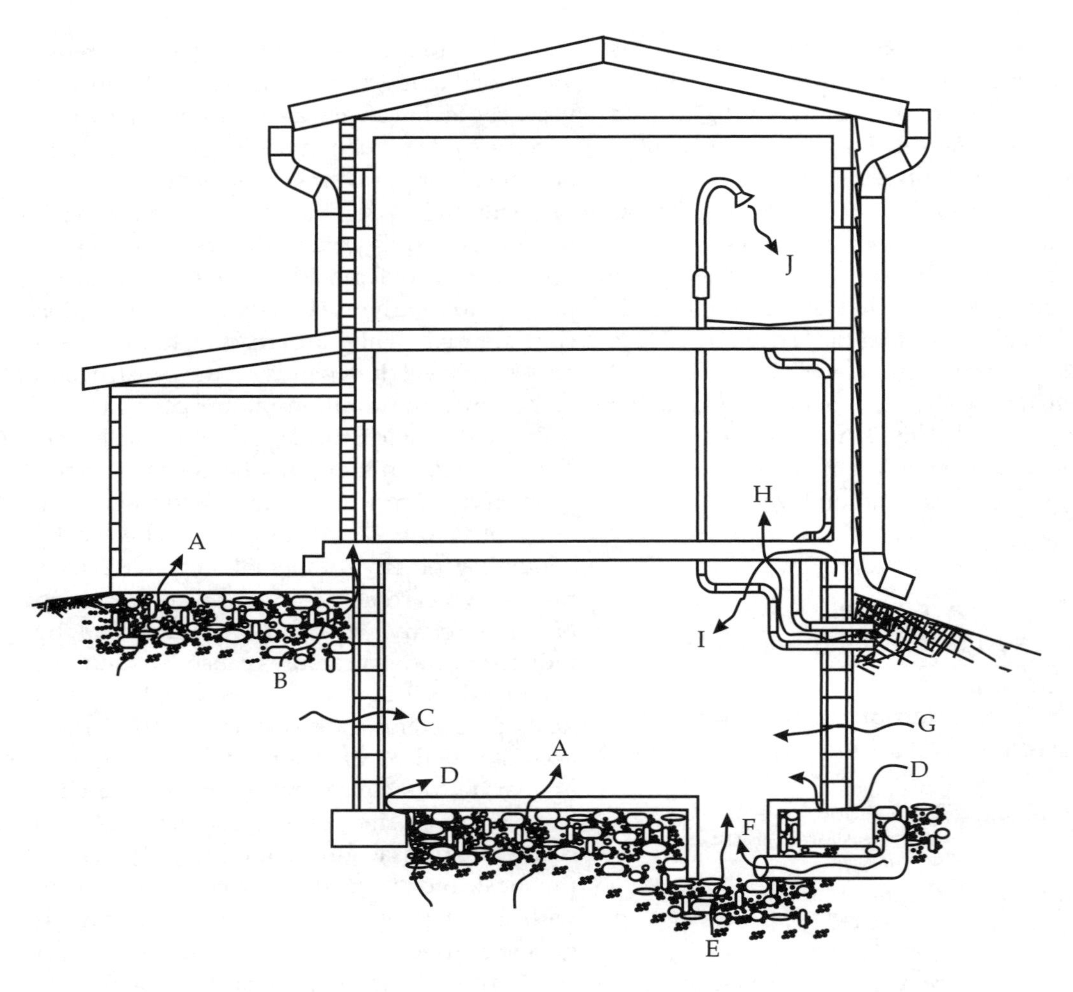

A. Cracks in concrete slabs
B. Spaces behind brick veneer walls that rest on uncapped hollow-block foundation
C. Pores and cracks in concrete blocks
D. Floor-wall joints
E. Exposed soil, as in a sump
F. Weeping (drain) tile, if drained to open sump
G. Mortar joints
H. Loose fitting pipe penetrations
I. Open tops of block walls
J. Shower

Fig. 20-1. *Major radon entry routes.*

A number of methods have been successful in reducing radon concentration levels in buildings to a point below the "action level." A typical mitigation procedure that is very effective is to use a 4-inch plastic pipe with an in-line fan that vents the radon gas from below the floor slab to above the roof. Correcting the problem is not a do-it-yourself task. It should be done by a radon-mitigation contractor that has completed the requirements for listing under the U.S. EPA's Radon Contractor Proficiency Program.

Radon can also enter the house through the domestic water supplied by a private well. It can be transferred into the air during a shower or when water is running in a sink. However, it takes relatively high levels of waterborne radon to produce a significantly elevated level of radon in a house; the large volume of air inside a house dilutes the radon being transferred to the air from the relatively small volume of water. It is estimated that the normal use of a water supply containing 10,000 pCi/l will produce a concentration of 1 pCi/l of radon in the air. It therefore takes a waterborne radon level of 40,000 pCi/l to reach the EPA's action level of 4 pCi/l.

In most parts of the country, radon gas emanating from the soil is the major contributor to indoor airborne radon. As a result, the water from a private well is normally not tested for radon during the initial screening test. However, in some areas of the United States, waterborne radon significantly contributes to the total radon concentration. You can check with your local health department to find out if testing the water is recommended for your area. As of this writing, the EPA has not set a maximum level for radon in drinking water. Nevertheless, if the radon level is greater than 10,000 pCi/l, water-supply mitigation should be considered. There are water-treatment methods available that can be used for removing radon at the point of entry. The specific method should be discussed with the radon-mitigation contractor. Incidentally, do not worry about drinking the water. The health effect of drinking waterborne radon is relatively insignificant compared to that associated with breathing in airborne radon.

Asbestos

A common concern of many home buyers is whether there is any asbestos in the house, and if there is, is it in a condition that would be considered a health hazard? Asbestos, which has been identified as a carcinogen, is a naturally occurring fibrous mineral found in certain types of rock formations throughout the world. Asbestos fibers are strong, won't burn, resist corrosion, and insulate well. These physical properties have made it a staple in architectural and construction applications. When the fibers are mixed during processing with a material which binds them together, they can be used in many construction products such as cement siding and roof shingles, vinyl floor tiles, ceiling tiles, textured paints or coatings, blown-in insulation, flexible fabric connections in ductwork, spackling compounds, boiler insulation, pipe insulation, caulking, putties, door gaskets on wood-burning stoves, and so on. (See FIG. 20-2.) The amount of asbestos in these products varies considerably, from approximately 1 percent to 75 percent.

Asbestos-containing material in the home doesn't necessarily pose a health risk. Asbestos materials become hazardous only when due to damage, disturbance, or deterioration over time, they release fibers into the air. Airborne asbestos fibers can be inhaled through the nose and mouth and lodge in the lungs. According to estimates by the EPA, every year 3,300–12,000 people die from cancer caused by exposure to asbestos. Of particular concern is asbestos-containing material that is sprayed or troweled or that has become friable. (Friable material can be crumbled, pulverized, or reduced to powder by hand pressure.) As long as the asbestos-containing material is intact, it does not pose a health hazard. If the asbestos material is not likely to be disturbed or is in an area where renovations will not occur, the EPA suggests that it is best left alone.

Many houses have old boilers that are insulated with asbestos. The insulation looks like a white plaster coating over the boiler shell. When the boiler is eventually replaced, it is

1. Some roofing and siding shingles are made of asbestos cement.
2. Houses built between 1930 and 1950 may have asbestos as insulation.
3. Asbestos may be present in textured paint and in patching compounds used on wall and ceiling joints. Their use was banned in 1977.
4. Artificial ashes and embers sold for use in gas-fired fireplaces may contain asbestos.
5. Older products such as stove-top pads may have some asbestos compounds.
6. Walls and floors around woodburning stoves may be protected with asbestos paper, millboard, or cement sheets.
7. Asbestos is found in some vinyl floor tiles and the backing on vinyl sheet flooring and adhesives.
8. Hot water and steam pipes in older houses may be coated with an asbestos material or covered with an asbestos blanket or tape.
9. Oil and coal furnaces and door gaskets may have asbestos insulation.

Fig. 20-2. *Locations of possible asbestos hazards in the home.*

necessary first to remove the asbestos insulation, even if the insulation is in good condition, before disassembling the boiler. Most heating contractors are not certified in asbestos removal. The job must be done by a qualified asbestos-removal contractor that has been trained and certified. Once removed, the asbestos cannot be thrown out like ordinary household garbage. A licensed industrial-waste hauler must take it, properly identified and contained, to a landfill that accepts asbestos.

Removal of damaged or deteriorating asbestos materials is not always necessary. In fact, it is the least desirable alternative because in the process it creates a considerable amount of airborne asbestos fibers that must be contained and removed. Depending on its condition, the asbestos material can be encapsulated by coating it with a sealant so that the fibers cannot be easily released. It can also be enclosed in airtight walls or ceilings that completely isolate and contain any fibers that become airborne. The decision whether to remove, encapsulate, or enclose deteriorating or damaged asbestos material, as well as the repair of damaged sections, should be made only by a certified trained professional.

Although most of the asbestos materials that had been used in construction are no longer being manufactured, various items can be found in many homes built prior to 1978. Identifying the more common types of asbestos material can generally be done in a visual inspection by home inspectors, asbestos-abatement personnel, or tradespeople who have frequently worked with asbestos material, and even by you. For example, in many older homes, the heating pipes in the basement or crawl spaces are covered with insulation that contains asbestos. The insulation on the straight sections of pipe, when viewed from an end, looks like corrugated cardboard. The angle fittings on the heating pipes are covered with an insulation coating that looks like plaster and is shaped around the fitting. If you see this type of insulation, look for torn, loose, crushed, or otherwise damaged sections. In many cases, it is not readily apparent whether building products and insulation materials contain asbestos. In those cases, positive identification of asbestos can be made by a qualified laboratory after analyzing representative samples of the material.

Drinking water

Domestic water is generally supplied to homes through private wells or public water companies. Water supplied by public water companies is usually safe to drink and does not pose a health risk. The quality of the water supplied by these companies is periodically checked because it must comply with rigid standards set by the U.S. EPA. Nevertheless, the EPA has indicated that some 40 million people have been using drinking water containing potentially hazardous levels of lead. The problem does not originate with the water supply but with distribution piping, solder used at the pipe fittings, and fixtures in the house. In some older homes, the inlet water pipe is made of lead; the solder used on pipe fittings in homes built before 1988 contained lead; and lead is contained in the metal alloy used in the manufacture of many faucets.

The most important factor causing a high concentration of lead in water is the contact time between the water and the lead. Water that is slightly acidic or soft (water that makes soapsuds easily) is corrosive and reacts with lead. When the water stands in pipes or faucets that contain lead for several hours without use, there is a potential for lead to leach, or dissolve, into the water. Also, hot water dissolves lead more quickly than cold water.

Dissolved lead in water has no odor and cannot be seen or tasted. Testing by an approved laboratory is the only way to determine if the drinking water has high levels of dissolved lead. The test is generally conducted in two parts, A "first draw" sample is collected—water that has been sitting in the pipes overnight or at least four hours. Then a "fully purged" sample is collected; the water is turned on and allowed to flow for at least one full minute before a sample is collected. According to preliminary studies at the University of North Carolina, about 30 percent of homes have a high lead concentration in the first draw, but purging corrects the problem more than 90 percent of the time. The current federal standards limit the amount of lead in water to 15 parts per billion (ppb), which is equivalent to 0.015 milligrams per liter (mg/l).

Another domestic water problem is excessive sodium. This condition is the result of using a water softener in the water supply. The softener replaces the calcium in the water with sodium. When the sodium concentration in the water is greater than 28 mg/l, people with high blood pressure and those with low-salt diets should be warned. When using a water softener, the sodium content is usually 100 mg/l or more. The problem can be corrected by having the pipe that supplies water to the kitchen sink bypass the water softener. All drinking and cooking water should then be taken only from the kitchen sink.

In addition to the possibility of high concentrations of lead and sodium, domestic water supplied by a private well can be contaminated by harmful bacteria resulting from faulty septic tanks, chemicals from a toxic spill that occurred years before, leaking underground storage tanks, or pesticides and fertilizers. The only way to tell whether the water is potable is to have it tested. The tests for pesticides and other chemicals are more complex and costly than the routine tests for bacteria or minerals. If you are concerned about pesticide and chemical contamination of your well, first contact your county officials and find out whether contamination problems have been reported in the area. As a precautionary measure, the water from a private well should be analyzed once a year for coliform bacteria to ensure that it is potable.

Lead

Lead poisoning is considered by many public-health officials the number-one environmental threat to children. It dwarfs radon and asbestos. High concentrations of lead in the body can cause permanent brain damage, even death. Low concentrations can result in reading disorders and hyperactivity and can affect a child's ability to perform in school. The government is estimating that one out of every nine children under the age of six has enough lead in his blood to place him at risk. The sad truth is that lead poisoning is entirely preventable.

Lead poisoning occurs in the home. It is not confined to children of low-income families living in the inner-city ghettos but has also been found in children of well-to-do families living in the suburbs. According to the EPA, about two-thirds of the homes built before 1940 and one-third to one-half of the homes built between 1940 and 1960 contain heavily leaded paint. A smaller percentage of homes built between 1960 and 1980 also had surfaces coated with lead-based paint. In 1978, the U.S. Consumer Product Safety Commission (CPSC) lowered the legal maximum lead content in most kinds of paint to trace amounts (0.06%), so houses built after 1978 should be relatively free of lead paint.

It has been known for years that children have been poisoned by eating chips of lead-based paint. However, it was only recently documented that they are also at risk of poisoning from exposure to lead dust in household air. Lead dust particles can become airborne when surfaces covered with lead-based paint are scraped, sanded, or heated with an open flame during paint stripping. Lead dust can also be created by the rubbing and sliding of lead-base-painted window sashes as they open and close.

Lead dust can be inhaled when airborne. When the dust settles on the floor, windowsills, or furniture, it can be ingested by children through their normal hand-to-mouth behavior. Also, settled lead dust particles can become airborne as a result of household cleaning. The particles, which are very fine, can penetrate the filter system of home vacuums and are recirculated in the exhaust airstream. Cleaning of lead dust should be done by a professional who specializes in lead abatement.

The only way to tell whether the paint in a home contains lead is to have samples from different areas tested, such as windowsills, door trim, radiators, banisters, and walls where the paint may be peeling and flaking. Testing by a qualified laboratory is considerably more accurate than using a do-it-yourself kit. Many of these kits aren't very precise, since other metals can cause false-positive results. They are not sensitive to low levels of lead, so that a sample might test negative and still be considered hazardous. Also, the kits cannot tell *how much* lead is in the paint. According to the U.S. Department of Housing and Urban Development (HUD), action should be taken to reduce exposure to lead when the lead content in the paint exceeds 0.5 percent. Measures to reduce exposure to lead are particularly important when the paint is deteriorating or when infants, young children, or pregnant women are present.

If you are planning on buying a house that has lead-painted walls and trim, you have several options not unlike those for asbestos. If it is in good condition and there is little possibility that it will be eaten by children, then leave it undisturbed. If there are damaged or deteriorating sections of paint on the walls or ceilings, you can have those areas covered over with gypsum wallboard or some other building material.

You can also have the lead-based paint removed. This task, however, must be done by professionals trained in removing lead-based paint, since each of the paint-removal methods (sanding, scraping, chemical paint stripping, and heat guns) can produce lead dust or fumes. If not done properly, this option will create a greater health hazard than the original one. In some cases, complete removal and replacement of items such as windows, doors, and wall and door trim might be the best approach because of the cost or difficulty of removing the paint. This task should also be done by professionals, who will control, contain, and remove the lead dust.

If a surface that has been painted with a lead-based paint is intact, painting it with a nonleaded paint is considered a viable method of reducing the hazard associated with lead paint. However, painting the surface is not considered a permanent or long-term solution because the lead-based paint below the top coat might eventually loosen and create lead flakes and dust. Before undertaking any abatement procedures, you should have a qualified lead inspector do a *lead hazard risk analysis* and have him develop an abatement strategy that considers all the options.

Because awareness of the pervasive nature of the lead-dust problem is relatively recent, there is currently a lack of qualified lead inspectors and abatement contractors. However, a number of states are in the process of implementing certification, licensing, or training requirements for abatement contractors. Also, HUD has prepared and published guidelines for identification and abatement of lead-based paint hazards. To find a qualified lead inspector or abatement contractor, check with your state department of health or environmental agency. In those states that have not implemented certification procedures, you might still be able to find qualified inspectors and contractors. Certified lead-abatement training courses are available in Massachusetts and Maryland. A number of companies, including asbestos-abatement contractors from other states, have taken these courses and listed their companies in the telephone yellow pages under the overall heading of "Lead Paint" with subheadings "Inspection" and "Abatement."

Formaldehyde

Although formaldehyde gas had been known for many years to be a powerful irritant that can contaminate the air in a house, it wasn't until the installation of urea formaldehyde foam insulation (UFFI) in houses that state and federal agencies began to focus their attention on it. Exposure to formaldehyde vapor can cause eye, nose, and throat irritation; coughing; skin rashes; headaches; dizziness; and nausea. It has been shown to cause cancer in laboratory animals.

After the oil embargo of 1973, many homeowners, realizing the need for energy conservation, had insulation or additional insulation installed in the exterior walls of their houses. It is estimated that about 500,000 homes were insulated with UFFI, most of which was done in the 1970s. With UFFI, the quality of formulation and installation determines the amount of formaldehyde released. The amount also increases under hot and humid conditions. By the end of 1980, the CPSC received more than 1,500 complaints of adverse health effects associated with the release of formaldehyde gas from UFFI. In 1982, the CPSC banned the use of UFFI in homes; however, in 1983, the ban was overturned by a U.S. court of appeals. By that time, most of the installation contractors were out of business.

At this time, the UFFI is not considered a major source of formaldehyde air contamination. Formaldehyde levels decline rapidly to below 0.1 parts per million (ppm) within the first year of the installation of UFFI. Although people vary in their susceptibility to formaldehyde, most healthy adults do not experience ill effects from exposure below 0.1 ppm. Also, since the UFFI was installed years ago, any vapors from the insulation would probably be negligible.

Notwithstanding the above, there might still be a high formaldehyde vapor concentration in the air. Formaldehyde is found widely in many household and construction products such as plywood, particle board, chipboard, plastic laminates, cosmetics, cleaners, paper products, drapes, carpets, and even tobacco smoke. Symptoms of exposure to formaldehyde generally resemble those associated with a cold or allergies; however, the symptoms usually cease once exposure is discontinued. If you are concerned about excessive formaldehyde concentration in the house, have the air analyzed by a private qualified testing laboratory.

Leaky oil tanks

Out of sight, out of mind. That's exactly how homeowners with oil-fired heating systems think about the buried tank that holds the heating oil. They could be in for a rude awakening. Most homeowners do not know that they are responsible if their oil tank leaks. If a tank leaks and contaminates the surrounding soil, the cost for the cleanup, which includes excavating around the tank, scooping up the oil-saturated soil, and carting it away by a licensed carter, must be paid by the homeowner. This effort can cost thousands of dollars. If the oil leak contaminates an aquifer or a stream, the cost for cleanup can run into tens of thousands.

Most of the buried tanks used to store home heating oil are steel and have a projected life of about twenty years, although the actual life depends on the extent to which water has been accumulating within the tank and on soil conditions. If the oil tank for the house that you want to purchase is that age or older, you have three options. You can ask the seller to provide you with a certification on the integrity of the tank; you can have the tank tested for leaks at your expense; or as a preventive measure you can replace the tank—the most expensive option.

Some homes have buried oil tanks that are no longer in use because their heating system has been converted from oil to gas. If this is the case, ask the seller if the tank has been properly put out of service. If it hasn't, it should be; otherwise, safety and environmental problems can develop. In many communities, properly putting a tank out of service is a legal requirement. An abandoned tank that has not been properly prepared will eventually deteriorate because of corrosion resulting in leakage, collapse, or both. Leakage, of course, contaminates the soil.

To put a tank out of service properly, the residual oil or sludge at the bottom of the tank must be pumped out and disposed of, using a licensed carter. The tank should then be removed entirely or filled with an approved filler such as sand. If the soil below the tank is contaminated, it must be removed. The exact abandonment procedure may vary depending on the municipality. Check with your local department of environmental conservation or health department for the approved method.

Electromagnetic fields

More and more attention by the news media is currently being devoted to electromagnetic fields. Whenever electricity is present, there are both electric fields and magnetic fields. These fields, when taken together, are referred to as electromagnetic fields, or EMF. They exist around electrical transmission lines, transformers on poles, house wiring, appliances, electronic equipment, and so on. An epidemiological study published in 1979 concluded that children living in Denver near certain power lines that were exposed to electromagnetic fields with a magnetic field strength of 2–3 milligauss (mG) were twice as likely to develop leukemia than children living in houses without such exposure. The conclusions of the above study were not universally accepted by the scientific community; some researchers felt that the study was flawed and not scientifically sound. In the early 1980s, the scientific community began looking extensively into whether exposure to electromagnetic fields is dangerous to a person's health. The complexity of the problem is compounded by the fact that the results vary according to the characteristics of the electromagnetic field, such as frequency, intensity, field orientation, period of exposure, and whether the field is continuous or pulsed.

The studies to date have not resulted in definite conclusions. They have shown that there is a statistical association between electromagnetic fields and cancer. However, they have not shown that the fields are involved in causing cancer. Based on all available information, many scientists have concluded that more research is needed before they can understand how EMF interacts with the human body to cause health effects and whether there is indeed a risk from exposure to EMF. Also, since not enough is known about which elements of the exposure are important to health, a safe exposure level has not been established. Both electric fields and magnetic fields are being studied. However, more attention is being focused on magnetic fields because they can penetrate through common objects as thick as concrete walls without losing their strength whereas electric fields can be shielded by materials like wood, aluminum, or insulation around the wires.

How does all of the above affect a person contemplating purchasing a home near power lines or an electrical substation? You can have measurements taken to determine the magnetic-field intensity in and around the house to see if the readings are within typical background levels. A few home inspectors provide this service. Unfortunately, as of this writing, no government regulation or scientific recommendation helps answer the question. Until the research is complete, it is premature to recommend any course of action. Nevertheless, prudent avoidance should be considered, because until the cloud of uncertainty is removed, this problem can affect the resale value of these homes.

Mold

Mold growth in a house is the latest environmental concern. *Mold* is the common name for simple plants or fungi, usually microscopic in size, which are naturally occurring and produce spores that can be found everywhere indoors and outdoors. It has been estimated that there are at least 40,000 known species of fungi. Molds thrive on materials such as cotton, wool, paper, leather, and wood. As a result mold, which will grow anywhere indoors where there is moisture, is quite a common occurrence in buildings. The moisture may be in the host material, on its surface, or in the form of humidity in the air. Although there have been known health problems associated with mold, the one problem that brought it to the attention of the media and therefore the public is a house in Austin, Texas. Because of a series of water leaks in the house, a toxic buildup of mold developed.

It is not unusual for a water leak in a house to cause the mold buildup. However, in this case the mold was the toxic species *Stachybotrys chartarum* (atra), which is black in color and produces airborne toxins. The spores of other molds can appear as black, brown, blue, orange, or white specks although they are not always visible to the naked eye. Because of the airborne toxins, the family living in the house suffered a number of very serious health problems, and the house was deemed uninhabitable.

In order not to overreact to finding a mold buildup in the house, it's important to separate the hype from reality. Fortunately, extensive *Stachybotrys* mold is relatively rare, and the molds usually found in homes are the less toxic species. Exposure to mold spores in a house does not always present a health problem. The testing for mold during a home inspection is not considered necessary even though there are a number of companies selling kits that test for mold. According to the U.S. Center for Disease Control, "It is not necessary to determine what type of mold you may have. All molds should be treated the same with respect to potential health risks and removal."

If there is a mold condition in the house that you are inspecting and you are allergic or otherwise sensitive to mold, you will usually know it because you may experience symptoms such as nasal stuffiness, eye irritation, or wheezing. During your inspection of the interior, check all the rooms, especially a basement, for indications of a possible mold buildup. In this effort your eyes and nose are the best tools. Large mold infestations can usually be seen or smelled. If you see a moldy buildup on a wall or ceiling, record the location on your worksheets for either cleanup or replacement. In most cases you can easily remove the mold by a thorough cleaning with bleach and water. *(Never combine bleach and ammonia because the mixture produces a toxic gas.)* However, if there is an extensive amount of mold, you may want to engage the services of a professional with experience in cleaning mold in homes. Eliminating the mold problem may include discarding moldy items and replacing sections of walls and/or ceilings.

Mold odors can be masked intentionally or unintentionally by a seller. If the house that you are inspecting has an odor because of incense burning, cinnamon water boiling on the stove, or some other pleasant odor, ask the seller if you can return for an additional inspection when there are no extraneous odors.

If you detect a moldy odor in a room, look for water stains, even those stains that are dry to the touch. It is possible that there is mold buildup on the backside of the drywall or paneling. In this case further investigation is required. However, this type of investigation is beyond the scope of a home inspection and should be performed by a company or consultant that specializes in "sick houses." Also, in those carpeted rooms that are vulnerable to occasional or more frequent flooding such as a bathroom or finished basement, if there is a moldy odor, the carpet should be suspect as the source of the mold buildup. If there is mold, the typical solution is to replace the carpet. The hidden dust in fibers and the underside of a wetted carpet are ideal locations for mold growth. In addition to the carpet, it is also possible that there is mold buildup behind the walls.

Since basements are below ground and are therefore vulnerable to water penetration, pay particular attention for any signs of past water problems because mold growth is almost always due to excessive moisture. A moldy odor in the basement during the warmer months is not uncommon and is the result of condensation because of high relative humidity.

The condition can be easily corrected with a dehumidifier. If the dehumidifier doesn't correct the problem, further investigation will be necessary.

21

Green home technology

In the residential real estate market, a "green" home is the latest buzz word. But what does green really mean? There is no single fixed definition. It is interpreted in many different ways. A green home encompasses a myriad of factors such as

- Its impact on the environment
- The site the house is located on
- The actual building design
- The material used for construction
- The use of nontoxic and renewable material for decorating
- The efficient use of energy, which includes heating and air conditioning equipment, appliances, and lighting
- The efficient use of water and its associated fixtures
- The quality of the indoor air
- Whether the project has a global impact

In order for a house to be certified as a green home, it must satisfy certain criteria. There are several shades of "green" depending on the extent to which the criteria have been satisfied. There are a number of organizations that provide guidelines for a green building certification program. However, the two main organizations that certify homes are the U.S. Green Building Council, which issues the Leadership in Energy and Environmental Design (LEED) for Homes certification, and the National Association of Home Builders (NAHB), which issues the National Green Building certification. In 2008, the American National Standards Institute (ANSI) approved the National Green Building Standard, which is used as part of the NAHB National Green Building program.

Each of the above items is subdivided into great detail, and each subitem is assigned a number of points. Based on the total number of points received, houses are certified at one of four levels of green. The green certification levels and the number of points required to achieve that level vary depending on the group that is doing the certifying. The LEED certification levels are Certified (45–59 points), Silver (60–74 points), Gold (75–89 points), and Platinum (90–136 points). The NAHB certification levels are Bronze (237 points), Silver (311 points), and Gold (395 points).

The following checklist is from the U.S. Green Building Council's *The Green Home Guide*. It lists those items in general terms that a person should look for and be aware of when looking for a green home.

- *Location*. New green homes and neighborhoods must not be built on environmentally sensitive sites like prime farmland, wetlands, and endangered species habitats. The greenest development sites are "in-fill" properties like former parking lots, rail yards, shopping malls, and factories. Look for compact development where the average housing density is at least six units per acre. The home should also be within easy walking distance of public transportation—like bus lines, light rail, and subway systems—so that you can leave your car at home. A green home should also be within walking distance of parks, schools, and stores.
- *Size*. No matter how many green building elements go into the home, a 5,000-square-foot green home still consumes many more natural resources than a 2,000-square-foot green home. The larger home will also require more heating, air conditioning, and lighting. If you really want a sustainable home, choose a smaller size.
- *Building design*. The home should be oriented on its site to bring abundant natural daylight into the interior to reduce lighting requirements and to take advantage of any prevailing breezes. Windows, clerestories, skylights, light monitors, light shelves, and other strategies should be used to bring daylight to the interior of the house. The exterior should have shading devices (sunshades, canopies, green screens, and—best of all—trees), particularly on the southern and western facades and over windows and doors, to block the hot summer sun. Dual-glaze windows reduce heat gain in summer and heat loss during cold winter months. The roof should be a light-colored, heat-reflecting Energy Star® roof, or a green (landscaped) roof, to reduce heat absorption.
- *Green building materials*. A green home will have been constructed or renovated with healthy, nontoxic building materials and furnishings, like low and zero volatile organic compound (VOC) paints and sealants and nontoxic materials like strawboard for the subflooring. Wood-based features should come from rapidly renewable sources like bamboo, but if tropical hardwoods are used, they must be certified by the Forest Stewardship Council. A green home uses salvaged materials like kitchen tiles and materials with significant recycled content.
- *Insulation*. A nontoxic insulation, derived from materials like soybean or cotton, with a high R (heat resistance) factor in a home's walls and roof will help prevent cool air leakage in the summer and warm air leakage in the winter.
- *Windows and doors*. Windows and exterior doors should have Energy Star ratings and they should seal their openings tightly to avoid heat gain in summer and heat loss in winter.
- *Energy efficiency*. A green home has energy-efficient lighting, heating,

cooling, and water-heating systems. Appliances should have Energy Star ratings.

- *Renewable energy*. The home should generate some of its own energy with technologies like photovoltaic (PV) systems.
- *Water efficiency*. A green home has a water-conserving irrigation system and water-efficient kitchen and bathroom fixtures. Look for a rainwater collection and storage system, particularly in drier regions where water is increasingly scarce and expensive.
- *Indoor environmental quality*. Natural daylight should reach at least 75 percent of the home's interior. Natural ventilation (via building orientation, operable windows, fans, wind chimneys, and other strategies) should bring plentiful fresh air inside the house. The heating, ventilation, and air conditioning (HVAC) system should filter all incoming air and vent stale air outside. The garage should not have any air handling equipment or return ducts, and it should have an exhaust fan.
- *Landscaping*. Vine-covered green screens, large canopy trees, and other landscaping should shade exterior walls, the driveway, patios, and other paved areas to minimize heat islands. Yards should be landscaped with drought-tolerant plants rather than water-guzzling plants and grass in most regions.

Along with green certification, homes are often rated by Energy Star, the government-backed symbol for superior energy efficiency. Products that can earn the Energy Star include windows, heating and cooling equipment, lighting, and appliances. Energy Star is a joint program of the U.S. Environmental Protection Agency (EPA) and the U.S. Department of Energy. To earn the Energy Star, a home must meet guidelines for energy efficiency set by the U.S. EPA. These homes are at least 15 percent more energy efficient than homes built to the 2004 International Residential Code (IRC), and include additional energy-saving features that typically make them 20 to 30 percent more efficient than standard homes.

Typical features to look for in Energy Star–qualified homes include

- An efficient home envelope, with effective levels of wall, floor, and attic insulation properly installed, comprehensive air barrier details, and high-performance windows
- Efficient air distribution, where ducts are effectively insulated
- Efficient equipment for heating, cooling, and water heating
- Efficient lighting, including fixtures that earn the Energy Star
- Efficient appliances, including Energy Star–qualified dishwashers, refrigerators, and clothes washers

In addition to the above, a green home is one that is environmentally sustainable. That is, the material used for the construction and the utilization of energy during the building's lifecycle will not compromise the health of the occupants or the ability of future generations to meet their own needs. Although a green home may be the ideal, in all probability you are more likely to find a home that is partially green than one that is certified green.

Every year there are millions of homes where one or more of the home's major components is being replaced or upgraded, or a new room addition to the house is being planned or constructed. To a large extent the degree to which these items will conform to green principles will depend on the homeowner's budget and finances.

During the construction phase of an addition, and in upgrading various home components, there are usually a number of different techniques, devices, materials, and equipment that can be used to comply with green principles. Discussing all of them is beyond the scope of this chapter. However, presented here is a sampling of some of the items that are used to green-up an existing house that you are likely to encounter when house hunting.

Roof

Asphalt shingles are the most common type of roof covering. Typically their lifespan has been between 15 to 25 years. When replacing a roof there are a number of "green" options. There are synthetic slate look-alikes that are now on the market. These products have the durability, texture, and appearance of natural slate and are lighter in weight. The materials used for synthetic slates are quite varied. One product is made from recycled steel reinforced rubber automobile tire treads. Others are made from recycled postindustrial rubber and plastic waste. There is also a synthetic slate made from ground natural slate, resin, and fiberglass, bonded under high pressure. Although synthetic slates are too new to comment on their life span, manufacturers estimate it at 40 to 60 years or more.

Metal shingles are becoming more popular for residential roofing. They are available in several different shapes and are primarily made from painted or coated aluminum or steel panels although they are also made from copper. The panels are formed to resemble wood shakes, slate or Spanish, Roman, and Mediterranean tile. The shingles are durable, lightweight, and fire-resistant. Some have embedded stone chips for additional texture. In addition to shingles, flat sheet metal roofing has also been gaining in popularity for residential pitched roofs.

Also, shake-like shingles have been made from 100 percent recycled vinyl and cellulose fibers, and reclaimed wood can be reformed into shakes, or wooden shingles. Photovoltaic (PV) roof shingles are also used. They are most effective when the roof slope faces a southern or southwestern exposure. These shingles convert sunlight into electricity.

Roof-mounted structures

Some homes have thermostatically controlled roof-mounted attic fans, which are wired to the household electrical system (see discussion on attic fans in chapter 9). These units are being replaced with PV solar-powered attic fans, which can be mounted on a roof pitch of 3/12 to 12/12. The PV panel on the unit can be adjusted from laying flat to a 45 degree angle. This unit qualifies for federal tax credits in the Solar Energy Systems category under Photovoltaic Systems as defined in the Energy Improvement and Extension Act of 2008.

Gutters and downspouts can be an important part in water conservation. A simple rainwater collection system can be made by connecting the base of the downspout to a barrel. The water in the barrel can then be used to water plants and shrubs, thereby saving the equivalent amount of potable water.

Paved areas around the structure

The driveways, walkways, and the patios in many homes have a surface area that is impervious to water. As a result, rain washing over those areas picks up contaminants, pollutants, and oils that dripped from the underside of the family auto and carry them to the storm drain where they flow to the local water supply. Some homes have minimized this problem with the use of permeable paving. With permeable paving, water drains naturally through the voids into the ground, where it helps to recharge the groundwater. For the most part, porous paving materials appear nearly indistinguishable from nonporous materials. Permeable paving can be made from asphalt, paving stones, bricks, or concrete. Even though the pavement may appear to be the same as traditional pavement, it is manufactured without the "fine" materials used in traditional pavement, and instead incorporates void spaces that allow for drainage.

Exterior walls—siding (cladding)

For hundreds of years wood, in the form of cedar shingles, shakes, and clapboard, had been the popular choice for siding a house. It was readily available, reasonably priced, and attractive. However, old growth forests have been depleted or are being depleted at a rapid rate, and as a result wood siding is no longer as readily available. Wood siding is still being used on houses, however; for a green house, the wood should be certified by the Forest Stewardship Council (FSC). This indicates that the siding came from a forest that was managed and harvested with sustainable methods. In fact, in addition to siding, all the lumber used for framing a green house should be FSC-certified.

Fiber-cement siding is a popular cladding for green homes. It is made from a mixture of cement, sand, cellulose fibers, and other components. The siding is popular because unlike wood siding it is fire-resistant and will not burn. It will not rot, warp, or delaminate, and is resistant to termites. It is also warranted to last 50 years. Fiber-cement siding can be textured to look like wood lap siding, shingles, and vertical siding.

Vinyl siding, which is commonly used throughout the country because of its low cost and low maintenance, does not have the blessings of many in the green building industry. Because of environmental concerns in the processing, manufacturing, and disposal of vinyl siding, there is controversy concerning its use. Dioxin, an environmental toxin, is a by-product of the manufacturing process. Also, in the event of a fire, when vinyl siding burns, it can release toxic fumes into the atmosphere.

Deck

In the past most residential decks were constructed entirely of wood, and to a large extent many still are. Wood is a renewable resource and can easily be recycled. In addition to using domestic wood for decking, exotic South American hardwoods such as Ipe, Cambara, and Meranti, normally used for furniture, are now also being used for decking. The hardwoods are beautiful, durable, and rot- and insect-resistant. Bamboo is also being used for decking. Of course, all wood decking should be FSC-certified.

In the last 20 years, composite decking has been displacing wood as the decking material of choice. Its popularity stems from the fact that it does not require waterproofing and resists damage from weather and insects, and is splinter-free. They are available in different colors and textures, which give the appearance of wood. The composites are generally made of recycled plastics such as shopping bags and wastewood fibers from reclaimed wood and sawdust. They are not intended for structural use. The structural support for those decks that use composite lumber, the joists, beams, and posts, will still be made of wood.

Windows

A big percentage of the heat loss in a house occurs at the windows. Most of the homes built before 1985 were erected with single glazed window panes. No doubt many windows have already been replaced, with many more due for replacement. A minimum replacement window for a green home is one that is wood framed and double glazed. A better window is one where the void between the panes is filled with an inert gas such as argon or krypton, which are better insulators than air. The best window is one where the glazing has a low-emissivity (low-e) coating. Depending on the climate that the house is located in, the low-e coating can be used to reflect the heat back into the house during the winter and reflect the sun's rays away during the summer.

Vinyl frame windows have also been used in green homes. The downside, however, is that vinyl manufacturing and incineration create dioxin, a highly carcinogenic chemical. Fiberglass frames are also energy-efficient but are more expensive than vinyl. Regardless of which windows are used they should have the Energy Star label indicating that they are high-performance windows.

Structural framing

Most existing homes were built with traditional 2 × 4″, 16″ on-center framing. With green homes the recommended framing is 2 × 6″, 24″ on-center. This will not only save wood, but the additional depth of the wall allows for increased insulation, which in turn reduces heat loss and saves on fuel. Instead of stick-built exterior walls, some green homes have walls built with structural insulated panels (SIPs). These panels consist of a foam core of expanded polystyrene (EPS) sandwiched between two layers of oriented strand board (OSB). Typically, panels are 4 to 24 feet wide, 8 or 9 feet high, and made in thicknesses of 4½ or 6½ inches. SIPs are 35 percent more energy-efficient than stick frame walls and are approximately 50 percent stronger than standard construction.

Framing of the interior portion of the house can be done with conventional lumber; however, engineered lumber has replaced solid-sawn lumber as the framing and structural support members of choice by builders in many newer homes and in homes that are currently being built. The decision is based on the fact that engineered lumber can be ordered to specific dimensions so that there is less wasted wood. Also, the lumber can be made from small-diameter, fast-growing trees that are grown in tree farms. The advantage of engineered lumber is that it is stronger, stiffer, and more dimensionally stable than solid-sawn dimensional lumber. Although the lumber is initially more expensive, the total installed costs are less than that of conventional lumber. The following products fall in the category of engineered lumber: Glued laminated beams (glulam), laminated veneer lumber (LVL), I-joists, and open web truss-joists. Glulam and wood I-joists can carry greater loads over longer spans than is possible with solid-sawn lumber of the same size. Because of their "I" cross-sectional shape, wood I-joists weigh up to 60 percent less than lumber joists, making them easier to handle. They generally do not shrink, warp, cup, crown, or twist. As with all green homes, the wood used for framing should have the Forest Stewardship Council certification.

Heating, ventilation, and air-conditioning (HVAC)

The heating and cooling equipment in a house provides a comfortable environment, but it is also a major consumer of energy. In homes that have been "greened up," the heating unit (furnace or boiler) is often replaced with a unit that has the Energy Star label. That is, the conventional heating unit, which had an operating efficiency of about 60 to 65 percent, has been replaced with a condensing boiler or furnace with an operating efficiency of about 90 percent. The increase in efficiency results from a redesign of the heat exchanger, which extracts the heat from the exhaust gases that normally flow up the chimney. During this process, the temperature of the exhaust gases drops to a point where the water vapor in the exhaust gases condenses, thereby releasing additional heat. With the conventional furnace, the temperature of the flue gases is about 450–550° F. With the high-efficiency units, it's about 120–130° F.

An HVAC system that you are more likely to find in new construction than in existing homes is a geothermal heating and cooling system. This system is basically a modified heat pump system. It is often referred to as a *ground source heat pump*. Unlike an air source heat pump where the compressor/condenser is located on the outside of the house, the geothermal heat pump is generally located in the basement or utility room. It is packaged in a single cabinet that includes the compressor, refrigerant heat exchanger, and various controls. The system uses the earth as a heat source in the winter, and heat sink in the summer. It takes advantage of the earth's relatively constant temperature, at depths below the frost line, of 40° to 55° F year round. The system circulates a water/antifreeze solution through a loop of high-density polyethylene pipe that is buried in the ground. There are two basic types of loops used with this system: closed and open. The simplest, but less common, open loop system circulates a constant source of ground water through the heat pump's heat exchanger, and discharges it back to the same aquifer. However, the supply and return lines must be placed far enough apart to ensure thermal recharge of the source. The open loop can also work with a stream, well, or pond, although in some areas this creates environmental concerns.

In a closed loop system, depending on the ground conditions and available space, the polyethylene pipe can be laid out vertically in holes 75 to 500 feet deep or horizontally in trenches dug below the frost line. Where there is a pond or stream that is deep enough and with enough flow, closed loop coils can be placed on the bottom of the pond or stream. Depending on which cycle the system is operating under (heating or cooling), the water/antifreeze solution circulating through the heat exchanger in the heat pump cabinet adds or extracts heat. It has been estimated that a homeowner may save anywhere from 20 to 60 percent annually on utilities by switching from an ordinary system to a ground source system.

Water heating

Many geothermal systems installed today are equipped with desuperheaters, which provide domestic hot water when the system is providing heat or air conditioning. A desuperheater is a small auxiliary heat-recovery system at the compressor outlet that uses superheated gases from the heat pump's compressor to heat water, which then circulates through a pipe to the home's water heater tank. In the summer, when the air conditioning runs frequently, a desuperheater can provide all the hot water needed by a household virtually for free. However, it provides less hot water during the winter months, and none during the spring and fall months if the system is not operating. The home's conventional water heater is then used to meet household hot water needs in winter if the desuperheater isn't producing enough hot water and in spring and fall when the geothermal heating and cooling system may not be operating at all.

As discussed above, there is no single fixed definition of a green home. It is interpreted in many different ways. It encompasses environmentally sustainable construction, which is the responsible use of natural resources, and how it affects the quality of life for both present and future generations. It's more than energy conservation; it includes the impact on the environment, the site the house is located on, the material used for construction, the use of nontoxic and renewable material for decorating, the efficient use of water and its associated fixtures, the quality of the indoor air, and whether the project has a global impact. At the present time, from a home inspection point of view, you are more likely to find an existing home that has been "greened up" than a home that has been certified as a green home.

22 Conclusion

For most families, the purchase of a home is an emotionally charged event filled with anxiety, hopeful dreams, and sometimes disappointments after moving in. The disappointment, however, can be avoided or at least minimized if you know the true condition of the house that you are buying. Prior to purchasing the house, you or a professional should perform a home inspection.

Professional home inspection

The purpose of this book is to provide sufficient background material and an overall procedure, so that you can perform a technical inspection of the house you are considering buying or the one you are currently living in. As previously pointed out, at times the services of a professional home inspector are necessary. A professional inspector, however, will not inspect in as detailed a fashion as suggested in this book. He or she will normally not operate every window; check every light switch; check the temperature of every radiator; indicate every cracked wall, floor, or ceiling (unless the crack represents a structural problem); check the caulking at every joint; and so on.

The main concerns of an inspector are the major problems or deficiencies, ones that will be costly to correct and ones that pose a health, safety, or fire hazard. Because of the inspector's training and experience, he or she is better able than a layperson to pick up problems through a more cursory inspection and relate problem conditions in one area to difficulties in another.

When selecting a professional home inspector, choose one whose sole endeavor is inspection. It is advisable not to use an inspector who is affiliated with an exterminating company, a real estate broker, or a contractor (plumber, electrician, roofer, and so on). The inspector must be completely unbiased and must not have a vested interest in finding problem conditions. Licensed professional engineers who specialize in home inspections have an advantage (because of their training) in that they can evaluate certain problems in depth.

Home inspection limitations

A home inspection is not perfect and will not necessarily reveal every problem, but it is the best thing available. A home inspection is basically a visual inspection as well as an operational check of the various visible elements and components of the house. If performed in the manner suggested herein, you should have a fairly good picture of the condition of the house at the time of the inspection. By taking into account the age and present condition of various items, predictions can usually be made about future problems.

However, a home inspection cannot reveal a deficiency or problem if the conditions causing it are not visible. The following are just a few examples of such possible situations:

- A house with a concrete block foundation may have termite infestation in the voids of the foundation walls. This condition would not be visible during an inspection. Nevertheless, a week later there may be visible termite shelter tubes or a swarm.
- A basement or crawl space can be "bone dry" for the first three years and then flood in the fourth year because of a new building development nearby.
- The operation of a septic system depends on the condition (porosity and degree of saturation) of the subsoil in the leaching field, an item that is not readily apparent during an inspection. Depending on the weather, the level of the water table, and the usage prior to the inspection, it is possible for a malfunctioning septic system to appear to operate properly during the time it takes to perform an inspection.
- If there is an inadequately insulated water pipe in an exterior wall, it can freeze and burst during a severe cold spell. Since the pipe would not be visible during an inspection, the condition would not be detected.

Even with such limitations, a proper home inspection will reveal considerable information about the house that very often even the owner does not know.

Real estate warranty program

Many homes throughout the United States are being sold with a warranty protection plan for the buyer. The warranty usually covers the following components:

- Electrical system
- Heating system
- Water heater
- Plumbing system
- Central air-conditioning system
- Built-in appliances

It is in effect for the first year that the buyer owns the house. The warranties are usually made available to the buyer through the real estate broker. Not all brokers, however, are affiliated with companies that offer home warranties.

The home warranties that are currently available when you buy a house are very limited in scope and should not be considered a replacement for a home inspection. The warranty is basically an insurance plan that will protect the new homeowner against defects or malfunctions in the covered components, and nothing more. When you are planning on buying a house, many questions about the house should be answered so that you can make an intelligent decision. These questions concern items that are not covered by the warranty, and their answers can have a significant impact on the true cost of buying the house. Remember, the true cost of a house is the purchase price plus the costs for upgrading substandard, deteriorated, or malfunctioning components.

For example, the electrical system is covered against defects under the warranty. The fact that the electrical service in the house might be totally inadequate (110 volts at

30 amps) and would have to be upgraded at a considerable expense is not covered by the warranty. Also not covered by the warranty but costly to correct are basements that are subjected to periodic water penetration, structural damage to the foundation and woodframing, termite infestation and deterioration, roof replacement, and improvements needed for energy conservation. In addition, items of concern are general and latent defects such as an inadequately ventilated attic. A home warranty is definitely of value as long as you realize that it is not the answer to all of your questions or potential problems.

Private home inspection warranty programs

For an additional fee, some private home inspection companies offer a warranty protection plan after they inspect the house. The warranty applies for a twelve-month period and is usually effective from the day of the inspection or the transfer of title. The warranty plan is basically an insurance program. In fact, in some states, inspection companies have been forced to drop their warranty programs because they were not licensed by the state insurance commission to sell insurance.

As with all types of insurance, you should know exactly what you are buying before you put your money down. The warranty plan protects the buyer against defects in most of the major housing components, providing that the components are in reasonably good condition to begin with. If during the inspection any covered component is judged to be in less than satisfactory condition, it is excluded from coverage, although this is not reflected in the warranty fee. Also, for an item such as the roof, there is usually a prorated repair schedule that is a function of age. For example, in one plan if the roof is nine years old (which is not considered old or even aging), only 30 percent of the repair or replacement costs are covered. If the deductible (which might be $250 or $500) is subtracted, there might not be any coverage.

The warranty agreement should be read very carefully. In addition to items excluded because of condition or age, many components that are costly to correct are often not included in the plan. As with real estate warranty plans, these warranties can also be of value as long as you know exactly what you are buying and understand the limitations.

Manufacturers' warranties

Many of the components and items in a house have the manufacturer's guarantee or warranty against defects for a specific number of years. Items that are usually covered are the furnace, water heater, roof shingles, well pump, air-conditioning compressor, and electrical appliances. If you are purchasing a relatively new house or are in an older house any of the above components are relatively new (because of replacement), you should obtain the warranties or bills of sale. Many warranties are transferable to a new owner. Also, if the house has been termite-proofed, do not forget to get a copy of the guarantee if it is still in effect.

Contract

Many lawyers indicate that the best time to inspect a house you are considering buying is *before* signing the contract. This avoids the problems that can develop if the contract has already been signed and the subsequent inspection reveals major deficiencies in the house. Often the buyer wants a price adjustment or wants to back out of the deal. By inspecting the house before signing the contract, the correction of problems and deficiencies can be written into the contract.

In a seller's market, however, the buyer might feel pressured into signing the contract prior to the inspection in fear of having the house sold to another buyer before the inspection has been performed.

If you do sign a contract before the house has been inspected, for your protection, the contract should include an "inspection contingency" clause.

It is important that the clause be broad enough to include all the major problems that can occur in a house. Otherwise, there might be legal problems in backing out of the deal because of a major defect that was not covered in the contract. For example, many years ago I received a telephone call from a buyer for whom I had done a home inspection. The buyer indicated that his contract had an inspection clause that provided him with the option of declaring the contract null and void if there were *structural* deficiencies. That was basically the extent of the clause; it was very limited. My inspection had revealed that the electrical service was inadequate, that the heating system had to be replaced, and that there was evidence of water penetration into the basement. After the inspection, the buyer decided that he no longer wanted to buy the house. He called to find out if the problem items were considered structural deficiencies. No, they were not, and unfortunately, because of the wording in the contract, the buyer lost part of his deposit.

One contingency clause that has been used states that the sales agreement is contingent upon having a home inspection within n days of the signing. The inspection is defined to include a physical check of the exterior and interior components of the house (including the structural integrity of the building) and the condition of the electrical and mechanical equipment. If the inspection reveals any defects for which the cost exceeds $\$n$, the buyer should have the following options: have defects corrected by the seller; negotiate the cost of the corrections with the seller; or declare the sales agreement null and void. The limitation of this clause is in defining the means by which the costs for correcting the defects are determined.

To avoid this type of problem, one attorney I know uses the following clause:

> *This contract is contingent upon the Purchaser's making, or having made on his behalf by another, a physical inspection of the entire premises. If such inspection reveals any condition or state of facts unsatisfactory to the Purchaser, in his sole discretion, the Purchaser shall receive back all monies paid hereunder. Upon receipt of said monies, this agreement shall be null and void, and neither party under any obligation whatsoever to the other.*

This clause might not be valid in your state. To avoid any legal entanglement, your best bet is to consult an attorney.

As mentioned in chapter 1, just before the contract closing you should perform a final "walk-through" inspection of the house. Although most sellers will not deny you the right to make this inspection, it is not ensured. Consequently, the right to a final walk-through inspection should be made part of the contract.

No two houses are alike. Just as a person changes with age, so does a house. Regardless of whether you are a home buyer or a homeowner, it is important to know the true condition of the house. With this book, you are now prepared to find out. Try doing the inspection yourself. It's really a lot of fun. Worksheets are provided on which you can record your observations. Good luck!

Appendix

Home inspector requirements

In a little over 30 years, home inspection as a service to home buyers has morphed from a part-time activity, by a number of individuals working in and around large metropolitan areas, into a mature profession that has been accepted and recommended by government agencies. When I started my home inspection company in 1971 there were no specific requirements covering the items in a house to be inspected, and there were no specific qualifications to be an inspector. At that time anyone could hang out a shingle claiming to be a home inspector. Things started to change when we formed the American Society of Home Inspectors (ASHI). In January 1976 a small group of independent inspectors met and discussed the need to form a nonprofit professional organization. We recognized that in order for inspectors to be accepted by the public as professionals, it would be necessary to establish an organization that has a set of standards regarding those items to be inspected, a code of ethics, and membership requirements.

Today, ASHI is the largest national professional organization of home inspectors with 88 local ASHI chapters servicing thousands of members throughout the United States and Canada. ASHI's Standards of Practice, Code of Ethics, and membership requirements are included at the end of this appendix. As the profession of home inspection matured, additional national organizations were formed to promote excellence and professionalism by also providing a set of standards and a code of ethics and providing educational opportunities such as technical seminars, continuing education courses, and monthly meetings at which there are relevant speakers and where members can exchange ideas and network.

National home inspection organizations and their date of establishment are

- American Society of Home Inspectors (ASHI)—1976
- National Association of Home Inspectors (NAHI)—1987
- National Association of Building Inspection Engineers (NABIE)—1989
- Housing Inspection Foundation (HIF)—1989
- National Association of Certified Home Inspectors (NACHI)—1990
- Canadian Association of Home & Property Inspectors (CAHPI)—1994
- National Society of Home Inspectors (NSHI)—2003

In addition to the national home inspection organizations above, there is also a number of independent state home inspection associations as listed below:

- California Real Estate Inspection Association (CREIA)—1976
- Texas Association of Real Estate Inspectors (TAREI)—1977
- Minnesota Society of Home Inspectors (MSHI)—1979

- Georgia Association of Home Inspectors (GAHI)—1989
- Connecticut Association of Home Inspectors (CAHI)—1992
- Kentucky Real Estate Inspection Association, Inc. (KREIA)—1992
- Arkansas Association of Real Estate Inspectors (AAREI)—1993
- Wisconsin Association of Home Inspectors, Incorporated (WAHI)—1995
- North Carolina Licensed Home Inspector Association (NCLHIA)—1997
- Southern Nevada Association of Professional Property Inspectors (SNAPPI)—1998

As of this writing, 32 states have enacted legislation that requires individuals performing home inspections to become licensed. This further ensures professionalism and provides consumer protection by preventing anyone who is not licensed by the state from holding oneself out as a home inspector, or from performing a home inspection. Although licensing requirements vary from state to state, most require the candidate to take the National Home Inspection Examination or equivalent, as a means of assessing their competence. Some states also require the candidate to successfully complete an exam of ASHI standards and ethics. The national examination is administered by Pearson VUE at more than 215 proctored test center locations throughout the United States, Canada, and Puerto Rico. In addition to the exam, depending on the state, there are requirements that the candidate must successfully complete 60 to 180 hours of classroom instruction, approved by the state, with some states also requiring 40 hours of field-based instruction; moreover, the candidate must have both liability insurance and errors and omission insurance, and must have completed 100 to 200 fee-paid inspections under the supervision of a licensed home inspector.

Included in this appendix is a 128-question quiz along with the answers. The quiz is based on the inspection information that is presented in this book. It is not intended as a study guide for the National Home Inspection Examination. However, it is a good check on your understanding of some of the basics of home inspection. If you are interested in taking the national exam, the national organizations or their local chapters can provide you with information on schools that provide training for the exam.

Once the candidate is approved for licensing, he or she must abide by the rules and regulations set out by the legislation. An example of an item that is included in the legislation of one state is the preinspection agreement. There is a requirement that the home inspector must send a preinspection agreement to the client no later than one business day after the appointment for the home inspection is made. The agreement must contain, among other things, the fee and a description of the systems and components that will be inspected, as well as those items that are not part of the home inspection. The preinspection agreement is also required to contain a description of any and all services not part of the home inspection for which the client is charged an additional fee. The agreement must be executed prior to the start of the home inspection.

In some states, these two areas of concern are not part of a home inspection and require the inspector to have a separate license in order to evaluate them: the presence of wood-destroying organisms and the condition of the septic system. A seperate license is also needed to perform a test or collect samples to check for environmental hazards, such as a high radon concentration in the house and the presence of mold, asbestos, and lead in the water or paint. Those tests or samples must then be analyzed by a certified laboratory.

Depending on the state, the licensing legislation may also contain a standards of practice section, which outlines in considerable detail those systems and components in a house that must be inspected, as well as those items that must be included in a written report to the client. The licensing law can be quite specific. One state, New Jersey, has included in its legislation a mandatory tools and equipment requirement. It states that all home inspectors during the performance of an inspection must be equipped with the following minimum tools and equipment; the inspector, however, is not limited to that list and can use additional tools and equipment.

1. A ladder, minimum 11 feet in length.
2. A flashlight, or another equivalent light source, with a minimum 15,000 candlepower illumination.
3. A flame inspection mirror.
4. An electrical outlet tester with ground fault circuit interrupter (GFCI) test ability.
5. Tools necessary to remove common fasteners on covers or panels that are required to be removed for inspection.
6. A measuring tape.
7. A probe.
8. Thermometers for testing air-conditioning.
9. Binoculars with a magnification between 8 × 42 and 10 × 50.
10. A moisture meter.
11. Combustible gas leak detection equipment.
12. A voltage detector.

Depending on the state requirements, when renewing the home inspection license, the inspector must attest or provide information showing successful completion of 10 to 40 hours of continuing education units (CEUs) per renewal period.

In most of those states that do not license home inspectors, there is usually a statute concerning the field of home inspection. In the statute there is a requirement that the inspector must be a member of a national or a state home inspection organization. This has helped the profession to be self-regulating. When there is a new inspector in town the more-experienced inspectors will check to determine whether or not the new inspector is a member of a national or state inspection organization. They check because the new inspector is more competition for them. If he or she is not a member of a national or state home inspection organization, which have strict membership requirements, the new inspector is encouraged to join one of the organizations.

Quiz based on *The Complete Book of Home Inspection*

1. A home inspection should begin on the
   a) exterior
   b) roof
   c) crawlspace
   d) interior
   e) electrical system

2. Roof sheathing is another term for
   a) wood shingles
   b) the roof deck
   c) thatched roofs
   d) roof runoff
   e) rain gutters

3. The most common form of roof shingle used in this country is
   a) wood shingles
   b) asphalt shingles
   c) concrete tile
   d) synthetic shakes
   e) tar and gravel

4. On a 17-year-old asphalt shingle roof, you would expect to find the shingles on the _________ slope more weathered than the shingles on the other slopes.
   a) easterly
   b) northerly
   c) southerly
   d) westerly
   e) none of the above

5. A Cedar Breather is
   a) a beetle found in cedar shingles
   b) a nylon matrix underlayment for cedar shingles
   c) a finish coating on cedar shingles
   d) used for regenerating the odor in a cedar closet
   e) a device for removing moss from cedar shingles

6. The most durable roof covering is
   a) modified bitumen
   b) treated shakes
   c) tile
   d) 40-year dimensional asphalt shingles
   e) slate

7. Ponded water on a flat roof that was not designed to hold standing water
   a) is desirable because it can reduce the heat load in summer
   b) is especially valuable in freezing weather
   c) must be maintained by hosing the roof in summer
   d) has more disadvantages than advantages and should be avoided
   e) is acceptable only on metal roofs

8. Single-ply membranes
   a) are used as an underlayment below shingle roofs
   b) are used on steeply sloped tile roofs
   c) can only be found on commercial roofs
   d) are vulnerable to leakage at seams
   e) cannot be covered with gravel

9. If the owner says the chimney has been leaning for 15 years
   a) it has withstood the test of time and is presumed safe
   b) corrective action is still needed
   c) it is because of uniform deterioration of the mortar on all sides of the chimney
   d) the chimney is not a potential hazard
   e) the owner probably removed an antenna from the chimney

10. A chimney cricket is
    a) an insect that nests in the chimney
    b) also known as a saddle
    c) found on every chimney

d) a code requirement
e) able to prevent downdrafts

11. Vent stacks
    a) are parts of the plumbing system that project through the roof surface
    b) should not go all the way through the roof
    c) are self-sealing and not vulnerable to roof leakage
    d) need not be checked unless doing a smoke bomb test of the septic tank
    e) should end below the eaves

12. The absence of rain gutters
    a) is a design defect, especially in snow country
    b) indicates that the soils are not expansive
    c) should be noted on your worksheet
    d) is normal in houses with eaves having sufficient overhang
    e) shows that the owner removed them rather than repaint them

13. Rain gutters without seams
    a) are always copper
    b) can be aluminum
    c) are not subject to corrosion
    d) do not require painting
    e) are an impossibility

14. The acceptable difference in height between stair risers
    a) should be less than one inch
    b) can be any amount when the steps are not level
    c) should be zero to avoid potential tripping hazards
    d) can be any amount as long as there is a handrail
    e) can be ½ inch

15. Sheathing on exterior walls
    a) is inappropriate and belongs only on the roof
    b) is installed to provide bracing and minimize air infiltration
    c) cannot also be the exterior siding material
    d) must be made of solid material, not fiberboard or gypsum
    e) is always used as the nailing base for exterior siding

16. When inspecting siding, it is important to pay special attention to
    a) the areas facing the north
    b) the area facing the prevailing winds
    c) the second story
    d) the areas facing east
    e) the sections that face south or southwesterly

17. If you find vines growing up an exterior wall
    a) you should consider removing them
    b) you are not responsible for inspecting the siding in that area
    c) the vines provide extra insulating value
    d) there is no problem as long as this is on a south-facing wall
    e) the paint is being protected from sunlight and will be OK

18. Aluminum siding
    a) is sometimes required to be grounded
    b) never needs repainting
    c) is durable and resistant to mechanical damage from impact
    d) is never used in new construction
    e) cannot incorporate insulation due to condensation problems

19. Stucco siding
    a) is a concrete sheet that is poured first then applied to a wall
    b) is made of recycled marble
    c) can be applied to curved or irregular-shaped surfaces

d) adheres directly to sheathing material
e) is built up in five or six coats

20. EIFS is the acronym for
a) Exterior Interior Flame Safety
b) Easily Installed Flue Shutoff
c) Exterior Insulation Finish System
d) Exterior Interior Final Siding
e) Every Interrupt Fault System

21. Fiber-cement wall shingles are popular because they
a) will not rot
b) will not warp
c) will not delaminate
d) are resistant to termites
e) all of the above

22. In a brick veneer wall, the weakest point is
a) the space between the bricks and the foundation
b) the metal ties to the wall frame
c) the mortar between the bricks
d) the weep holes
e) the area over window or door openings

23. Exterior trim
a) includes all portions of the exterior finish other than the siding
b) serves a structural function
c) is installed in addition to the protection for joints and edges
d) refers only to wood and not to metal
e) cannot be exposed to the weather

24. The most common type of window in older homes is
a) awning window
b) horizontal sliding window
c) casement style
d) Craftsman style
e) double-hung

25. Jalousie refers to a style or type of
a) soffit
b) window
c) porch
d) cornice
e) tango

26. A swale is a
a) measuring stick to determine the water table
b) type of marine mammal
c) leach field for a septic tank system
d) junction box for an underground drainage system
e) depression in the ground to intercept and redirect surface runoff

27. A high water table
a) is especially good when there is a septic tank
b) has nothing to do with local rainfall
c) can be lowered by puncturing the ground
d) will likely lead to flooding in basements
e) is never within 20 feet of the ground surface

28. A dry retaining wall is one constructed
a) without weep holes to let water out
b) only in areas where the soils percolate moisture readily
c) without mortar
d) with the footing at least two feet below the frost line
e) in one piece

29. A retaining wall must be designed or approved by a licensed engineer if it is taller than
a) 3 feet
b) 4 feet
c) 5 feet
d) 6 feet
e) 7 feet

30. Where deck joists attach to a house, they should be secured by
    a) being toe-nailed into a header
    b) toe-nails and a ledger, or by metal brackets
    c) notches in the siding material
    d) diagonal braces
    e) posts set into the siding

31. Wood decks
    a) that show any signs of deterioration should be replaced completely
    b) should have proper drainage slope on the surface
    c) do not require guardrails except on second stories and balconies
    d) should have repairs or replacement of deteriorated planks as needed
    e) must always be constructed with pressure-treated pine

32. Composite decks
    a) are constructed entirely of wood
    b) are maintenance free
    c) do not expand and contract with temperature changes
    d) can be used for structural members
    e) use wood for the structural support members

33. Free-standing decks
    a) require two or more additional posts
    b) are as safe and secure as attached decks
    c) need diagonal bracing between the posts and the beams
    d) all of the above
    e) have posts that do not extend below the frost line

34. A door between the house and the garage
    a) should have a metal surface on the house side
    b) must not be on a higher elevation than the garage floor
    c) should be self-closing
    d) cannot have a tight seal
    e) may not be solid wood

35. A wall between the garage and the house
    a) may not have the attic entrance hatch, even if covered
    b) should have vents to the crawlspace or basement
    c) need not be inspected if covered with storage shelving
    d) may have exposed wood framing if the rest of the garage is also open
    e) must be covered with plaster, stucco, or other fire-resistant material

36. Garage doors with automatic openers
    a) should be checked for automatic reversal of the door
    b) need not be tested if the garage is detached from the main house
    c) should be tested only by disconnecting them from the automatic opener
    d) should only be tested by operating the remote control
    e) cannot be single-piece doors

37. Garage door restraining cables are used to
    a) prevent the door from closing too fast
    b) prevent the spring from whipping around in the event it breaks
    c) prevent the rollers from slipping out of the channels
    d) prevent fingers from getting crushed in door sections
    e) prevent the door from reversing automatically

38. The insect that causes the most damage to houses in the United States is the
    a) powder post beetle
    b) subterranean termite
    c) drywood termite

d) carpenter ant
e) dampwood termite

39. In general, subterranean termites
  a) are very fast working and immediately destructive
  b) are likely to colonize within the home near the heating plant
  c) can be permanently eradicated by pesticides and termite barriers
  d) are social insects with a rigid caste structure
  e) are rare in California and the Southeast

40. Termite baits are
  a) considerably less toxic that most liquid termiticides
  b) slow-acting poisons combined with a termite food material
  c) usually installed in plastic tubes or boxes
  d) used by pest control operators
  e) all of the above

41. Fiberglass batt insulation in an unfinished attic
  a) should be placed between rafters with the foil side showing
  b) should be placed between joists with the foil facing toward the attic
  c) is not necessary in warm climates such as California
  d) should be placed between joists with the foil facing toward the heated portion of the structure
  e) should be slit to allow water vapor to escape

42. Attic ventilation
  a) is not necessary in climates with no snowfall
  b) is very difficult to upgrade
  c) is important in both summer and winter
  d) should not be provided by openings in gable walls
  e) is typically provided by ducts from the kitchen range hood

43. An air admittance valve (AAV)
  a) is a pressure-actuated one-way mechanical vent
  b) is a code requirement
  c) allows sewer gases to vent to the attic
  d) negates the need for vent stacks
  e) is only effective on rainy days

44. Defective Chinese drywall
  a) is a gypsum wallboard made in China
  b) produces a rotten egg odor
  c) causes air-conditioning coils to corrode
  d) causes copper plumbing and electrical wiring to corrode
  e) all of the above

45. Squeaks in flooring
  a) indicate a significant structural defect
  b) are generally difficult to eliminate
  c) can be easily repaired, especially on the second floor
  d) indicate problems with water vapor
  e) can be corrected by carpeting

46. Roof truss uplift causes
  a) cracks in door frames
  b) cracks in joint between floor and walls
  c) cracks in joint between walls and ceiling
  d) cracks in window frames
  e) permanent sag in roof ridge

47. If a thermal pane window appears cloudy
  a) it indicates poor housekeeping on the part of the owner
  b) the hermetic seal should be repaired
  c) water has leaked into the window from the exterior

d) there is no insulating value to the window
e) the window requires replacement

48. Poor water flow is typically caused by
 a) too many fixtures on the second floor
 b) poor water pressure
 c) constriction in the supply pipes
 d) inadequate meter sizing
 e) clogged aerators

49. S-type traps are
 a) used only in second floor bathrooms
 b) required in two-pipe steam systems
 c) installed on a/c condensate lines
 d) not permitted by most plumbing codes
 e) installed on water heaters

50. Improper venting at sink traps
 a) can be determined by filling the sink with water and listening as it drains
 b) is not an important enough problem to be worth fixing
 c) is the term for lack of siphoning in a drain trap
 d) can be seen by tape on the drain trap
 e) will cause water hammer in the drain pipes

51. A rule of thumb for the minimum height of the chimney above the fireplace floor is
 a) 8 feet
 b) 10 feet
 c) 12 feet
 d) 15 feet
 e) 18 feet

52. The best location for a smoke detector is
 a) in the return air duct of the heating system
 b) over the stove, since most fires start there
 c) on the ceiling of the hallway leading to the bedroom
 d) on a hallway wall but no closer than 6 inches from the ceiling
 e) in the upper corners of each bedroom

53. Soils with a high clay content can swell or shrink up to
 a) 10 percent
 b) 20 percent
 c) 30 percent
 d) 40 percent
 e) 50 percent

54. Deteriorated mortar joints in foundation walls
 a) do not represent a weakened structural condition
 b) should be repointed to improve the appearance of the foundation
 c) should be repointed to restore load-bearing capacity
 d) should be repointed to prevent basement water entry
 e) can be painted to prevent further deterioration

55. Superior walls are
 a) walls made from superior materials
 b) less likely to crack under pressure
 c) insulated, precast, ribbed concrete wall panels
 d) made from a mix of concrete and fiber
 e) concrete blocks with insulation in the voids

56. Engineered lumber
 a) can only be used for roof framing
 b) is termite-resistant
 c) is stronger and more dimensionally stable than solid-sawn lumber
 d) will prevent squeaks in flooring
 e) is most effective when installed in cold weather

57. The inspection for water seepage into a basement or crawlspace begins
   a) after you see standing water in the crawlspace
   b) on the roof
   c) on the exterior
   d) by consulting maps available from the U.S. Geological Survey (USGS)
   e) by smelling damp conditions

58. Efflorescence is
   a) a significant structural defect
   b) an enthusiastic spontaneous response from realtors reading a favorable inspection report
   c) a chalky residue of sand and dissolved concrete created by the action of acidic soils
   d) an indication of effective damp proofing
   e) an accumulation of mineral salts created by moisture in masonry or concrete

59. The minimum FHA standards for crawlspace clearance are
   a) 24 inches between any wood and soil
   b) 18 inches soil-to-joist, 12 inches soil-to-girder
   c) 18 inches between any wood and soil
   d) 12 inches between the subfloor and soil
   e) any distance if pressure-treated lumber is used

60. In a conditioned crawl space
   a) the foundation walls are insulated
   b) wall vents are closed
   c) a dirt floor, if there is one, is covered with large plastic sheets
   d) the area is heated or air-conditioned
   e) all of the above

61. 110-volt electrical services
   a) are adequate for all modern appliances
   b) are adequate in most cases
   c) will suffice for all houses except those with air-conditioning
   d) are inadequate by current standards
   e) require that you use a gas refrigerator

62. Circuit breakers
   a) are an inconvenience when compared to fuses
   b) are more reliable than fuses and always provide greater safety
   c) cannot be used to provide overcurrent protection
   d) prevent overloading by allowing just enough voltage to pass through to meet demand
   e) are somewhat less reliable than fuses

63. The capacity of a branch circuit fuse
   a) can be increased by inserting a penny behind it
   b) should be as large as possible for added safety
   c) should match the current-carrying capacity of the branch circuit
   d) is electrically interchangeable with any other Edison fuse
   e) is increased by replacing the fuse with a fusestat

64. Removal of the interior cover of an electrical panel box
   a) is necessary to locate the main circuit breaker
   b) cannot be done without an electrical permit
   c) will enable determination of the true electrical capacity
   d) is not necessary when determining proper sizing of circuit breakers
   e) should be part of routine homeowner maintenance

65. If a house is wired with aluminum
   a) it is normal and acceptable for switches and outlets to be warm to the touch

b) the wiring is unsafe and should be completely replaced
c) there is no problem provided that outlets marked AL/CU were used
d) you should scan all the walls with an infrared heat detector to check for bad wiring
e) the connections should all be checked by a qualified electrical contractor

66. A ground fault circuit interrupter (GFCI) is an electronic device that
a) can be used instead of a fuse
b) is retroactively required for all bathroom receptacles
c) will not work properly if the main grounding is faulty
d) opens the circuit when it senses a potentially hazardous condition
e) can be reached when touching another ground source such as a shower head

67. An arc fault circuit interrupter (AFCI) is an electronic device that
a) is a backup if the GFI fails to operate
b) does not require periodic testing
c) disconnects the power if it detects arcing in electrical wiring
d) is an important part of a fire alarm system
e) is a retroactive building code requirement

68. Knob-and-tube wiring is
a) obsolete but usually safe if unmodified
b) acceptable when it includes grounding conductors
c) dangerous because the wires are not insulated
d) exposed and readily accessible in walls and ceilings
e) found only in Canada, Europe, and the East Coast

69. Systems with light fixture switches operating at 24 volts
a) indicate a significant loss of current and waste of electricity
b) can be operated from only one location, and do not work for three-way switches
c) can be repaired by removing the transformer
d) should be provided with spare replacement relays
e) are borrowed from automobile wiring systems

70. The house service main water pipe
a) is maintained by the water company
b) must always be at least 1 inch in diameter
c) would be the homeowners' expense if in need of repair or replacement
d) must not be made of metal or a dangerous electrical ground would result
e) should be fully visible and accessible to inspection

71. Hot water piping
a) should be no further than 4 inches from the parallel cold water piping
b) cannot be a part of the heating system of the house
c) is not visually accessible due to insulation
d) is required for all plumbing fixtures
e) should be separated from cold water piping by at least 6 inches

72. Drainage systems
a) are simpler than supply systems
b) consist of horizontal runs of piping
c) include traps, drainpipes, and vents
d) cannot operate properly by gravity alone
e) must be the same diameter as supply pipes

73. The gases that are generated in a septic tank are usually
    a) explosive and must be kept out of the house drainage-vent system
    b) must be dissipated through a fresh air inlet at the house trap
    c) dissipated into the earth through the leach field
    d) dissipated to the atmosphere through the house drainage-vent system
    e) an indication that the septic tank is sterilized and not functioning

74. A mound septic system
    a) is most effective in areas with hard water
    b) is normally found on top of a mountain ridge
    c) is used when the drainfield has a problem such as inadequate percolation rates
    d) can have a mound 6 to 8 feet high
    e) uses gravity to move the effluent

75. Vent pipes in drainage systems
    a) allow air to flow in and out of drain systems
    b) are not necessary when connected to a public sewer
    c) carry water and wastes
    d) should not connect near fixtures
    e) are not allowed to discharge into the atmosphere

76. The presence of a septic tank
    a) cannot always be determined during an inspection
    b) can be determined by whether or not there is a house trap
    c) can be ruled out if there are sewers for other houses on the street
    d) cannot be determined from the building department or other public records
    e) is a significant defect that must always be reported

77. The appropriate size for a septic leaching field is determined by
    a) the number of seepage pits
    b) the size of the water supply piping
    c) the elevation above sea level
    d) whether or not a garbage disposer is installed
    e) the percolation rate of the ground

78. Septic tank leaching field
    a) must be supplied with fresh leaches each year
    b) works best in clay soils
    c) will receive nitrogen and other natural fertilizers from the septic effluent
    d) definitely failed if the vegetation over them is lush and green
    e) are a less likely area for failure to be visible as compared to the area over the tank

79. An atmospheric vacuum breaker
    a) cannot be used in residential plumbing without the approval of the Department of Air Quality
    b) is required for all lawn sprinkler systems
    c) may not be more than 6 inches above grade
    d) is required at each sprinkler head
    e) prevents dirty water from backing up from the drains

80. A wall hatch providing access to the fixture side of a tub
    a) is a violation of most plumbing codes
    b) indicates that leakage occurred requiring the wall to be opened for repairs
    c) must be distinctly noticeable and not blocked by furniture
    d) is common in many older homes
    e) should never be found above the first floor

81. Inlet service pipes made of lead
   a) can sometimes be identified by a large bulge around a wiped joint
   b) are still installed in modern homes
   c) pose no hazard after prolonged use as interior corrosion prevents direct water contact with lead
   d) are the responsibility of the water supply company not the homeowner
   e) are common, but testing of the effects on water is very expensive and usually unnecessary

82. PEX piping (tubing)
   a) should only be used for cold water
   b) should only be used for hot water
   c) is flexible, and won't corrode or develop pin holes
   d) uses a special adhesive to seal the joints
   e) reacts with water-soluble oxidants

83. Leaching zinc is the cause of
   a) encrustation around fittings of copper pipe
   b) pinhole leaks in brass pipe
   c) rusting in galvanized pipe
   d) galvanic corrosion
   e) a green color at copper pipe fittings

84. A pressure tank
   a) can be located near a well pump house or in a basement
   b) is an indication that the well pump is inadequate
   c) compensates for lack of water in the well
   d) is illegal in most building jurisdictions
   e) is only needed for wells that yield less than 5 gallons per minute

85. Central heating
   a) can be easily extended to room additions
   b) always includes the basement
   c) has fewer moving parts than area heaters
   d) can usually be designed to provide distribution to separate zones
   e) automatically balances itself

86. Every heating system should have
   a) a humidistat
   b) a setback thermostat with timer
   c) multiple zones
   d) a master shutoff switch
   e) a filter

87. A condition under which a furnace must be replaced is
   a) a cracked heat exchanger
   b) introduction of carbon monoxide into the air return
   c) blocking of the chimney
   d) being older than the anticipated service life of the furnace
   e) being over 30 years old

88. Examples of a warm air heating system are
   a) single-pipe steam systems
   b) forced-air heating systems with ducts
   c) two-pipe steam heating system
   d) electric baseboard heating system
   e) all of the above

89. A multizoned forced-air heating system
   a) provides truly independent operation of one zone from another
   b) can only be controlled by a single thermostat
   c) requires motorized duct dampers
   d) bypasses potential dangers of failed heat exchangers
   e) has no moving parts

90. An "octopus" heating system
   a) is another term for a system that smells like dead fish
   b) requires air filters to function properly

c) will not work in thin atmospheres above 1,000 feet
d) requires frequent lubrication of the blower motor
e) is another term for a large gravity warm air central furnace

91. Condensing furnaces are
a) furnaces with reduced clearances that fit into a condensed space
b) the new generation of high efficiency furnaces
c) furnaces that extract the heat from a refrigeration system
d) furnaces without internal air-conditioning coils
e) types with a single heat exchanger cell

92. An extended plenum is
a) a type of warm air distribution system
b) the same as a radial duct system
c) a platform to provide clearance above a garage or basement floor
d) a vibration damper in a duct system
e) always composed of square or rectangular ducts

93. The optimum location for heating registers is
a) high on the wall or on the ceiling opposite from windows
b) in the center of each room's ceiling
c) along the outside walls and below windows
d) as close as possible to the return air inlet
e) where it will produce a "short cycle"

94. All hot water heating systems must be supplied with
a) a pump
b) an expansion tank
c) radiators
d) storage-tank boilers
e) an oil filter

95. Older coal-fired gravity-type boilers
a) cannot be converted to oil or gas
b) should be replaced with domestic water heaters
c) are typically cast iron and very large
d) produce condensation and require drain piping
e) operate at pressures in excess of 100 PSI

96. A gravity hot water heating system does not need an expansion tank
a) True
b) False

97. An advantage of a forced hot-water heating system is
a) immunity from problems with freezing
b) easy adaptability to air-conditioning
c) lack of any moving parts requiring maintenance
d) the absence of a need for any other energy source other than water pressure
e) less heat fluctuation and adaptability to be multiple zones

98. When a boiler is also used to provide heat for domestic hot water, it requires a
a) flow-control valve
b) second pressure relief valve
c) seasonal setting on the thermostat
d) sight glass
e) supplementary circulating pump

99. A pressure-temperature gauge is
a) a necessary safety control on a hot-water heating system
b) an operational control
c) illegal, as there must be separate gauges for each function
d) only necessary on steam heating
e) sometimes provided with a scale in altitude

100. A quick means to determine whether a boiler is for steam or hot water is
   a) whether the system was installed before 1900
   b) if it is hooked up to the municipal steam supply
   c) if it includes a water-level gauge (sight glass)
   d) if the tank has welded pressure boiler plates
   e) the presence of a steam whistle

101. A Hartford loop can be found on
   a) gas-fired water heaters
   b) steam boilers
   c) hot water boilers
   d) oil-fired water heaters
   e) air-conditioning piping

102. The purpose of a Hartford loop is to
   a) provide a more efficient balancing of steam to the radiators
   b) prime the system and prevent freezing
   c) allow expansion and contraction of piping with changes in temperature
   d) prevent water from draining out of the boiler in the event of a leak
   e) qualify the house for insurance by a certain company in Connecticut

103. A blow-off valve is
   a) an automatically self-cleaning valve
   b) a part of the fuel delivery system in an oil burner
   c) part of an externally mounted low water cutoff valve
   d) a phrase to be used when disclaiming any knowledge about steam heating
   e) a type of pressure relief valve

104. A one-pipe steam heating system can be easily converted to an induced hot-water heating system by changing the radiators and adding a circulator pump and expansion tank.
   a) True
   b) False

105. Oil burners
   a) burn cleaner than gas burners
   b) do not require maintenance
   c) must bear an approval stamp from the American Gas Association
   d) are often maintained by the companies that supply the fuel oil
   e) do not require a source of combustion air

106. Condensing furnaces
   a) can be vented into masonry chimneys
   b) are noisier than conventional furnaces
   c) have less air flow than conventional furnaces
   d) must have vertical or near-vertical exhaust piping
   e) can be combined with vents from a gas-fired water heater

107. Gas burners
   a) should have a blue flame with little or no yellow
   b) should have a flame that lifts well above the burner
   c) should have an orange flame
   d) will normally vary in height depending on the barometric pressure
   e) do not produce soot or water vapor when functioning properly

108. A water heater temperature and pressure relief valve
   a) is not required for modern efficient water heaters
   b) can be placed in either the cold or hot water piping near the tank
   c) must always be placed in the lower third of the tank

d) should connect to the inlet with the dip tube
e) should be provided with a discharge pipe ending in a visible location

109. A rumbling noise from a water heater tank
a) is an indication of imminent explosion
b) means that the pressure-relief valve is blocked
c) is a result of loose gas or oil burners
d) results from accumulated sediment
e) means that the water heater will leak soon

110. A dripping water heater relief valve could be an indication of
a) excessive high pressure
b) defective valve
c) thermal expansion
d) all of the above
e) none of the above

111. Recovery rate most often refers to
a) air-conditioning compressor
b) warm air furnace capacity
c) tank-type water heaters
d) hot water heating systems
e) water pressure reducing valves

112. A water heater should be replaced when
a) the burners malfunction
b) the flue is worn out
c) the sacrificial anode contains magnesium
d) it is more than 10 years old
e) it leaks

113. Electric water heaters
a) are usually instantaneous, with no storage tank
b) use much more energy and are not allowed by most cities for energy conservation reasons
c) have lower recovery rates than gas- or oil-fired water heaters
d) have higher recovery rates than gas, so the tanks are usually smaller
e) always require a mixing valve to prevent scalding

114. One ton of cooling capacity
a) means that the air conditioner will produce one ton of ice per hour
b) is needed for approximately every 700 square feet of living space
c) is a measure of pressure in an average size compressor
d) is the weight of the air cooled by a one-ton air conditioner each hour
e) is equal to 2,000 Btu of cooling capacity

115. An air-conditioning condenser
a) draws approximately 7 amps per ton of cooling capacity
b) must be provided with a condensate drain pipe
c) should be located in a basement or furnace room
d) must be protected from rainfall
e) will weigh approximately one ton for each ton of cooling capacity

116. An evaporator coil is
a) part of a hydronic heating system used for cooling interior air
b) a ring on the outside of an air-conditioning compressor
c) an adaptation often found on air-conditioning when distilling spirits
d) located on the building exterior
e) the place where coolant changes state from a liquid to a gas

117. An electrical disconnect switch
a) is needed to deactivate the evaporator coils

b) must be turned on to energize the condensing coils
c) is not needed except for the circuit breaker controlling the furnace
d) is required for a light near the condenser
e) should be found near the portion of the air-conditioning system containing the compressor

118. A condensate trap in air-conditioning
a) is to prevent any condensate from escaping
b) means that the system has trapped too much cold air
c) prevents the blower from sucking air up through the drain line
d) must be vented to dissipate sewer gases
e) recycles the coolant to the condenser

119. When a split-system air conditioner is also part of a heating system
a) the blower should have at least two speeds
b) there should be a separate set of smaller ducts for the air conditioner
c) the split should be healed before operating the heating system
d) the coolant can be run through the furnace heat exchanger
e) the compressor should be next to the furnace plenum

120. The temperature of air discharging from the registers during the cooling cycle
a) should be 10 to 20 degrees lower than the outside air
b) should be 15 degrees lower than the temperature during the heating cycle
c) should be no more than 60 degrees
d) should be about 15 degrees lower than the temperature of the air in the room
e) should be at least 20 degrees lower than the temperature of the return air

121. A heat pump should
a) not ever need supplementary electric-resistance heating
b) be sized based upon the cooling load, not the heating load
c) be tested in both the cooling and heating modes to ensure that the compressor works
d) be located entirely within the building
e) not be tested when the outside air temperature is 40 degrees

122. Evaporative coolers are most effective where?
a) East Coast
b) Florida
c) Southwest
d) California
e) Washington and Oregon

123. If attic insulation does not have a vapor barrier
a) one should be laid over the top of it in an attic
b) one cannot be added to it or the manufacturer's warranty is void
c) one can be laid under the insulation so that it is adjacent to the heated ceiling
d) one can be added to the underside of the roof deck
e) it is because a vapor barrier is never necessary

124. The average water loss in one week in a swimming pool due to evaporation is
a) less than 1 inch
b) 1 inch
c) 1½ inches
d) 2 inches
e) 2½ inches

125. Radon gas
   a) is one of the most significant forms of manmade pollution
   b) is odorless, tasteless, invisible, and formed by natural uranium deposits
   c) is a result of high frequency radio waves generated for communications lines
   d) can be prevented from forming when a house is properly insulated
   e) most often enters a home through shower heads

126. Asbestos
   a) kills thousands of people annually from asbestos-related cancers
   b) must always be removed prior to sale of a home
   c) should never be encapsulated, as this only increases its dangers
   d) can be removed only by a licensed heating or insulating contractor
   e) in any level is a health risk within a home

127. Mold growth in a house thrives on moisture and
   a) cotton
   b) wool
   c) paper
   d) wood
   e) all of the above

128. A desuperheater
   a) is used to steam clean clothing
   b) eliminates static electricity in a house
   c) is a device on a geothermal system that is used to heat water
   d) uses gas as an energy source
   e) is a device on newly designed oil burners

Quiz answers

1. a
2. b
3. b
4. c
5. b
6. e
7. d
8. d
9. b
10. b
11. a
12. c
13. b
14. c
15. b
16. e
17. a
18. a
19. c
20. c
21. e
22. b
23. a
24. e
25. b
26. e
27. d
28. c
29. b
30. b
31. d
32. e
33. d
34. c
35. e
36. a
37. b
38. b
39. d
40. e
41. d
42. c
43. a
44. e
45. b
46. c
47. e
48. c
49. d
50. a
51. d
52. c
53. e
54. d
55. c
56. c
57. c
58. e
59. b
60. e
61. d
62. e
63. c
64. c
65. e
66. d
67. c
68. a
69. d
70. c
71. e
72. c
73. d
74. c
75. a
76. a
77. e
78. c
79. b
80. d
81. a
82. c
83. b
84. a
85. d
86. d
87. a
88. b
89. c
90. e
91. b
92. a
93. c
94. b
95. c
96. b
97. e
98. a
99. e
100. c
101. b
102. d
103. c
104. b
105. d
106. b
107. a
108. e
109. d
110. d
111. c
112. e
113. c
114. b
115. a
116. e
117. e
118. c
119. a
120. d
121. b
122. c
123. c
124. a
125. b
126. a
127. e
128. c

ASHI Standards of Practice

1. INTRODUCTION

The American Society of Home Inspectors®, Inc. (ASHI®) is a not-for-profit professional society established in 1976. Membership in ASHI is voluntary and its members are private home *inspectors*. ASHI's objectives include promotion of excellence within the profession and continual improvement of its members' inspection services to the public.

2. PURPOSE AND SCOPE

2.1 The purpose of the Standards of Practice is to establish a minimum and uniform standard for home *inspectors* who subscribe to these Standards of Practice. *Home inspections* performed to these Standards of Practice are intended to provide the client with objective information regarding the condition of the *systems* and *components* of the home as *inspected* at the time of the *home inspection*. Redundancy in the description of the requirements, limitations, and exclusions regarding the scope of the *home inspection* is provided for emphasis only.

2.2 Inspectors shall:

A. adhere to the Code of Ethics of the American Society of Home Inspectors.

B. *inspect readily accessible*, visually observable, *installed systems* and *components* listed in these Standards of Practice.

C. report:

1. those *systems* and *components inspected* that, in the professional judgment of the *inspector*, are not functioning properly, significantly deficient, *unsafe*, or are near the end of their service lives.
2. recommendations to correct, or monitor for future correction, the deficiencies *reported* in 2.2.C.1, or items needing *further evaluation*. (*Per* Exclusion 13.2.A.5 *inspectors* are NOT required to determine methods, materials, or costs of corrections.)
3. reasoning or explanation as to the nature of the deficiencies *reported* in 2.2.C.1, that are not self-evident.
4. *systems* and *components* designated for inspection in these Standards of Practice that were present at the time of the *home inspection* but were not *inspected* and the reason(s) they were not *inspected*.

2.3 These Standards of Practice are not intended to limit *inspectors* from:

A. including other inspection services or *systems* and *components* in addition to those required in Section 2.2.B.

B. designing or specifying repairs, provided the *inspector* is appropriately qualified and willing to do so.

C. excluding *systems* and *components* from the inspection if requested by the client.

3. STRUCTURAL COMPONENTS

3.1 The *inspector* shall:

A. *inspect*:

1. *structural components* including the foundation and framing.
2. by probing a *representative number* of *structural components* where deterioration is suspected or where clear indications of possible deterioration exist. Probing is NOT required when probing would damage any finished surface or where no deterioration is visible or presumed to exist.

B. *describe:*
1. the methods used to *inspect under-floor crawl spaces* and attics.
2. the foundation.
3. the floor structure.
4. the wall structure.
5. the ceiling structure.
6. the roof structure.

3.2 The inspector is NOT required to:

A. provide any *engineering* or architectural services or analysis.
B. offer an opinion as to the adequacy of any *structural system* or *component*

4. EXTERIOR

4.1 The inspector shall:

A. *inspect:*
1. *siding*, flashing and trim.
2. all exterior doors.
3. attached or adjacent decks, balconies, stoops, steps, porches, and their associated railings.
4. eaves, soffits, and fascias where accessible from the ground level.
5. vegetation, grading, surface drainage, and retaining walls that are likely to adversely affect the building.
6. adjacent or entryway walkways, patios, and driveways.

B. *describe:*
1. *siding*.

4.2 The *inspector* is NOT required to *inspect*

A. screening, shutters, awnings, and similar seasonal accessories.
B. fences.
C. geological and/or soil conditions.
D. *recreational facilities*.
E. outbuildings other than garages and carports.
F. seawalls, break-walls, and docks.
G. erosion control and earth stabilization measures.

5. ROOFING

5.1 The inspector shall:

A. *inspect:*
1. roofing materials.
2. *roof drainage systems*.
3. flashing.
4. skylights, chimneys, and roof penetrations.

B. *describe:*
1. roofing materials.
2. methods used to *inspect* the roofing.

5.2 The inspector is NOT required to inspect:

A. antennae.
B. interiors of flues or chimneys that are not *readily accessible*.
C. other *installed accessories*.

6. PLUMBING

6.1 The inspector shall:

A. *inspect:*
1. interior water supply and distribution *systems* including all fixtures and faucets.
2. drain, waste, and vent *systems* including all fixtures.
3. water heating equipment and hot water supply system.
4. vent *systems*, flues, and chimneys.
5. fuel storage and fuel distribution *systems*.
6. drainage sumps, sump pumps, and related piping.

B. *describe:*
1. water supply, drain, waste, and vent piping materials.
2. water heating equipment including energy source(s).
3. location of main water and fuel shut-off valves.

6.2 The Inspector is NOT required to:

A. *inspect:*

1. clothes washing machine connections.
2. interiors of flues or chimneys that are not *readily accessible*.
3. wells, well pumps, or water storage related equipment.
4. water conditioning *systems*.
5. solar water heating *systems*.
6. fire and lawn sprinkler *systems*.
7. private waste disposal *systems*.

B. determine:

1. whether water supply and waste disposal *systems* are public or private.
2. water supply quantity or quality.

C. operate *automatic safety controls* or manual stop valves.

7. ELECTRICAL

7.1 The inspector shall:

A. *inspect*:

1. service drop.
2. service entrance conductors, cables, and raceways.
3. service equipment and main disconnects.
4. service grounding.
5. interior *components* of service panels and sub panels.
6. conductors.
7. overcurrent protection devices.
8. a *representative number* of *installed* lighting fixtures, switches, and receptacles.
9. ground fault circuit interrupters.

B. *describe:*

1. amperage and voltage rating of the service.
2. location of main disconnect(s) and sub panels.
3. presence of solid conductor aluminum branch circuit wiring.
4. presence or absence of smoke detectors.
5. *wiring methods*.

7.2 The inspector is NOT required to:

A. *inspect:*

1. remote control devices.
2. *alarm systems* and *components*.
3. low voltage wiring *systems and components*.
4. ancillary wiring *systems* and *components* that are not a part of the primary electrical power distribution *system*.

B. measure amperage, voltage, or impedance.

8. HEATING

8.1 The inspector shall:

A. open *readily openable access panels*.

B. *inspect:*

1. *installed* heating equipment.
2. vent *systems*, flues, and chimneys.

C. *describe:*

1. energy source(s).
2. heating *systems*.

8.2 The inspector is NOT required to:

A. *inspect:*

1. interiors of flues or chimneys that are not *readily accessible*.
2. heat exchangers.
3. humidifiers or dehumidifiers.
4. electronic air filters
5. solar space heating *systems*.

B. determine heat supply adequacy or distribution balance.

9. AIR CONDITIONING

9.1 The inspector shall:

A. open *readily openable access panels*.

B. *inspect:*

1. central and through-wall equipment.
2. distribution *systems*

C. *describe:*

1. energy source(s).
2. cooling *systems*.

9.2 The inspector is NOT required to:

- **A.** *inspect* electronic air filters.
- **B.** determine cooling supply adequacy or distribution balance.
- **C.** *inspect* window air conditioning units.

10. INTERIORS

10.1 The *inspector* shall *inspect:*

- **A.** walls, ceilings, and floors.
- **B.** steps, stairways, and railings.
- **C.** countertops and a *representative number* of *installed* cabinets.
- **D.** a *representative number* of doors and windows.
- **E.** garage doors and garage door operators.

10.2 The *inspector* is NOT required to *inspect:*

- **A.** paint, wallpaper, and other finish treatments.
- **B.** carpeting.
- **C.** window treatments.
- **D.** central vacuum *systems.*
- **E.** *household appliances.*
- **F.** *recreational facilities.*

11. INSULATION & VENTILATION

11.1 The inspector shall:

- **A.** *inspect:*
  1. insulation and vapor retarders in unfinished spaces.
  2. ventilation of attics and foundation areas.
  3. mechanical ventilation *systems.*
- **B.** *describe:*
  1. insulation and vapor retarders in unfinished spaces.
  2. absence of insulation in unfinished spaces at conditioned surfaces.

11.2 The *inspector* is NOT required to disturb insulation.

See 13.2.A.11 and 13.2.A.12.

12. FIREPLACES AND SOLID FUEL BURNING APPLIANCES

12.1 The *inspector* shall:

- **A.** *inspect:*
  1. *system components.*
  2. chimney and vents.
- **B.** *describe:*
  1. fireplaces and *solid fuel burning appliances.*
  2. chimneys.

12.2 The inspector is NOT required to:

- **A.** *inspect:*
  1. interiors of flues or chimneys.
  2. fire-screens and doors.
  3. seals and gaskets.
  4. automatic fuel feed devices.
  5. mantles and fireplace surrounds.
  6. combustion make-up air devices.
  7. heat distribution assists (gravity fed and fan assisted).
- **B.** ignite or extinguish fires.
- **C.** determine draft characteristics.
- **D.** move fireplace inserts and stoves or firebox contents.

13. GENERAL LIMITATIONS AND EXCLUSIONS

13.1 General limitations:

- **A.** The *inspector is* NOT required to perform any action or make any determination not specifically stated in these Standards of Practice.
- **B.** Inspections performed in accordance with these Standards of Practice:
  1. are not *technically exhaustive.*
  2. are not required to identify concealed. conditions, latent defects, or consequential damage(s).
- **C.** These Standards of Practice are applicable to buildings with four or fewer dwelling units and their garages or carports.

13.2 General exclusions:

A. *Inspectors* are NOT required to determine:

1. conditions of *systems* or *components* that are not *readily accessible.*
2. remaining life expectancy of any *system* or *component.*
3. strength, adequacy, effectiveness, or efficiency of any *system* or *component.*
4. the causes of any condition or deficiency.
5. methods, materials, or costs of corrections.
6. future conditions including but not limited to failure of *systems* and *components.*
7. the suitability of the property for any specialized use.
8. compliance with regulatory requirements (codes, regulations, laws, ordinances, etc.).
9. market value of the property or its marketability.
10. the advisability of purchase of the property.
11. the presence of potentially hazardous plants or animals including, but not limited to, wood destroying organisms or diseases harmful to humans including molds or mold-like substances.
12. the presence of any environmental hazards including, but not limited to, toxins, carcinogens, noise, and contaminants in soil, water, and air.
13. the effectiveness of any *system installed* or method utilized to control or remove suspected hazardous substances.
14. operating costs of *systems* or *components.*
15. acoustical properties of any *system* or *component.*
16. soil conditions relating to geotechnical or hydrologic specialties.

B. *Inspectors* are NOT required to offer:

1. or perform any act or service contrary to law.
2. or perform *engineering* services.
3. or perform any trade or any professional service other than *home inspection.*
4. warranties or guarantees of any kind.

C. *Inspectors* are NOT required to operate:

1. any *system* or *component* that is *shut down* or otherwise inoperable.
2. any *system* or *component* that does not respond to *normal operating controls.*
3. shut-off valves or manual stop valves.

D. *Inspectors* are NOT required to enter:

1. any area that will, in the opinion of the *inspector,* likely be dangerous to the *inspector* or other persons or damage the property or its *systems* or *components.*
2. *under-floor crawl spaces* or attics that are not *readily accessible.*

E. *Inspectors* are NOT required to inspect:

1. underground items including but not limited to underground storage tanks or other underground indications of their presence, whether abandoned or active.
2. items that are not *installed.*
3. *installed decorative* items.
4. items in areas that are not entered in accordance with 13.2.D.
5. detached structures other than garages and carports.

6. common elements or common areas in multi-unit housing, such as condominium properties or cooperative housing.

F. *Inspectors* are NOT required to:

1. perform any procedure or operation that will, in the opinion of the *inspector,* likely be dangerous to the *inspector* or other persons or damage the property or its *systems* or *components.*
2. describe or report on any *system* or *component* that is not included in these Standards and was not *inspected.*
3. move personal property, furniture, equipment, plants, soil, snow, ice, or debris.
4. *dismantle* any *system* or *component, except* as explicitly required by these Standards of Practice.

ASHI Standards of Practice glossary of italicized terms

Alarm Systems
Warning devices *installed* or free-standing including but not limited to smoke detectors, carbon monoxide detectors, flue gas, and other spillage detectors, and security equipment

Automatic Safety Controls
Devices designed and *installed* to protect *systems* and *components* from unsafe conditions

Component
A part of a *system*

Decorative
Ornamental; not required for the proper operation of the essential *systems* and *components* of a home

Describe
To identify (in writing) a *system* or *component* by its type or other distinguishing characteristics

Dismantle
To take apart or remove any *component*, device, or piece of equipment that would not be taken apart or removed by a homeowner in the course of normal maintenance

Engineering
The application of scientific knowledge for the design, control, or use of building structures, equipment, or apparatus

Further Evaluation
Examination and analysis by a qualified professional, tradesman, or service technician beyond that provided by the *home inspection*

Home Inspection
The process by which an inspector visually examines the *readily accessible systems* and *components* of a home and which *describes* those *systems* and *components* in accordance with these Standards of Practice

Household Appliances
Kitchen, laundry, and similar appliances, whether *installed* or free-standing

Inspect
To examine any *system* or *component* of a building in accordance with these Standards of Practice, using *normal operating controls* and opening *readily openable access panels*

Inspector
A person hired to examine any *system* or *component* of a building in accordance with these Standards of Practice

Installed
Attached such that removal requires tools

Normal Operating Controls
Devices such as thermostats, switches, or valves intended to be operated by the homeowner

Readily Accessible
Available for visual inspection without requiring moving of personal property, *dismantling*, destructive measures, or any action that will likely involve risk to persons or property

Readily Openable Access Panel
A panel provided for homeowner inspection and maintenance that is *readily accessible*, within normal reach, can be removed by one person, and is not sealed in place

Recreational Facilities
Spas, saunas, steam baths, swimming pools, exercise, entertainment, athletic playground or other similar equipment, and associated accessories

Report
Communicate in writing

Representative Number
One *component* per room for multiple similar interior *components* such as windows, and electric receptacles; one *component* on each side of the building for multiple similar exterior *components*

Root Drainage Systems
Components used to carry water off a roof and away from a building

Shut Down
A state in which a *system* or *component* cannot be operated by *normal operating controls*

Siding
Exterior wall covering and cladding; such as: aluminum, asphalt, brick, cement/asbestos, EIFS, stone, stucco, veneer, vinyl, wood, etc.

Solid Fuel Burning Appliances
A hearth and fire chamber or similar prepared place in which a fire may be built and that is built in conjunction with a chimney; or a listed assembly of a fire chamber, its chimney, and related factory-made parts designed for unit assembly without requiring field construction

Structural Component
A *component* that supports non-variable forces or weights (dead loads) and variable forces or weights (live loads)

System
A combination of interacting or interdependent *components*, assembled to carry out one or more functions.

Technically Exhaustive
An investigation that involves *dismantling*, the extensive use of advanced techniques, measurements, instruments, testing, calculations, or other means

Under-floor Crawl Space
The area within the confines of the foundation and between the ground and the underside of the floor

Unsafe
A condition in a *readily accessible, installed system* or *component* that is judged to be a significant risk of bodily injury during normal, day-to-day use; the risk may be due to damage, deterioration, improper installation, or a change in accepted residential construction standards

Wiring Methods
Identification of electrical conductors or wires by their general type, such as non-metallic sheathed cable, armored cable, or knob and tube, etc.

ASHI Code of Ethics

For the home inspection profession

Integrity, honesty, and objectivity are fundamental principles embodied by this Code, which sets forth obligations of ethical conduct for the home inspection profession. The Membership of ASHI has adopted this Code to provide high ethical standards to safeguard the public and the profession.

Inspectors shall comply with this Code, shall avoid association with any enterprise whose practices violate this Code, and shall strive to uphold, maintain, and improve the integrity, reputation, and practice of the home inspection profession.

1. **Inspectors shall avoid conflicts of interest or activities that compromise, or appear to compromise, professional independence, objectivity, or inspection integrity.**

   A. Inspectors shall not inspect properties for compensation in which they have, or expect to have, a financial interest.
   B. Inspectors shall not inspect properties under contingent arrangements whereby any compensation or future referrals are dependent on reported findings or on the sale of a property.
   C. Inspectors shall not directly or indirectly compensate realty agents, or other parties having a financial interest in closing or settlement of real estate transactions, for the referral of inspections or for inclusion on a list of recommended inspectors, preferred providers, or similar arrangements.
   D. Inspectors shall not receive compensation for an inspection from more than one party unless agreed to by the client(s).
   E. Inspectors shall not accept compensation, directly or indirectly, for recommending contractors, services, or products to inspection clients or other parties having an interest in inspected properties.
   F. Inspectors shall not repair, replace, or upgrade, for compensation, systems or components covered by ASHI Standards of Practice, for one year after the inspection.

2. **Inspectors shall act in good faith toward each client and other interested parties.**

   A. Inspectors shall perform services and express opinions based on genuine conviction and only within their areas of education, training, or experience.
   B. Inspectors shall be objective in their reporting and not knowingly understate or overstate the significance of reported conditions.
   C. Inspectors shall not disclose inspection results or client information without client approval. Inspectors, at their discretion, may disclose observed immediate safety hazards to occupants exposed to such hazards, when feasible.

3. **Inspectors shall avoid activities that may harm the public, discredit themselves, or reduce public confidence in the profession.**

   A. Advertising, marketing, and promotion of inspectors' services or qualifications shall not be fraudulent, false, deceptive, or misleading.
   B. Inspectors shall report substantive and willful violations of this Code to the Society.

ASHI membership categories and requirements

All ASHI members are required to abide by the ASHI Code of Ethics, which specifically forbids home inspectors from active brokerage or sale of real estate, or performing repairs on homes that they inspect. ASHI members also agree to perform and report inspections in accordance with the ASHI Standards of Practice. Not only is the ASHI logo widely regarded as the mark of a professional home inspector, its use by the ASHI membership is also carefully controlled.

There are three designations of active professional home inspector membership in ASHI. These ASHI membership categories are as follows:

ASHI Certified Inspectors—Inspectors in this category have:

- Passed the National Home Inspector Examination and ASHI's Standards and Ethics module.
- Had inspection reports successfully verified for compliance with ASHI's Standards of Practice.
- Submitted valid proof of performance of at least 250 fee-paid home inspections that meet or exceed the ASHI Standards of Practice.

ASHI Associates w/ Logo—Inspectors in this category have:

- Passed the National Home Inspector Examination and ASHI's Standards and Ethics module.
- ASHI has verified performance of 50 fee-paid inspections in substantial compliance with the Standards of Practice.
- Had inspection reports successfully verified for compliance with ASHI's Standards of Practice.

ASHI Associates—Inspectors in this category have:

- Just joined ASHI and may be new to the inspection profession or may be a seasoned inspector who has not yet completed ASHI's requirements to move up in membership and logo use.

Glossary

alligatoring Extensive surface cracking in a pattern that resembles the hide of an alligator.

areaway An open subsurface space around a basement window or doorway; provides light, ventilation, and access.

backfill The gravel or earth replaced in the space around a building wall after the foundation is in place.

bottom plate The bottom horizontal member of a frame wall.

Btu (British thermal unit) The amount of heat required to raise the temperature of 1 pound of water by 1 degree Fahrenheit.

cantilever A structural member that projects beyond its supporting wall or column.

Cedar Breather A three-dimensional nylon matrix that is stiff enough to resist crushing, thereby allowing air movement to the underside of wood shingles and shakes.

Conditioned crawl space A crawl area that has insulated foundation walls, closed vents, is heated and/or air conditioned and has a dirt floor that is covered over with large plastic sheets or cement. Basically an area that is dry and comfortable all year long.

control joint A groove that is formed, sawed, or tooled in a concrete or masonry structure to regulate the location and amount of cracking.

cornice Horizontal projection at the top of a wall or under the overhanging portion of the roof.

cross connection A direct arrangement of a piping line that allows the potable water supply to be connected to a line that contains a contaminant.

cupping An inward-curling distortion at the exposed corners of asphalt shingles.

downdraft A downward current of air in a chimney, often carrying smoke with it.

dry well A covered pit, either with open-jointed lining or filled with coarse aggregate, through which drainage from downspouts or foundation footing drains may seep into the surrounding soil.

eaves The lower edge of a roof that projects beyond the building wall.

efflorescence A white powdery substance appearing on masonry wall surfaces, composed of soluble salts that have been brought to the surface by water or moisture movement.

fascia (or facia) A horizontal board nailed vertically to the ends of roof rafters; sometimes supports a gutter.

feathering The tapering of one surface into another.

flashing Sheet-metal or other thin, impervious material used around roof and wall junctions to protect joints from water penetration.

flue A passageway in a chimney for conveying smoke, gases, or fumes to the outside air.

frost line The depth of frost penetration in soil; varies in different parts of the country.

gable roof A double-sloped roof from the ridge to the eaves; the end section appears as an inverted V.

girder The main structural support beam in a wood-framed floor; a girder supports one end of each joist.

glulam (glued laminated lumber) An engineered structural timber product composed of several layers of dimensioned timber glued together.

By laminating several smaller pieces of timber, a single large, strong, structural member is manufactured.

grade The ground level at the outside walls of a building or elsewhere on a building site.

header A framing member that crosses and supports the ends of joists.

hip roof A roof that slopes upward from all four sides of a building.

hydro-air Basically a hot water/warm air system, which essentially consists of an air handler that looks like the blower compartment of a furnace, a hot water boiler, and a duct system.

joist One of a series of parallel beams used to support floor and ceiling loads, supported in turn by larger beams (*girders*) or bearing walls.

knee wall A wall that acts as a brace by supporting roof rafters at an intermediate position along their length.

Lally column A steel tube, filled with concrete, used to support girders and other floor beams.

lath A building material of wood, metal, gypsum, or insulating board fastened to the frame of a building to act as a plaster base.

lintel A horizontal structural member that supports the load over an opening such as a door or window.

nosing The rounded edge of a stair tread that projects over the riser.

parapet wall The part of a wall that extends above the roofline.

pier A masonry column, usually rectangular in horizontal cross section, used to support other structural members.

pigtail (1) A flexible conductor attached to a light fixture that provides a means of connecting the fixture to a circuit. (2) A short length of copper conductor that is attached to the end of an aluminum branch circuit by a special fastener, then fastened to the terminal of a switch or outlet.

pilaster A pier-type projection of the foundation wall used to support a floor girder or stiffen the wall.

plenum A chamber or large duct above a furnace that serves as a distribution area.

ply A sheet in a layered construction, such as plywood, roofing.

pointing (repointing) Filling open mortar joints; removing deteriorated mortar between joints of masonry units and replacing it with new mortar.

rafter One of a series of inclined structural roof members spanning from an exterior wall to a center ridge beam or ridge board.

refractory A material, usually nonmetallic, used to withstand high temperatures, as in the combustion chamber of an oil-fired heating system.

relay An electromechanical switch; a device in which changes in the current flow in one circuit are used to open or close electrical contacts in a second circuit.

resilient tile A manufactured interior floor covering that is resilient, such as vinyl or vinyl-asbestos tile.

ridge beam The beam or board placed on edge at the ridge (top) of the roof into which the upper ends of the rafters are fastened.

riser The vertical height of a stair step; also the vertical boards that close the space between the treads of a stairway.

sheathing The structural covering, usually wood boards or plywood, over a building's exterior studs or rafters.

sheave A wheel with a grooved rim (pulley).

shelter-tube Mud-type tube (tunnel) built by termites as a passageway between the ground and the source of food (wood).

sill plate The lowest member of the house framing resting on top of the foundation wall. Also called *mud sill*.

soffit The visible underside of a roof overhang or eaves.

stringer (step) One of the enclosed sides of a stair supporting the treads and risers.

stud One of a series of slender wood or metal vertical structural members placed as supporting elements in walls and partitions.

subfloor Boards or plywood laid on joists, over which a finished floor is to be laid.

swale A shallow depression in the ground to form a channel for storm-water drainage.

thermocouple A device consisting of two junctions of dissimilar metals; when the two junctions are at different temperatures, a voltage is generated. Used in controlling gas valves.

toenail Driving a nail at an angle into the corner of one wood-frame member to penetrate into a second member.

tread The horizontal board in a stairway on which feet are placed.

truss A combination of structural members usually arranged in triangular units to form a rigid framework for spanning between load-bearing walls.

weep hole A small opening at the bottom of a retaining wall or the lower section of a masonry veneer facing on a wood-frame exterior wall, that permits water to drain.

Inspection worksheets

Address:______________________________

EXTERIOR

Front Exposure: N E S W

Condition

1. **Sidewalks:** none_____ concrete_____ asphalt_____ stone_____ cracked_____ uneven_____ _______
2. **Exterior walls:** brick_____ stucco_____ stone_____ wood siding_____ fiber-cement siding_____
   wood/asbestos shingle_____ aluminum/vinyl siding_____ other_____ vines growing_____ _______
3. **Trim:** cracked_____ rotting_____ broken_____ loose_____ needs paint touch-up_____ _______
4. **Paths:** concrete_____ stone_____ brick_____ asphalt_____ cracked_____ settled_____
   overgrown_____ _______
5. **Steps:** concrete_____ stone_____ brick_____ wood_____ need repair_____
   Handrail:_____ _______
6. **Porch:** front/rear/left/right: cracked_____ rotting_____ chipped_____ loose_____ _______
7. **Windows:** wood_____ aluminum_____ cracked_____ broken_____ reputty_____
   broken cords_____ needs lubrication/adjustment_____ do not close properly_____
   corroding/rotting frames_____ thermal panes_____ faulty seals_____ _______
8. **Storms/screens:** none_____ partial_____ full_____ cracked_____ broken_____
   torn screens_____ _______
9. **Roof:** asphalt_____ wood_____ slate_____ tile_____ roll roofing_____ other_____
   Shingles: missing_____ loose_____ cracked_____ aging_____ eroded_____ _______
10. **Gutters/downspouts:** none_____ partial_____ full_____ built-in_____ aluminum_____
    copper_____ galvanized_____ wood_____ loose/sagging_____ open joints_____
    need splash plates/elbows_____ need cleaning_____ _______
11. **Chimney:** brick_____ masonry_____ prefabricated_____ needs repair_____ _______

Condition

12. **Garage:** none_____ attached_____ detached_____ car occupancy_____

    Doors: operational_____ need lubrication/adjustment_____ automatic opener_____ _______

13. **Driveway:** cracked_____ uneven_____ broken_____ heaving_____ drain_____ _______

14. **Patio:** yes_____ no_____ concrete_____ stone_____ brick_____ needs repair_____

    cracked_____ settled_____ _______

15. **Deck:** attached_____ free-standing_____ composite decking_____ wood_____

    cracked_____ rotting_____ broken_____ needs additional bracing_____

    loose railings/handrails_____ loose steps_____ loose posts_____ needs CO_____ _______

16. **Landscaping:** grass needs recultivation_____ shrubs need pruning_____

    Trees_____ overhanging roof_____ need pruning_____ dead/dying_____ _______

17. **Retaining walls:** brick_____ stone_____ concrete block_____ wood_____ need repair_____ _______

18. **Fencing:** wood/metal_____ broken_____ rotting/rusting_____ needs painting_____ _______

19. **Drainage/grading:** satisfactory_____ poor_____ low spots_____ needs regrading_____

    stairwell_____ window well_____ _______

20. **Termites:** no visible evidence:_____ visible evidence_____ where_____ _______

21. **Caulking:** needed around exterior joints: yes_____ no_____ _______

    Comments: ____

    ____

ATTIC AREA

❍ **Type:** full_____ crawl_____ finished_____ unfinished_____

❍ **Accessible:** yes_____ no_____ **Evidence of condensation:** yes_____ no_____

❍ **Insulation:** floor_____ roof_____ walls_____ partial_____ adequate_____ inadequate_____

none visible_____ loose sections_____ improperly installed_____

Vapor barrier: yes_____ no_____

❍ **Ventilation:** adequate_____ inadequate_____ none_____ need roof/gable vents_____ ridge vent_____

power ventilator_____ Attic fan: _____ operational: yes_____ no_____

❍ **Past leakage:** chimney_____ plumbing vent stacks_____ rafters_____ other_____

❍ **Firestopping:** needed around chimney and attic floor: yes_____ no_____

CRAWL SPACE (BASEMENT LEVEL)

❍ **None:** _____ **Accessible:** yes_____ no_____ Conditioned space: yes_____ no_____

❍ **Floor:** dirt_____ cement_____ asphalt_____ other_____

Vapor barrier needed for floor: yes_____ no_____

Past water seepage: yes_____ no_____

❍ **Insulation needed:** yes_____ no_____ loose section(s)_____

❍ **Ventilation:** adequate_____ inadequate_____ Dehumidifier needed: yes_____ no_____

❍ **Termites:** evidence of_____ no visible evidence_____ conditions conducive to activity_____

Comments: ____

____

BASEMENT AND/OR UTILITY ROOM(S)

❍ **Evidence of water seepage:** yes_____ no_____

water stains_____ efflorescence: on walls_____ on floor_____

swelled floor tiles/joints_____

peeling and flaking paint on walls_____ corrosion below radiators_____

Seepage condition extensive: yes_____ no_____

Sump pump: yes_____ no_____ Dehumidifier needed: yes_____ no_____

	Condition
❍ **Foundation walls:** concrete_____ block_____ stone_____ brick_____ other_____	
cracked_____ reseal_____ cracked or loose mortar joints_____ repoint_____	_______
❍ **Girders:** wood_____ steel_____ none_____ not visible_____	_______
❍ **Joists:** (partially) covered_____ sagging_____	_______
❍ **Columns:** wood_____ steel_____ screw-jack type_____	_______
❍ **Mold:** evidence of: yes_____ no_____	_______

Comments: ____

____

ELECTROMECHANICALS

❍ **Electricity**_____ amps 110-120 volts circuit breakers_____ fuses_____

Service: adequate_____ marginal_____ inadequate_____

Wiring/outlets: adequate_____ marginal_____ inadequate_____ knob-and-tube wiring_____

aluminum branch circuits_____

❍ **Heating system:** forced_____ gravity_____; hot water_____ hot air_____ steam_____

high-efficiency condensing type_____ hydro-air_____ geothermal_____

oil_____ gas_____ electric_____; zones_____

Condition: good_____ fair_____ aging_____ poor_____

Buried oil tank_____ Gas meter adequately sized_____

Furnace room ventilation: adequate_____ inadequate_____

Fire-code plasterboard needed: yes_____ no_____

Asbestos insulation on heating pipes and/or boiler_____

❍ **Domestic hot water:** tankless coil thru heating system_____ tankless wall unit_____

separate tank unit_____ capacity_____ recovery_____ age_____

oil_____ gas_____ electric______

Capacity: adequate_____ marginal_____ inadequate_____ recirculating system_____

Relief valve adequately sized_____

Indirect fired storage type_____

Condition: good_____ fair_____ aging_____ poor_____

❍ **Plumbing:** Pipes and fittings: copper_____ brass_____ iron_____ PEX_____ other_____

Water flow: adequate_____ low_____

Drain lines: cast iron_____ galvanized_____ copper_____ lead_____ plastic_____

Condition: good_____ fair_____ aging_____ poor_____; leakage_____ corrosion_____

Water inlet: copper_____ iron_____ lead_____ plastic_____ inaccessible_____

Septic system: yes_____ no_____

❍ **Well pump:** none_____ jet_____ submersible_____ other_____

Storage tank: yes_____ no_____ Insulation needed: yes_____ no_____

Relief valve needed: yes_____ no_____

Condition: good_____ fair_____ aging_____ poor_____

SWIMMING POOL

In-ground _______ Aboveground ________

❍ **Safety fence:** Y___ N___

Needs repair: Y___ N___ Gate self-closing and self-latching: Y___ N___

❍ **Masonry deck:** cracked___ chipped___ uneven joints___ open joints___

Grab rail___ adequately secured___ Ladder___ adequately secured___

Slide___ adequately secured___ Diving board___ cracked___ warped___

❍ **Wood deck:** cracked___ rotting___ broken___ Handrail: Y___ N___ condition___

❍ **Vinyl lined pool:**

❍ **Lining:** stained___ tears___ discolored___

Lining pulled out of edge retainer: Y___ N___ Replacement recommended: Y___ N___

❍ **Concrete pool side walls:**

Tiles___ cracked___ loose___ chipped___ missing___ Plaster___ flaking___ chipped___ cracked___

Paint___ faded___ flaking___ cracked___

❍ **Skimmer:** Weir: moves freely___ broken___ missing___ Strainer basket: damaged___ missing___

❍ **Pump:** noisy___ hot to the touch___ water dripping around pump___

Strainer basket: needs cleaning___

❍ **Heater:** operational Y___ N___ Cycles on/off: Y___ N___

Rust dust and flakes around base of heater: Y___ N___

❍ **Filter:** Pressure gauge operational: Y___ N___ Filter needs cleaning: Y___ N___

❍ **Cover:** worn___ torn___ damaged___ Repairs needed___ Replacement recommended___

Comments:____

____

Room____________

Condition

1. **Ceiling/walls:** plaster_____ sheetrock_____ Chinese wallboard_____

   Cracked: yes_____ no_____ Broken: yes_____ no_____

   Evidence of past leakage: wall_____ ceiling_____ peeling and flaking paint/paper_____

   old layers of paint_____ lead-based paint_____ _______

2. **Floor:** wood_____ bamboo_____ laminate_____ tile_____ concrete_____ carpet_____

   other_____squeaks_____ slopes_____ sagging_____ refinish_____ torn/chipped_____

   loose_____ _______

3. **Trim:** cracked_____ chipped_____ missing_____ lead-based paint_____ _______

4. **Electrical outlets:** not observable_____ number_____ should have_____ _______

5. **Doors:** Exterior:_____ weatherstripped: yes_____ no_____ sliding glass_____

   Interior:_____ cracked_____ chipped_____ broken_____ pocket door_____ _______

6. **Hardware** (locks, hinges, knobs, handles): missing_____ needs repair_____ _______

7. **Closets:** yes_____ no_____ number_____ _______

8. **Heating:** radiators_____ baseboard_____ air registers_____ convectors_____

   radiation panels_____ none_____ _______

9. **Fireplace:** yes_____ no_____ cracked/chipped mortar joints_____ needs cleaning_____

   Damper: missing: yes_____ no_____ needs repair: yes_____ no_____

   indication of backsmoking_____ _______

10. **Cabinets/counter:** none_____ kitchen_____ medicine/vanity_____

    cracked_____ chipped_____ loose_____ _______

11. **Plumbing fixtures:** yes_____ no_____ leaks_____ Operational: yes_____ no_____

    Water flow: adequate_____ low_____ Drainage: adequate_____ blockage_____

    Regrouting/caulking needed at tub/shower joints: yes_____ no_____ _______

12. **Exhaust fan:** yes_____ no_____ operational_____ _______

    Comments:____

    ____

Room____________

Condition

1. **Ceiling/walls:** plaster_____ sheetrock_____ Chinese wallboard_____
   Cracked: yes_____ no_____ Broken: yes_____ no_____
   Evidence of past leakage: wall_____ ceiling_____ peeling and flaking paint/paper_____
   old layers of paint_____ lead-based paint_____ _______
2. **Floor:** wood_____ bamboo_____ laminate_____ tile_____ concrete_____ carpet_____
   other_____squeaks_____ slopes_____ sagging_____ refinish_____ torn/chipped_____
   loose_____ _______
3. **Trim:** cracked_____ chipped_____ missing_____ lead-based paint_____ _______
4. **Electrical outlets:** not observable_____ number_____ should have_____ _______
5. **Doors:** Exterior:_____ weatherstripped: yes_____ no_____ sliding glass_____
   Interior:_____ cracked_____ chipped_____ broken_____ pocket door_____ _______
6. **Hardware** (locks, hinges, knobs, handles): missing_____ needs repair_____ _______
7. **Closets:** yes_____ no_____ number_____ _______
8. **Heating:** radiators_____ baseboard_____ air registers_____ convectors_____
   radiation panels_____ none_____ _______
9. **Fireplace:** yes_____ no_____ cracked/chipped mortar joints_____ needs cleaning_____
   Damper: missing: yes_____ no_____ needs repair: yes_____ no_____
   indication of backsmoking_____ _______
10. **Cabinets/counter:** none_____ kitchen_____ medicine/vanity_____
    cracked_____ chipped_____ loose_____ _______
11. **Plumbing fixtures:** yes_____ no_____ leaks_____ Operational: yes_____ no_____
    Water flow: adequate_____ low_____ Drainage: adequate_____ blockage_____
    Regrouting/caulking needed at tub/shower joints: yes_____ no_____ _______
12. **Exhaust fan:** yes_____ no_____ operational_____ _______

    Comments:____

    ____

Index

D

E

F

G

H

I

K

L

M

O

P

R

S

T

V

W

Y